Software-Defined Wide Area Network Architectures and Technologies

Data Communication Series

Cloud Data Center Network Architectures and Technologies
Lei Zhang and Le Chen

Campus Network Architectures and Technologies
Ningguo Shen, Bin Yu, Mingxiang Huang, and Hailin Xu

Enterprise Wireless Local Area Network Architectures and Technologies
Rihai Wu, Xun Yang, Xia Zhou, and Yibo Wang

Software-Defined Wide Area Network Architectures and Technologies
Cheng Sheng, Jie Bai, and Qi Sun

SRv6 Network Programming:
Ushering in a New Era of IP Networks
Zhenbin Li, Zhibo Hu and Cheng Li

For more information on this series, please visit: https://www.routledge.com/Data-Communication-Series/book-series/DCSHW

Software-Defined Wide Area Network Architectures and Technologies

Cheng Sheng
Jie Bai
Qi Sun

CRC Press
Taylor & Francis Group
Boca Raton London New York

CRC Press is an imprint of the
Taylor & Francis Group, an **informa** business

人民邮电出版社
POSTS & TELECOM PRESS

First edition published 2021
by CRC Press
6000 Broken Sound Parkway NW, Suite 300, Boca Raton, FL 33487-2742

and by CRC Press
2 Park Square, Milton Park, Abingdon, Oxon, OX14 4RN

Translated by Wanda Wang

CRC Press is an imprint of Taylor & Francis Group, LLC

Library of Congress Cataloging-in-Publication Data
Names: Sheng, Cheng, author.
Title: Software-defined wide area network architectures and technologies / Cheng Sheng, Jie Bai, Qi Sun.
Other titles: SD-WAN jia gou yu ji shu. English
Description: First edition. | Boca Raton, FL : CRC Press, 2021. | Includes bibliographical references. |
Summary: "Starting with problems and challenges faced by enterprise WANs, this book provides a detailed description of SD-WAN's background and basic features, as well as the system architecture, operating mechanism, and application scenarios of the SD-WAN solution based on the implementation of Huawei SD-WAN Solution. It also explains key SD-WAN technologies and analyzes real SD-WAN deployment cases, affording readers with design methods and deployment suggestions for the SD-WAN solution. The information presented in this book is easy to understand and very practical. This book enables you to become adept in the SD-WAN solution's implementation and design principles, and it is intended for ICT practitioners, such as network technical support engineers, network administrators, and network planning engineers, to use in studying theory. Furthermore, it serves as material that network technology enthusiasts can reference"— Provided by publisher.
Identifiers: LCCN 2020051430 (print) | LCCN 2020051431 (ebook) |
ISBN 9780367695774 (hbk) | ISBN 9781003144038 (ebk)
Subjects: LCSH: Software-defined networking (Computer network technology) |
Wide area networks (Computer networks)
Classification: LCC TK5105.5833 S64613 2021 (print) |
LCC TK5105.5833 (ebook) | DDC 621.39/8—dc23
LC record available at https://lccn.loc.gov/2020051430
LC ebook record available at https://lccn.loc.gov/2020051431

ISBN: 978-0-367-69577-4 (hbk)
ISBN: 978-0-367-69967-3 (pbk)
ISBN: 978-1-003-14403-8 (ebk)

Typeset in Minion
by codeMantra

Contents

Summary

STARTING WITH PROBLEMS AND CHALLENGES FACED by enterprise WANs, this book provides a detailed description of SD-WAN's background and basic features, as well as the system architecture, operating mechanism, and application scenarios of the SD-WAN solution based on the implementation of Huawei SD-WAN Solution. It also explains key SD-WAN technologies and analyzes real SD-WAN deployment cases, affording readers with design methods and deployment suggestions for the SD-WAN solution.

The information presented in this book is easy to understand and very practical. This book enables you to become adept in the SD-WAN solution's implementation and design principles, and it is intended for ICT practitioners, such as network technical support engineers, network administrators, and network planning engineers, to use in studying theory. Furthermore, it serves as material that network technology enthusiasts can reference.

Introduction

Starting with the service challenges faced by enterprise wide area networks (WANs), this book describes the background and basic features of SD-WAN. It then explains the system architecture, technical implementation, planning, and design of the SD-WAN solution. Finally, this book introduces design principles and deployment suggestions for the SD-WAN solution. This book is a useful guide for network architecture planning and design, as well as network deployment, for ICT practitioners such as network engineers. This book can also be used by network technology enthusiasts and students as a reference for learning and understanding the latest network technologies.

HOW IS THE BOOK ORGANIZED

This book consists of 11 chapters.

CHAPTER 1: WAN DEVELOPMENTS

This chapter describes the basic concepts of WAN as well as the development history of and problems and challenges faced by enterprise WANs.

CHAPTER 2: EMERGENCE OF SD-WAN

This chapter combines SDN and WAN to introduce SD-WAN, including the background, basic concepts, and industry viewpoints of SD-WAN. It then goes into detail about the basic features and core values of SD-WAN.

CHAPTER 3: INTRODUCTION TO THE SD-WAN SOLUTION

This chapter describes the system architecture, basic components, mechanisms, and panorama of the SD-WAN solution, making it easier for you to understand the SD-WAN solution before deployment.

CHAPTER 4: STARTING FROM SITES

This chapter describes sites and their various categories, as well as forms, functions, and key capabilities of customer-premises equipment (CPE). It also describes zero-touch provisioning (ZTP) for CPEs, including USB-based deployment, email-based deployment, DHCP-based deployment, and deployment through the registration center, making CPEs plug-and-play.

CHAPTER 5: SITE INTERCONNECTION

This chapter describes the interconnection between sites through SD-WAN. It begins by describing the design principles of SD-WAN networking and then proceeds with overlay networking, network orchestration and automation, and reliability. It further illustrates the implementation of services such as Internet access, NAT traversal, interconnection with legacy sites, Point of Presence (POP) networking, and connection to the public cloud in the SD-WAN solution.

CHAPTER 6: GUARANTEED APPLICATION EXPERIENCE

This chapter describes the measures that the SD-WAN solution takes to assure application experience. Such measures include application identification (premise for ensuring application experience), traffic steering, quality of service (QoS), and WAN optimization technologies.

CHAPTER 7: SECURITY: TOP PRIORITY

This chapter describes security risks on SD-WAN networks and details security protection measures employed by the SD-WAN solution from system security and service security perspectives.

CHAPTER 8: EASY O&M

This chapter describes the SD-WAN solution's O&M capabilities. Leveraging the SDN controller, the O&M architecture's intelligent brain, the SD-WAN solution significantly eases O&M. The main topics in this chapter include the O&M mode and management roles, network-wide monitoring, fault demarcation and locating, routine device maintenance, log management, and intelligent O&M, as well as the solution design and operation principles for reconstructing a legacy site into an SD-WAN site.

CHAPTER 9: SD-WAN BEST PRACTICES

This chapter describes the SD-WAN solution's application scenarios and lays out the best practices of the SD-WAN solution in typical industry scenarios, such as finance, carriers, and managed service providers (MSPs).

CHAPTER 10: SD-WAN COMPONENTS

This chapter describes the application scenarios, product architectures, and main functions and features of major components in Huawei's SD-WAN Solution: NetEngine AR routers and iMaster NCE-WAN.

CHAPTER 11: SD-WAN OUTLOOK

This chapter describes the impact of new technologies and trends such as Segment Routing over IPv6 (SRv6), 5G, and artificial intelligence (AI) on SD-WAN from the perspectives of new technology development and industry landscape changes, and provides an outlook into the future of SD-WAN.

ICONS USED IN THIS BOOK

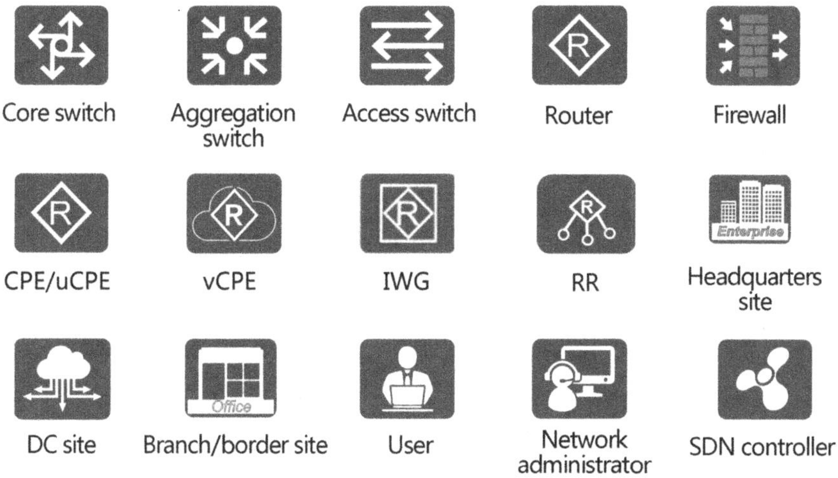

Acknowledgments

THIS BOOK HAS BEEN JOINTLY WRITTEN by the Information Digitalization and Experience Assurance (IDEA) Department and Switch & Enterprise Gateway Design Department of Huawei Technologies Co., Ltd. During the process of writing the book, high-level management from Huawei's Data Communication Product Line provided extensive guidance, support, and encouragement. We are sincerely grateful for all their support.

The following is a list of participants involved in the preparation and technical review of this book.

Editors-in-Chief: Cheng Sheng

Deputy Editors-in-Chief: Jie Bai, Qi Sun

Editorial board: Chunning Wang, Yonglong Fang, Hongwei He, Jianqiang Hou, Penghe Tang, Xu Tian, Tao Ye, Zhifeng Yin, Lingling Yu, Hui Li, Hua Long

Technical reviewers: Xiongfei Gu, Junhui Li, Jun Guo, Jiaofeng Li, Xinfeng Yang, Hao Zhang, Yong Chen, Qi Yu, Fangli Li, Bayaer Dahu

Translators: Wanda Wang, Haiyan Tang, Baishun Yu, Fang He, Rene Okech, Hannah OCallaghan

While the writers and reviewers of this book have many years of experience in ICT and have made every effort to ensure accuracy, it may be possible that minor errors have been included due to time limitations. We would like to express our heartfelt gratitude to the readers for their unremitting efforts in reviewing this book.

Acknowledgments

Authors

Cheng Sheng is the Chief Architect of Huawei's SD-WAN Solution. He has nearly 20 years of experience in network product and solution design, as well as extensive expertise in product design and development, network planning and design, and network engineering project implementation.

Jie Bai is an Architect of Huawei's SD-WAN Solution. He is well versed in Huawei security products and SD-WAN Solution, and wrote books such as *Huawei Firewall Technology Talk* as well as *Huawei Anti-DDoS Technology Talk*.

Qi Sun is a Senior Information Architect of Huawei, and he is well versed in Huawei SD-WAN Solution, CloudVPN Solution, and Cloud Management Solution. He also participated in the information architecture design and delivery of multiple solutions.

CHAPTER 1

WAN Developments

From the early days of pigeon post and telegram to today's telephones and instant messaging, society has always pursued new ways to communicate and collaborate seamlessly beyond geographical boundaries. Over time, technological advances have made it possible for us to interact as if we were face to face, no matter how far apart we are. Such immersive interactions are no longer limited to science fiction — they are already part of our daily lives.

Similarly, enterprises have high demands for truly productive communication and collaboration. As economic globalization and digital transformation take shape, the scale of enterprises is expanding with more branches being dispersed throughout different regions. But despite being so spread out, enterprise headquarters and branches need to be able to seamlessly communicate and collaborate with each other. This is where Wide Area Networks (WANs) come in. WANs are used by enterprises to connect branches that are far apart so as to better roll out services. This is especially true in today's information society where networks are the root and connections are the foundation.

WANs enable enterprise headquarters and branches thousands of miles apart to connect, share information, and communicate seamlessly with each other. But more importantly, WANs inspire and motivate network vendors to constantly improve connection quality and to create a better network experience. Next, let's delve into the world of enterprise WANs.

1.1 EVOLUTION OF ENTERPRISE WANs

WAN is not a new concept for practitioners in the communications industry. Along with Local Area Network (LAN) and Metropolitan Area Network (MAN), it is one of the first concepts learners of computer networking come across. As Confucius said, "One will get something new in looking over one's old studies." In this spirit, let's first review the development of WANs.

As its name suggests, a WAN is a wide area interconnection network used for long-distance communication between enterprises, organizations, or individuals, and can span multiple countries, regions, or cities. With coverage often ranging from tens to thousands of kilometers, WANs enable information and resource sharing over vast distances. Due to costs and construction difficulties, WANs are generally provided by carriers. The Internet, for example, which is now deeply rooted in our daily lives, is actually a global WAN built by different carriers around the world.

Benefiting from WANs, enterprises can easily connect their geographically dispersed branches and roll out various services, which in turn greatly advance economic globalization.

You may be wondering: How do enterprises get their own WANs? Most enterprises set up their own WANs by leasing WAN private lines from carriers. For example, an enterprise may lease multiple peer-to-peer private lines provided by a carrier to connect its dispersed branches. Alternatively, the enterprise can rent the public network provided by the carrier to establish its own WANs. With the public network, each branch of the enterprise can connect to every other branch through a single link.

As network technologies are developing at a remarkable pace, enterprise WANs also keep evolving, from the earliest Time Division Multiplexing (TDM) era to the IP/Multi-Protocol Label Switching (MPLS) era, and finally to today's cloud era. Correspondingly, private lines provided by carriers for constructing enterprise WANs are constantly transforming, as shown in Figure 1.1.

Private lines provided by carriers are primarily classified into two types:

- The first type refers to such private lines as TDM-based Synchronous Digital Hierarchy (SDH) and Multi-Service Transport Platform (MSTP), as well as Wavelength Division Multiplexing (WDM)-based Optical Transport Network (OTN). Private lines of this type are traditional point-to-point physical private lines and are highly secure.

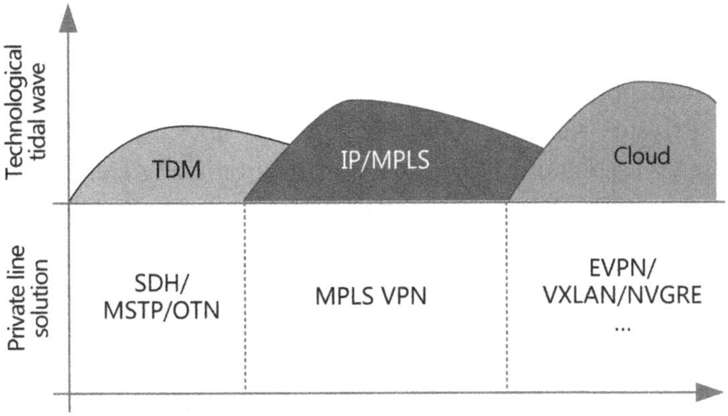

FIGURE 1.1 Evolution of enterprise WANs.

Users can exclusively occupy private line resources, with high guaranteed bandwidth and Quality of Service (QoS), and therefore costs are inevitably very high. Organizations in the financial services industry prefer to use this type of private line, as they have high requirements on WAN line quality as well as communications security and privacy.

- The second type is typified by MPLS Virtual Private Network (VPN) private lines based on Packet Switched Network (PSN) technology. Private lines of this type achieve reachability from any point on the network to any other point. In essence, MPLS VPN private lines are the same as the abovementioned first type of private lines, in the sense that both are provided by carriers. Where they differ is that MPLS VPN private lines offer more networking convenience and lower costs. In particular, MPLS VPN private lines are noted for better bandwidth and reliability assurance, as well as security isolation, to name a few. It is these merits that have led to MPLS VPN private lines being so widely used in the enterprise WAN interconnection market.

With the rapid development of enterprise IT digitization and economic globalization, enterprises are now stepping into the cloud era. In this new era, enterprises will be looking for a more convenient, intelligent, and simplified network connection mode to easily implement ubiquitous, high-quality WAN interconnection anytime, anywhere. In particular, driven by the development of the Internet, IP networks have made great progress in terms of coverage and network quality, making it possible to interconnect enterprise WANs through the Internet.

Against this backdrop, overlay VPN technologies — such as Ethernet Virtual Private Network (EVPN), Virtual eXtensible Local Area Network (VXLAN), and Network Virtualization using Generic Routing Encapsulation (NVGRE) — are emerging as next-generation technologies. Due to their advantages in service provisioning agility, networking flexibility, and interoperability, these up-and-coming technologies are gradually becoming the mainstream for building the next-generation enterprise WAN interconnection.

It should be noted that it isn't a case of choosing just one of the three phases of enterprise WAN private lines. Rather, in the long run, they should be used in combination to get the best of each. For example, MPLS VPN private lines will be around for a long time due to advantages in networking and reliability. The emerging overlay VPN private lines — which are often regarded as networking options for flexible scheduling — will not replace MPLS VPN private lines. Instead, they will run on top of MPLS VPN private lines, providing the ability to schedule the Internet and offering other WAN capabilities.

1.2 CHALLENGES FACING ENTERPRISE WANs

Traditionally, enterprises use WANs to interconnect their organizations that are widely dispersed throughout different regions, as shown in Figure 1.2.

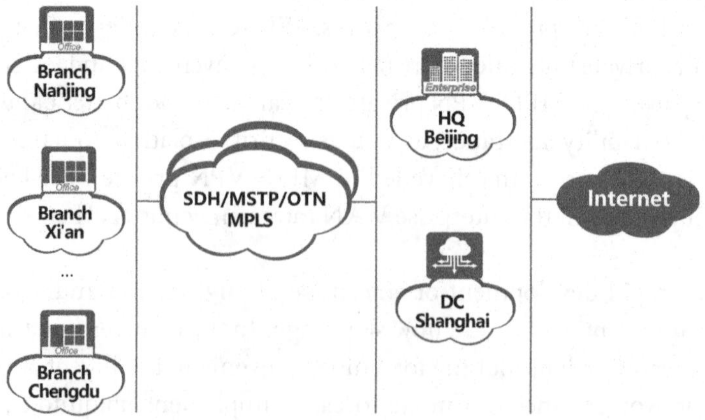

FIGURE 1.2 Traditional enterprise WAN interconnection.

This traditional enterprise WAN interconnection scenario has the following characteristics:

- An enterprise typically includes the headquarters, branches, and data centers (DCs).
- Enterprises build WANs by purchasing SDH, MSTP, OTN, and MPLS private lines from carriers.
- Enterprise branches implement services by communicating with the headquarters or DC, through which they access the Internet.

In this traditional enterprise WAN interconnection scenario, the enterprise WAN architecture is relatively closed due to the following reasons:

- Key applications, information, and data are stored inside the enterprise. The demand for WAN bandwidth is low, and services do not change frequently.
- Carriers provide only limited choices of private lines. The network topology seldom changes.
- Security policies are centrally implemented in the headquarters or DC to ensure security.

For quite some time, this traditional WAN architecture has played an important role in enterprise branch interconnection and, to some extent, met enterprise service requirements.

However, as enterprises further their digital transformation, we are seeing the emergence of new external business and technical factors, including cloud computing, network virtualization, and Network as a Service (NaaS). These new drivers are exerting a profound impact on the service model and network architecture of traditional enterprise WANs. As such, enterprise WANs are being presented with the following new challenges.

1.2.1 Service Cloudification in Full Swing

The rise of cloud computing catalyzes the booming public cloud market. More and more enterprises are choosing to build their IT systems on the public cloud in order to reduce construction costs while accelerating construction. They are also choosing to move their traditional applications to the cloud. In most cases, enterprises are using WANs to access Software as a Service (SaaS) applications, such as office software and databases.

As enterprise services are migrated to the cloud, enterprise WANs are carrying more and more cloud-related application traffic, resulting in WAN traffic surges and therefore far higher demand for WAN bandwidth. Due to this, the traffic transmission quality of enterprise WANs directly impacts enterprises' cloud application experience. Because of the large latency and high packet loss rate over Internet links, as well as difficult access to the cloud through MPLS private lines, enterprises still find it difficult to implement cloud-network synergy, preventing enterprises from fully enjoying the unprecedented speeds and performance that cloud computing offers.

1.2.2 Network Virtualization Taking Off

Network Functions Virtualization (NFV) is the use of Virtual Machine (VM) or container technology to virtualize traditional network devices, such as routers and firewalls, into software to be run on general-purpose servers. The result is network functions that are decoupled from network devices.

NFV uses software to implement multiple network functions, which has obvious advantages such as reducing hardware spending as well as Operations and Maintenance (O&M) costs and deploying services more flexibly and rapidly. Network virtualization technologies like NFV are accelerating the transformation of enterprise WAN infrastructure deployment and O&M modes toward rapid service provisioning and on-demand simplified deployment.

1.2.3 NaaS as an Inevitable Trend

The fundamental building blocks of today's enterprise WANs are twofold: network devices (such as routers and switches) and enterprise-class WAN private lines (such as MSTP and MPLS). In essence, both are products. They are sold by equipment vendors or resold by carriers to enterprise customers, with commitments extending only to product functions and performance specifications. That is, vendors and carriers are not responsible for enterprise customers' IT application experience. As such, if the experience of enterprise WAN applications is poor after enterprises have integrated purchased products, they are left to deal with the problem by themselves.

In stark contrast to product sales, the purpose of service sales is to meet the final service requirements of customers. Services innately offer more direct consumption patterns than products. For example, NaaS enables enterprises to purchase business services and commitments instead of separate product components. Quality is measured according to how satisfied customers are with the obtained services.

Currently, we are seeing network resale evolving from product-oriented to service-oriented. This evolution radically changes carriers' product sales modes, meaning that enterprises only need to propose their IT requirements as opposed to constructing IT networks themselves, as before. In this way, enterprises are freed from much workload and can redirect their resources to core services, thereby improving production efficiency.

1.2.4 Internet-Based Enterprise Communications on the Rise

In recent years, the Internet improved drastically in terms of coverage and performance, and it is now able to offer network quality closer to that of private lines. Thanks to these advances, more enterprises are able to use the Internet as a network transmission medium in order to realize more effective utilization of network resources. In addition to providing traditional Internet access services, the Internet is playing an ever-more prominent role in the interconnection between enterprise headquarters, branches, and DCs. As such, it has become a viable alternative to traditional private lines provided by carriers in enterprise WAN interconnection scenarios.

In short, enterprise WAN interconnection has changed tremendously, taking on a completely new look. Such change is largely driven by network technology advances and business model transformations.

Figure 1.3 shows the current state of enterprise WAN interconnection.

In addition to the headquarters, branches, and DCs that are typical of traditional enterprise WAN interconnection, today's enterprise WAN interconnection includes public cloud and SaaS applications. The Internet has also become an element that should not be ignored.

Today's enterprise WAN interconnection scenarios are now more complex than ever, and are facing the following challenges regarding cloud-network synergy, application experience, network performance, and O&M:

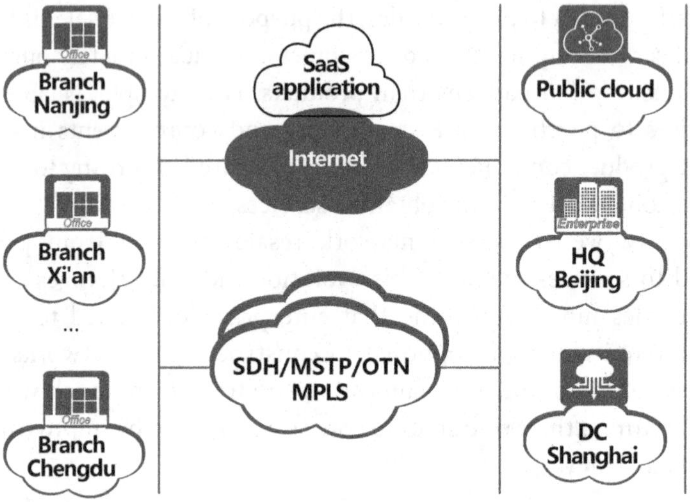

FIGURE 1.3 Current state of enterprise WAN interconnection.

1. **Connecting multiple clouds and networks amid the trends of globalization and cloudification**

   Within the next few years, most enterprises will migrate their services to the cloud. Meanwhile, we will see increasingly widespread use of emerging services such as 5G, 4K/8K video, Virtual Reality(VR)/Augmented Reality (AR), and the Internet of Things (IoT).

   Driven by these, enterprise WAN traffic will experience explosive growth, and enterprise interconnection scenarios will become increasingly complex, involving enterprise headquarters, branches, hybrid clouds, Infrastructure as a Service (IaaS), SaaS, mobile offices, and more. In addition, different network environments from multiple carriers across regions will need to be accommodated, including optical fibers, Digital Subscriber Lines (DSLs), Long-Term Evolution (LTE), and 5G.

   Against this backdrop, an important concern for enterprise WAN construction is how to build a scalable and cost-effective network for the cloud era.

   Such concern will be of particular importance to multinational companies. Take Huawei as an example. As a globalized enterprise, Huawei has business operations spanning more than 170 countries and has established over 900 branches, 15 R&D centers, and 36 joint innovation centers. Also, Huawei has more than one million partners and suppliers. Huawei's connection requirements vary in different scenarios, such as for the headquarters, branches, research centers, Huawei public cloud, third-party IaaS/SaaS, private cloud, suppliers, and partners. For Huawei, efficiently implementing multi-cloud and multi-network interconnection to support such a large and complex organization has become the key to its successful digital transformation.

2. **Ensuring the experience of key applications despite the sharp increase in application quantities**

   With the advent of the cloud and digital era, numerous emerging enterprise services are being closely integrated with cloud computing, which is driving their rapid development. For this reason, the number and types of enterprise applications, such as voice, video, file transfer, email, and SaaS applications, are increasing explosively. These varied applications have different requirements on link quality. For example, cloud desktop services require a latency of less than 20 ms in order to deliver an optimal user experience. Video conferences — which are sensitive to packet loss and latency — must be free from freeze frame and artifacts, allowing users to clearly see their colleagues when holding video conferences using cloud desktops.

   Traditional enterprise private lines cannot detect services or obtain the status of applications. As a result, when link congestion or quality deterioration occurs due to traffic bursts, key services cannot be guaranteed. Overcoming such issues requires new methods for quickly identifying which of the many applications are key ones and optimizing user experience accordingly. Doing so is the key to improving internal efficiency and customer satisfaction.

3. **Eliminating network congestion even with low-performance branch network devices**

   In traditional enterprise WAN scenarios, carriers sell private lines to customers while offering hardware network devices free-of-charge. These devices, however, have only basic routing and forwarding capabilities. With the emergence of services such as cloud, video, and VR, enterprise WAN bandwidth demands are increasing quickly. To ensure application experience, devices must be upgraded to include application identification and WAN optimization capabilities.

In short, as branch services become ever-more complex, traditional devices whose core capabilities are basic routing and forwarding will inevitably form performance bottlenecks. Especially after value-added services (VASs) are enabled, key services will often be congested in case of traffic bursts during peak hours, ultimately affecting enterprise WAN services. As such, a key consideration for enterprise WANs in the new era is how to further optimize and improve device performance.

4. **Minimizing manual configuration to facilitate deployment and O&M of branch networks**

   As enterprises ride the tidal wave of digitalization and globalization, more and more geographically dispersed enterprise branches need to be interconnected. As well as this, branch networks need to go online more quickly and flexibly, and maintaining them must be simpler and more convenient. Only then can we fully support rapid service development.

   The difficulty is that branches have complex and changing network demands; many branch networks are scattered and require innumerable configurations; and professional network engineers are required to perform network provisioning, service commissioning, and fault locating onsite. Inevitably, the result is high network deployment and O&M costs.

   Indeed, the companies that are able to quickly roll out services will be those that stay ahead of the competition. Let's take the insurance industry as an example: A large insurance company usually has thousands of branches, and many branch networks often need to be newly built or migrated. If the company decides to create a branch through traditional private lines, it has to go through multiple steps, such as private line application, resource allocation, and network admission construction, with the entire process possibly taking several months. Besides this, service provisioning relies heavily on onsite manual configuration by highly trained network engineers, resulting in a low provisioning efficiency.

   Given all of these factors, it is key that enterprises focus on how to reduce network deployment and O&M costs while improving customer satisfaction.

CHAPTER 2

Emergence of SD-WAN

Enterprise WANs have already undergone decades of development. One perpetual fact proven against such a long-enough development history is that we must embrace transformation if we want enterprise WANs to remain in step with industry development.

Recent years have witnessed the momentum of software-defined networking (SDN) as it has created a new order in the networking industry. At the same time, SDN has given birth to a new networking philosophy — integrating SDN into WANs, paving a way through the challenges facing legacy enterprise WANs. This philosophy signifies the very beginning of enterprise WAN transformation.

2.1 WHEN WAN AND SDN MEET…

2.1.1 What Is SDN?

A conventional network architecture adopts distributed control where each network device is a closed system consisting of hardware, an operating system (OS), and network applications. These devices are able to independently deliver tightly coupled network control and data forwarding functions, but the network architecture has plenty of shortcomings, such as poor flexibility, complex network protocols, heavy dependency on network device vendors, and difficult network O&M and management. As such, this network architecture, even in its full play, is still unable to meet user requirements for flexible service deployment.

SDN is an ideal solution to address these issues. SDN enables applications to participate in network control and management to satisfy upper-layer service requirements. It also achieves automated network deployment, enhancing service agility and multiplying network operations efficiency.

You can think of SDN as either a brand-new network architecture or a network design philosophy. In other words, SDN is ideal for resolving the issues currently confronted by networks and meeting user expectations for network transformation. Many world-leading institutions have shed light on what is SDN. For example, the Open Network Foundation (ONF) has defined SDN as illustrated in Figure 2.1.

ONF defines SDN as a network architecture that separates network control from data forwarding and implements programmable control. As shown in Figure 2.1, the SDN architecture consists of three layers: the application layer, control layer, and infrastructure layer.

2.1.1.1 Application Layer

The application layer is the uppermost interaction interface of the SDN architecture and is responsible for communication with external systems.

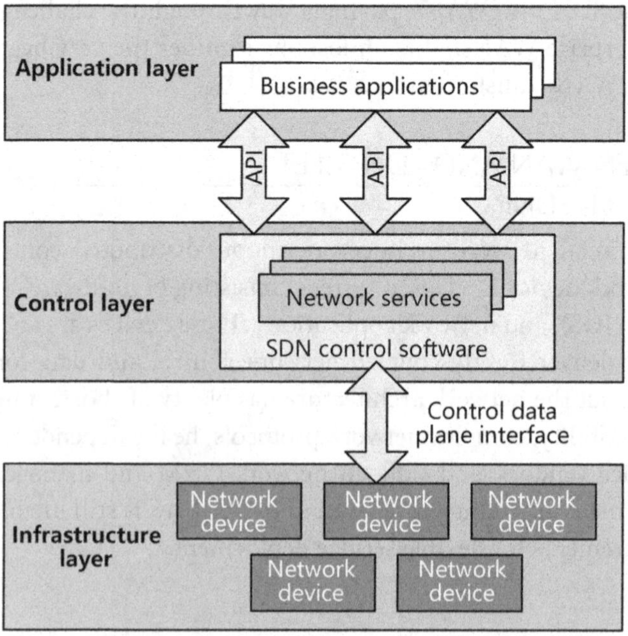

FIGURE 2.1 ONF-defined SDN architecture.

This layer consists of various business applications and is responsible for interpreting users' service requirements and defining and orchestrating network services based on user requirements.

2.1.1.2 Control Layer

The control layer is the brain of the SDN architecture. It bridges the application layer (northbound) and the infrastructure layer (southbound). To elaborate, it provides open and abstracted network functions and services for the application layer through northbound Application Programming Interfaces (APIs), and controls data forwarding behaviors of network devices at the infrastructure layer through southbound interfaces.

2.1.1.3 Infrastructure Layer

The infrastructure layer is the body of the SDN architecture. It contains a wide range of common network devices that forward traffic according to policies delivered by the control layer.

ONF's definition of SDN paints it as a framework for network design, not a technology for specific use cases. The framework adheres to three criteria in its pursuit to reconstruct conventional networks: separation of the forwarding plane from the control plane, centralized control, and an open network with high levels of programmability.

SDN came to life in a university research project and has changed the momentum of network development after more than a decade's development. SDN is not an intricate concept making sense only at the theoretical level or just a gimcrack and infeasible network design approach. Instead, it is a solid, feasible approach to network design and has made a significant difference in the data center network (DCN) domain, innovatively redefining the DCN framework to better satisfy ever-changing service requirements.

2.1.2 SD-WAN Emerges at the Right Time

As SDN is setting a new trend for DCNs, a question emerges: Can SDN be applied in the WAN domain? Open Networking User Group (ONUG) is the closest to answering this question.

ONUG is an influential, IT user-focused user organization led by large enterprises. The organization was established by IT executives from renowned enterprises in North America, and its members come from a wide range of industries, such as financial services, insurance, medical care, and retail. The organization is dedicated to promoting IT implementation and network technology transformation in large enterprises, and functions as a platform for customers in North America to discuss and exchange IT requirements.

As SDN takes the network industry by storm, we can see it used in a broad range of enterprise LAN applications, such as cloud DCN. ONUG, with its keen insight into network development, conducted a thorough discussion among its members, sparking the application of SDN in enterprise WANs so as to capitalize on its technical strengths and resolve the many pressing challenges enterprise WANs are confronted with.

ONUG elaborated on this discussion in a 2014 conference by unveiling a brand-new concept: Software Defined Wide Area Network (SD-WAN). SD-WAN, just like its name, applies the innovative and field-proven architecture and philosophy of SDN to redefine WANs.

In addition, ONUG defined ten requirements for SD-WAN[1], as illustrated in Table 2.1.

On the basis of the pioneering definition of SD-WAN made by ONUG, an increasing number of standards organizations have actively participated in this future-proof wave, extending on ONUG's definition with

TABLE 2.1 Ten SD-WAN Requirements Defined by ONUG

No.	Description
1	Hybrid WAN featuring flexible interworking of public and private WANs between enterprise sites
2	Able to deploy the CPE in a physical or virtual form with no dependence on hardware
3	A secure hybrid WAN architecture that allows for dynamic load balancing among MPLS links, enabling enterprise WAN traffic to be dynamically adjusted among WAN links upon a link fault or deterioration of application quality, thereby ensuring service convergence performance
4	Visualization, preferential scheduling, and QoS guarantee for key applications
5	Hybrid WAN with high reliability and elasticity, ensuring optimal experience of applications
6	Compatibility and interoperability with Layer 2 and Layer 3 devices on live networks
7	Able to visualize the performance of sites, applications, and VPNs
8	Open northbound APIs
9	Able to perform ZTP at sites to simplify service provisioning
10	Federal Information Processing Standard (FIPS) 140-2 validation certification for security authentication

SD-WAN characteristics and technical standards. The following takes two world-renowned organizations as typical examples to showcase the ongoing, popular pursuit for SD-WAN.

2.1.2.1 Gartner

As the world's most authoritative IT research and advisory company, Gartner spreads its research scope to as wide as all IT industries, including IT research, development, evaluation, application, and market. It aims to provide objective, fair demonstrations and market surveys for its customers, assisting them in market analysis, technology selection, project demonstration, and investment decision-making.

The Magic Quadrant research tool launched by Gartner is well-known in the industry. It presents a comprehensive, graphic description of the market at given intervals. Particularly, Gartner released the Magic Quadrant for WAN Edge Infrastructure in 2018, in which it analyzed SD-WAN vendors and market conditions.

Gartner had clearly defined key requirements for SD-WAN, involving both lower-layer links and upper-layer applications. For example, SD-WAN shall provide dynamic and policy-based application traffic steering across WAN links, and support a broad variety of VASs. In addition, Gartner believes that SD-WAN has tangible advantages in enhancing application experience, shortening configuration duration, improving network availability, and reducing deployment costs.

2.1.2.2 MEF

Metropolitan Ethernet Forum (MEF) is a nonprofit organization that focuses on resolving technical issues of metro Ethernet, aiming to use Ethernet technologies for data switching and transmission over MANs. With this objective, MEF dedicates itself to promoting the implementation of existing and new network standards, Ethernet service definitions, test procedures, and technical specifications, ultimately turning Ethernet-based WANs into carrier-class networks.

MEF also focuses on providing Lifecycle Service Orchestration (LSO) solutions and architectures for carriers' managed service market, and defining northbound APIs for interworking between devices of various vendors.

A paper published by MEF in 2017[2] introduced fundamental characteristics of SD-WAN as well as the components involved in a typical

TABLE 2.2 SD-WAN Characteristics Defined by MEF

No.	Description
1	Secure, IP-based overlay network
2	Independent underlay network that operates over any type of wired or wireless access networks
3	Service assurance of each SD-WAN tunnel
4	Application-based traffic steering and forwarding
5	Hybrid access using multiple types of WAN links, featuring high reliability
6	Policy-based packet forwarding
7	Automated service provisioning through centralized management, control, and orchestration, such as ZTP
8	WAN optimization

SD-WAN solution, and defined the framework and service specifications for all APIs of these components. Table 2.2 presents a brief description of the SD-WAN characteristics defined by MEF.

If we compare the two, Gartner and MEF have many similarities in their SD-WAN characteristics, such as multi-link access, application-based traffic steering, WAN optimization, and security. However, MEF's definition is much more fine-grained, and MEF considers centralized management, control, and orchestration of WANs as essential to achieving automated service deployment.

Other companies and organizations in the industry, just like Gartner and MEF, may have different approaches to defining SD-WAN. However, there seems to be a consensus on SD-WAN fundamentals, illustrated as follows:

- SD-WAN shall implement rapid deployment and rollout of branches through Zero-Touch Provisioning (ZTP), improving service deployment efficiency.
- SD-WAN shall dynamically adjust traffic paths by application type, making flexible, convenient traffic scheduling a reality.
- SD-WAN shall implement centralized network management and control, network-wide status visualization, and automated, intelligent network O&M.
- SD-WAN shall support plenty of VASs which exemplify WAN optimization and security to achieve rapid service provisioning.

In addition to the preceding companies and organizations, proactive and future-oriented communications between all parties in the industry also boost the development of SD-WAN. A national or international SD-WAN event is held almost every year, which gathers industry experts together for in-depth brainstorming about SD-WAN, promoting sustainable development of the SD-WAN industry. The following describes several examples of SD-WAN events:

- **ONUG conference**: ONUG gave birth to and maintains great enthusiasm for SD-WAN. SD-WAN remains a hot topic at ONUG conferences, where they present use cases, proof of concept (POC) demonstrations, and provider technology showcases to jointly promote SD-WAN innovation and improvement.
- **Global SD-WAN Summit**: Since 2017, SD-WAN Summit — a world-renowned SD-WAN event — has been held every year. This event attracts test organizations, consulting companies, standards organizations, vendors, and carries from around the world, enabling comprehensive, in-depth communication and discussion between these participants so as to boost SD-WAN application and development.
- **China SD-WAN Summit**: The first-ever China SD-WAN summit was successfully held in 2018. As the first of its kind in China, this summit attracted numerous experts and scholars from the Internet industry, carriers, and SD-WAN suppliers, and provided them with a platform to discuss SD-WAN development.

Plenty of vendors proactively contributed to the sustainable development of the SD-WAN ecosystem by providing differentiated SD-WAN products and solutions. This vigorous development of SD-WAN has incubated huge opportunities for all parties in the industry. For example, network equipment providers have been proposing their own innovative SD-WAN solutions by capitalizing on their mature devices and inventory markets. Emerging companies, or startups, concentrate on providing a variety of unique SD-WAN solutions. Security and WAN acceleration device vendors have also been seizing the opportunities brought by SD-WAN to enter the SD-WAN domain by leveraging their advantages in firewalls and WAN acceleration devices.

Yet another example comes to carriers and cloud service providers. Specifically, carriers integrate SD-WAN technology and private line services to spark service growth, and cloud service providers attempt to remain in-step with SD-WAN development by connecting enterprise branches to high-performance, intelligent cloud-based backbone networks with global connectivity. These changes give enterprise branches better access to the cloud and each other.

Further Reading

There is a term similar to SD-WAN in the industry — SDN WAN. These two terms are easy to confuse, but could not be any more different. The following will compare the two terms from the perspectives of application scenarios and networking technologies.

The first perspective comes from their application scenarios. SD-WAN is evolved from enterprise WAN interconnection and focuses on providing intelligent enterprise branch interconnection as a service. In contrast, SDN WAN is typically used for carrier service solutions, is evolved from carrier-built WANs (IP backbone network+MAN), and focuses on "IP+optical" backbone networks and MANs. This solution capitalizes on SDN to optimize data forwarding paths on WANs, simplifying network deployment and multiplying utilization of network bandwidth.

In terms of networking technologies, both SD-WAN and SDN WAN require network controllers but traffic steering is implemented differently. In particular, SD-WAN establishes an overlay network on top of an underlay network. The overlay network houses a control plane that is built via Border Gateway Protocol (BGP) EVPN, and a forwarding plane that is set up via Internet Protocol Security (IPsec). These practices help achieve multi-path intelligent traffic steering between enterprise sites. This solution mainly involves low-end and mid-range enterprise routers. In contrast, the SDN WAN solution centrally computes optimal paths throughout the network via a network controller. In addition, Border Gateway Protocol-link state (BGP-LS) or Path Computation Element Communication Protocol (PCEP) is deployed on devices throughout the network to collect network-wide topology information and distribute paths. This solution mainly involves high-end routers oriented to the carrier backbone network.

2.2 INTERPRETATION OF SD-WAN

As described above, different standards organizations have different approaches to defining and interpreting the functions of SD-WAN. As shown in Figure 2.2, the industry holds different opinions on what SD-WAN is, and has yet to reach a consensus on a unified definition.

Against this backdrop, this section focuses on figuring out what is SD-WAN. To put it simply, SD-WAN is a type of network service that applies SDN technology to the interconnection of WANs in enterprises. Leveraging the SDN controller — a centralized network control system, this type of network service boasts automated configuration for WANs, centralized control and management, and high openness and programmability.

SD-WAN deeply integrates conventional enterprise WAN technologies, such as routing, QoS, security, and WAN acceleration, as well as future-proof brand-new technologies, including SDN, NFV, and service orchestration. As such, using the SDN controller, SD-WAN achieves centralized orchestration, control, and management of WAN interconnections.

2.2.1 Fundamental Characteristics of SD-WAN

Having now interpreting SD-WAN, in this section we will proceed to present ten fundamental characteristics of SD-WAN in terms of basic enterprise WAN networking, cloud-network synergy, application experience, and network O&M.

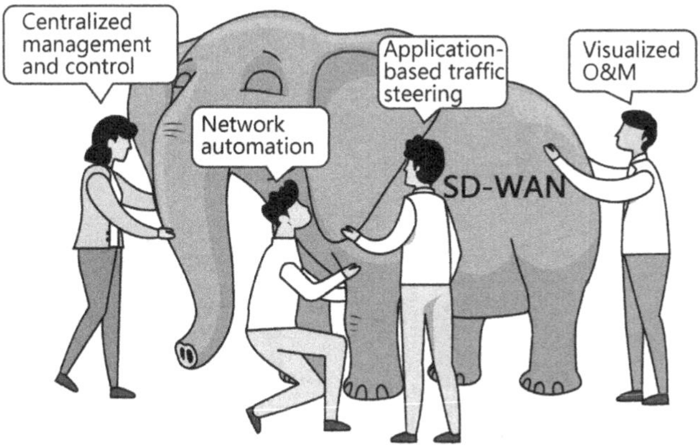

FIGURE 2.2 Various opinions on SD-WAN in the industry.

2.2.1.1 Hybrid WAN Links, Achieving Flexible Overlay Networking

This section first delves into the development of WAN networking. If we view an enterprise WAN like a nation-wide logistics system, enterprise branches would be logistics stations. Goods can be transported between these stations in different ways, such as by air, rail, and road. Traditionally, goods are usually transported by air or rail throughout the entire process, which outperforms other logistics methods in terms of speed and reliability. However, this speed and reliability require higher costs. As the national highway network is rapidly developing, goods transportation by road has also become a rapid and cost-effective approach to logistics, as illustrated in Figure 2.3.

Traditional WANs are interconnected through private lines provided by carriers. As the quality and coverage of the Internet improve, the Internet turns to be a new choice for WAN interconnection. In addition to the MPLS private lines provided by carriers, enterprises can also use the Internet to connect WAN branches, as shown in Figure 2.4. This achieves hybrid interconnections, slashes deployment costs, and multiplies utilization of WANs.

As illustrated in Figure 2.5, to achieve a hybrid WAN, some key network technologies, such as overlay and VPN, are required.

Even though the overlay technology delivers outstanding performance in transparent traffic transmission, it requires that the user overlay network be securely isolated from the carrier's underlay network. This is where the VPN is needed.

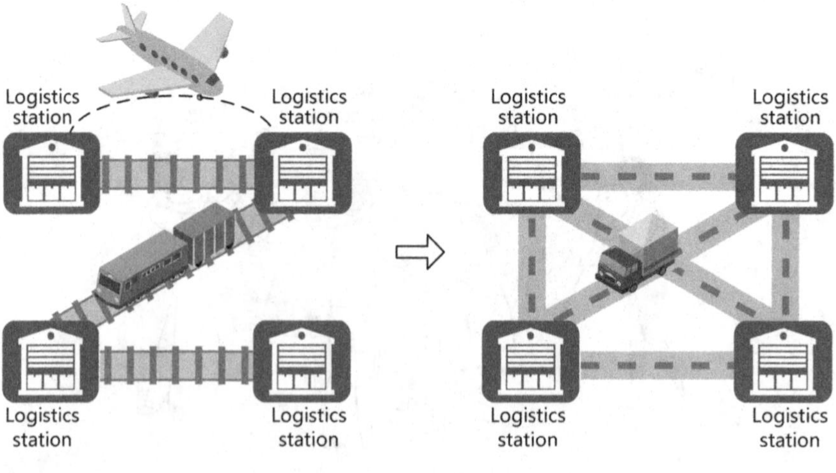

FIGURE 2.3 Development history of the logistics system.

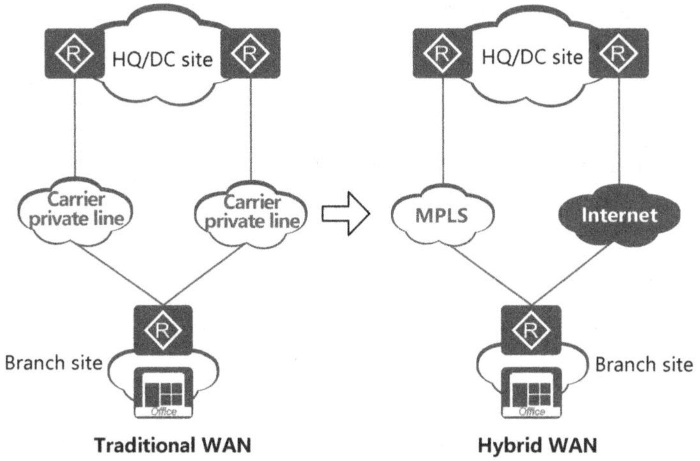

FIGURE 2.4 WAN networking transition.

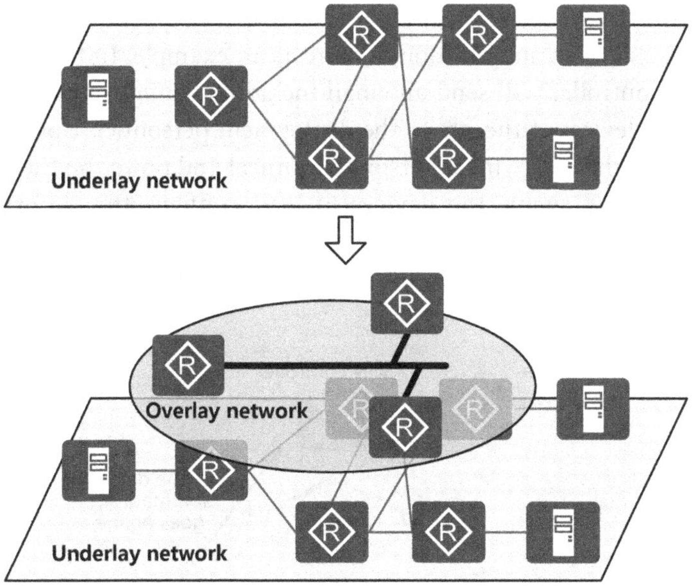

FIGURE 2.5 Overlay technology.

VPN technology adds additional VPN fields into IP packets to differentiate network domains. This ultimately isolates enterprises' private networks from carriers' public networks, and even isolates different private networks of a single enterprise.

Essentially, overlay and VPN are a pair of technologies that are deployed together.

2.2.1.2 *Device Plug-and-Play, Achieving Fast Service Rollout*

Typically, enterprises have the following requirements when it comes to network deployment:

- New branches need to be provisioned quickly.
- Network devices need to be easily deployed by most personnel without the need for professional network skills.

The plug-and-play function can help meet these requirements. In addition to this function, the SDN controller — a centralized network control system — is required for registering and managing devices. This system can deliver network configurations to devices automatically and remotely on demand. In this manner, branch services can be quickly rolled out.

We take device configuration at a site as an example. In this scenario, the SDN controller will send an email including configuration information about devices at the site to the deployment personnel. Upon receipt of the email, the deployment personnel connect and power on the devices at the site, and enable the devices to load configuration information contained in the email just with several simple operations. The network devices then automatically go online and register with the SDN controller, as illustrated in Figure 2.6.

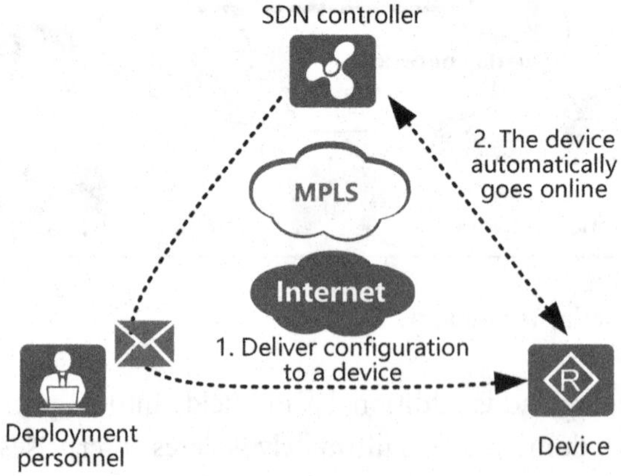

FIGURE 2.6 Plug-and-play.

The plug-and-play function can effectively reduce the skill requirements for deployment personnel, as well as shorten the branch service rollout and adjustment duration from days, or even months, to just hours, significantly improving the rollout flexibility of branch network services.

2.2.1.3 High-Performance Branch Devices, Achieving Application-Centric Full-Service Processing

To adapt to the ever-changing requirements of services, devices at enterprise branches must center around applications and provide a wide range of functions, such as VPN, QoS, application identification, network monitoring, traffic steering, security, and application optimization, posing higher requirements on device performance than ever before. At legacy branches, Customer Premise Equipment (CPE) performs well in forwarding packets at Layer 1–Layer 3 but are unable to process all services oriented to Layer 3–Layer 7 applications.

SD-WAN can resolve all these issues. In this solution, high-performance branch devices innovatively adopt a "multicore central processing unit (CPU)+network processor (NP)" forwarding architecture and are equipped with professional NP chips. These highlights make them ideal for forwarding Layer 2–Layer 4 traffic at an ultra-high speed and implementing efficient QoS. In addition, leveraging the highly open and programmable multicore CPU, the NP chip integrates a variety of hardware-based intelligent acceleration engines, including hierarchical quality of service (HQoS), application identification, and IPsec acceleration, to provide high-performance full-service processing capabilities at Layer 3–Layer 7.

2.2.1.4 Service Intent-Oriented Network Orchestration and Automated Service Rollout

Traditionally, network services are rolled out manually and statically, which complicates deployment and poses high requirements on the network personnel's skills. For example, the following operations can be performed only by highly professional, experienced network engineers: implementing network planning, configuration, and O&M, as well as inputting and provisioning the service configurations as planned on each device through the Command Line Interface (CLI) or network management system (NMS). Such high skill requirements on personnel result in low service rollout efficiency and high operating expense (OPEX).

Against this backdrop, rolling out WAN services more conveniently and quickly becomes the major aim for SD-WAN.

SDN enables the abstraction, orchestration, and automated rollout of network services on demand through the SDN controller. The SDN controller abstracts and models network services to shield the specific technical details of the network. In this way, only service-oriented interfaces and parameters are exposed. In addition, the system provides external graphical user interfaces (GUIs) or programmable APIs that are tailored for services, with which end users can use the system to orchestrate and automatically roll out network services according to their service requirements.

Figure 2.7 shows the evolution lineup of service rollout approaches.

Taking a scenario where an enterprise wants to connect its branch sites through the headquarters site, for example, the traditional approach is to manually configure the relevant network parameters, such as interfaces, IP addresses, routing protocols, security, and VPNs, on each branch device. Assuming that an enterprise has as many as 100 branch sites, each device at each site needs to be configured. Furthermore, specific routing configurations are required to enable the traffic of all branch sites to be centrally forwarded through the headquarters site, which is a complex

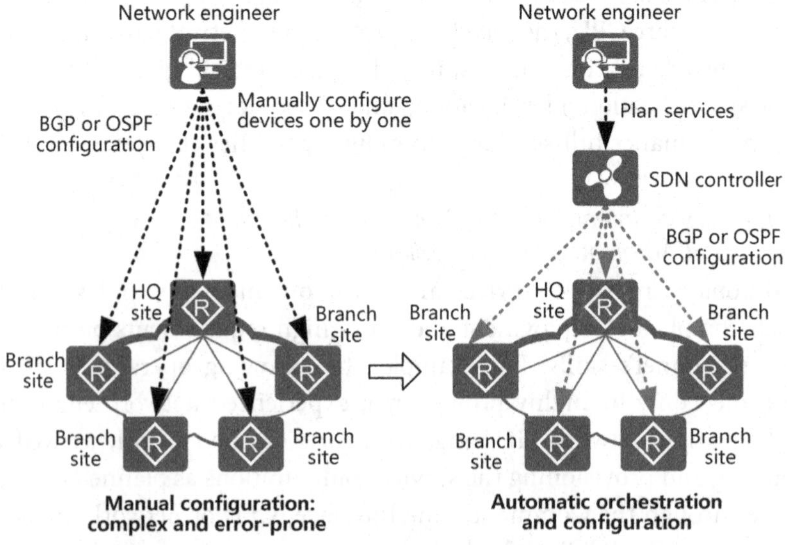

FIGURE 2.7 Evolution lineup of service rollout approaches.

configuration process and requires operators to be familiar with the technical details of the network, such as switching and routing configurations. As a result, services are rolled out at a low speed, and the configuration is prone to errors.

All these issues can be resolved with SD-WAN, which uses the SDN controller as the network control center to manage devices at all branch sites.

As the brain of the network, the SDN controller externally presents network connection operation primitives that can be understood even by users who are not specialized in networking. If SD-WAN is deployed in the preceding example, the enterprise only needs to enter an instruction — to provision 100 branch sites and connect the branch sites through the headquarters site — and does not need to enter any network parameters. In this process, the SDN controller can automatically translate this instruction into conventional network configuration operations, such as routing, VPN, and other network operations that are manually executed in traditional approaches. The system then delivers these operations to the 100 branch sites, implementing automated network service rollout.

In SD-WAN, the SDN controller automatically performs the entire process of network service rollout, reducing the network skill requirements on users and simplifying network operations. As the process is automated, errors caused by manual operations are effectively prevented, the WAN service rollout efficiency is increased, and the operation experience for users is significantly optimized.

2.2.1.5 On-Demand, Efficient Cloud Connectivity

With the advent of the cloud era, enterprise WANs need to open up their traditionally closed network architecture and flexibly connect various cloud resources. The cloud resources closely related to enterprise WANs include IaaS and SaaS.

As public clouds pick up pace among enterprises of all sizes, a broader range of enterprises are considering migrating their IT systems onto public clouds. An IT system on a public cloud can be seen as a special branch site, namely, a cloud site, as illustrated in Figure 2.8.

Similar to a legacy branch site, a cloud site also requires a device to function as a gateway to connect enterprise branches and the public cloud. Since the device must reside on the cloud, NFV is needed. In addition, such a device needs to be created quickly and connected to an enterprise

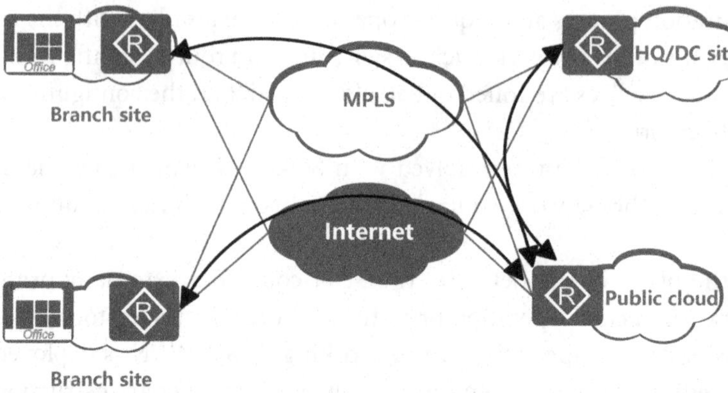

FIGURE 2.8 Connecting to the public cloud.

branch site in real time. To this end, the SDN controller remotely schedules public cloud APIs and resources, and automatically deploys cloud devices to enable branch sites to communicate with the public cloud.

To increase the efficiency of access to SaaS applications, SD-WAN must be able to provide optimal paths for accessing these applications.

In Figure 2.9, an enterprise attempts to access a remote SaaS application on the cloud through a WAN. Many paths are available for the enterprise to access the application, including the Internet link, MPLS link, or diversion through the headquarters or DC site. This requires branch sites to be able to detect the service level agreement (SLA) of each path in real time. SD-WAN not only meets this requirement but also selects and adjusts the optimal SaaS access path through the SDN controller in real time.

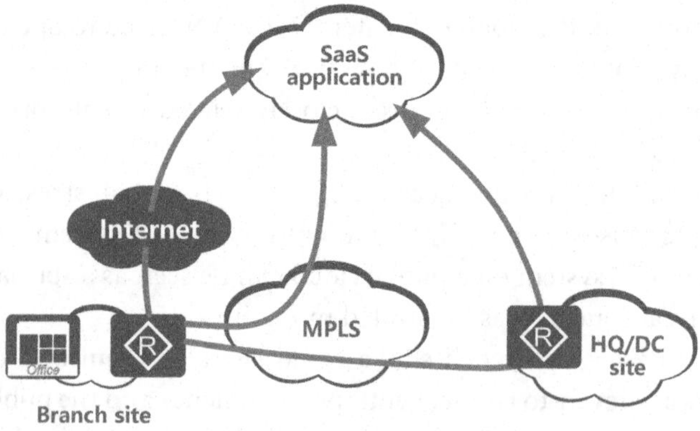

FIGURE 2.9 Accessing SaaS applications.

2.2.1.6 Intelligent Traffic Steering, Guaranteeing Application Experience

The introduction of hybrid WANs enables enterprise service traffic to be transmitted over a variety of WAN links.

The network quality varies according to WAN link, including the packet loss rate, delay, and jitter. For example, MPLS private lines can guarantee high-quality link SLAs, though they are expensive. In contrast, even though Internet links provide better network quality than before and can support bandwidth-hungry applications, they still are prone to large latency and packet loss, failing to offer the required SLA levels. For this reason, Internet links are not suitable for audio and video services, which are latency-sensitive.

To eliminate such limitations, traffic needs to be steered in accordance with SLA requirements of specific applications.

In particular, the quality of different WAN links needs to be measured, and network quality requirements, including packet loss rate, delay, and jitter, of specific applications need to be customized. Additionally, among all WAN links that meet the SLA requirements of specific applications, enterprise users can customize traffic steering policies to select the optimal WAN link for transmitting traffic of high-value applications. This delivers optimal user experience for such high-value applications.

As shown in Figure 2.10, high-value audio traffic is preferentially transmitted over the MPLS link. If the quality of the MPLS link deteriorates

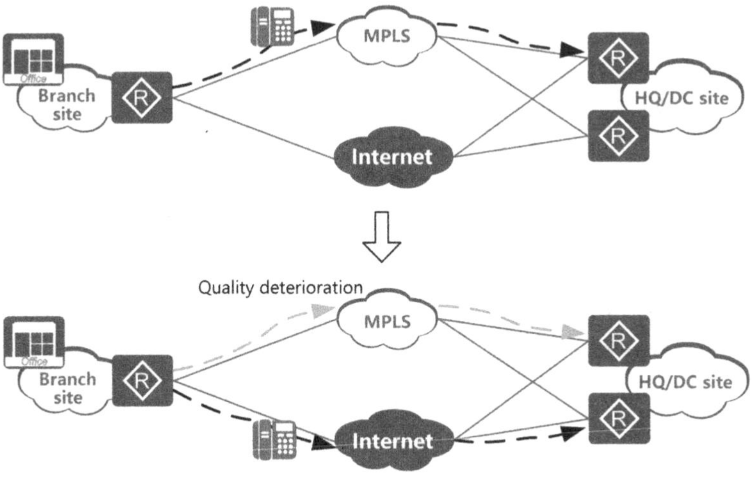

FIGURE 2.10 Intelligent traffic steering.

and can no longer meet the SLA requirements of the audio service, the SD-WAN network automatically switches the traffic to another qualified WAN link through the dynamic traffic steering capability.

2.2.1.7 WAN Optimization, Improving WAN Transmission Quality

WAN links that meet the SLA requirements of an application must be available before traffic steering can be implemented for the application. If the quality of the WAN links deteriorates due to packet loss or large latency, WAN optimization technologies are required to improve the fault tolerance on the network and ensure high-quality data transmission.

Typical WAN optimization technologies include transmission optimization, data optimization, and packet loss mitigation.

Taking Forward Error Correction (FEC), a typical packet loss mitigation technology, as an example to describe the benefits of WAN optimization for customers. FEC focuses on improving applications' tolerance to deterioration of link quality by reconstructing or optimizing data transmission protocols. With this technology, redundancy packets are transmitted together with normal data flows. This way, even if some service packets are lost due to packet loss during data transmission, the receive end can properly restore the lost packets based on the redundancy packets, ensuring integrity of data.

Figure 2.11 illustrates the FEC implementation.

To sum up, the core objective of WAN optimization is to utilize computing or storage resources to improve the performance of data transmission, thereby optimizing application experience. WAN optimization provides an effective technical means to ensure a smooth, optimal service experience for users even upon link quality deterioration.

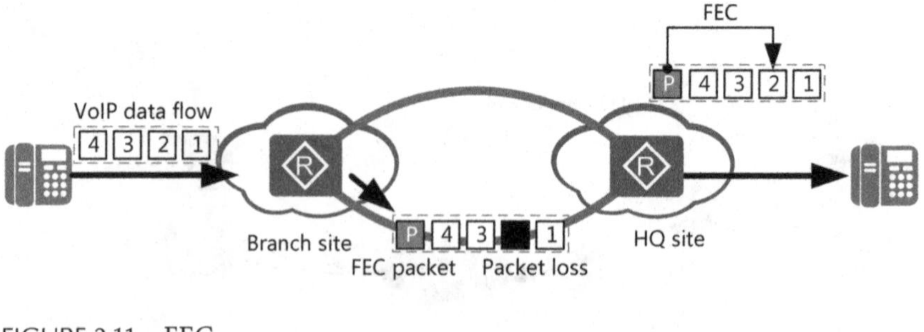

FIGURE 2.11 FEC.

2.2.1.8 Secure, Reliable Connectivity

Transmitting application data over enterprise WANs requires a comprehensive security assurance mechanism, which covers both system and service security.

To ensure the security of a network system, the network devices and SDN controller must first be secured. In most cases, NEs need to be connected to a WAN, or even to the Internet, and are faced with a variety of network intrusion and attacks. To prevent them, NEs must be equipped with basic security protection capabilities, such as attack defense.

Service security focuses on the secure transmission of enterprises' service data in and out of their networks. Typically, in order to prevent data leakage, enterprise application data needs to be authenticated and encrypted prior to transmission to the WAN. In addition, different access policies are required to schedule access between departments in an enterprise and between an enterprise's internal network and the external network. Ensuring service security in such a complex network environment requires SD-WAN to provide a wide range of security measures, such as Access Control Lists (ACLs), firewalls, and attack defense.

2.2.1.9 Centralized, Visualized Management and Control, Facilitating O&M

Enterprise WAN branches are usually geographically dispersed and large in number, thereby requiring a centralized management, control, and O&M system. Such a system must have the following capabilities:

1. Remotely managing branch devices, visually displaying WAN topologies, and remotely monitoring alarms, logs, and other key event information of each branch device in real time.

2. Receiving network performance data, including the packet loss rate, delay, jitter, and traffic statistics of key WAN links, from devices in real time.

   Enterprise users can use such data to get insights into the WAN network performance and quality of applications. In the SDN controller, the dashboard displays key network performance data, the bandwidth usages of key applications, and the health of applications, as shown in Figure 2.12.

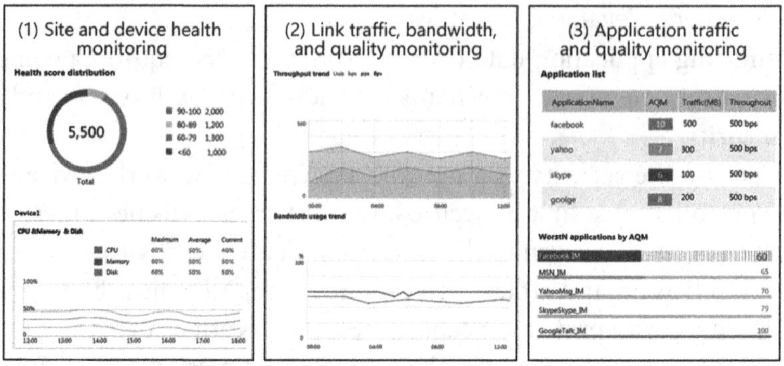

FIGURE 2.12 Visualized O&M.

3. Intelligently locating and preventing faults.

   The SDN controller can analyze and locate the root causes of key events on devices and networks, enhancing the O&M efficiency of networks. Capitalizing on professional network analysis components, the system provides advanced intelligent analysis capabilities such as application quality analysis and fault locating, predict network faults and events, as well as provide warnings or suggestions for troubleshooting accordingly.

2.2.1.10 Open Northbound APIs, Implementing Programmable SD-WAN

SD-WAN is a coherent extension of SDN that stands out for openness and programmability.

As the brain of SD-WAN, the SDN controller provides open, programmable northbound RESTful APIs to interwork with third-party business support systems (BSSs) and operational support systems (OSSs). In addition, the controller invokes its northbound APIs to create policies through software-defined methods. Third-party systems then can integrate and customize SD-WAN.

2.2.2 Core Benefits of SD-WAN

SD-WAN provides enterprise customers with the following four benefits:

1. Efficient multi-cloud and multi-network interconnection, meeting connectivity requirements of different services

2. Intelligent application identification and traffic steering, ensuring superb experience of enterprises' key applications
3. Application-centric full-service processing via high-performance branch devices, preventing congestion in services
4. Plug-and-play, centralized and visually displayed network management and control, slashing network deployment and O&M costs

Overall, SD-WAN mainly benefits enterprises in terms of helping them flexibly and efficiently build a high-quality WAN featuring robust connectivity, superior experience, high performance, easy O&M, and convenient user access anytime, anywhere.

2.2.2.1 Robust Connectivity: Flexible Networking, Building On-Demand Multi-cloud and Multi-network Connectivity

SD-WAN can fully utilize hybrid link resources. No matter where enterprise branches are located, they can access the network through a wide range of available link types including optical fibers, DSLs, and LTE links, for fast network provisioning and lower link costs. In addition, SD-WAN provides a wealth of networking models, including hub-spoke, full-mesh, partial-mesh, and hierarchical networking, to better adapt to enterprises' services.

With SD-WAN deployed, enterprises simply need to orchestrate devices and specify the topology model on the SDN controller based on the network construction intent. The SDN controller then automatically generates the corresponding network model based on the specified topology model, and controls route advertising and receiving at different sites based on the routing policy. In this manner, SD-WAN achieves flexible user isolation and mutual access based on the topology model and interconnects multiple clouds and networks on demand.

2.2.2.2 Superior Experience: Application-Based Traffic Steering and Optimization, Ensuring High-Quality Experience of Key Applications

SD-WAN can monitor the quality of links in real time, detect link connectivity, and record information including the packet loss rate, delay, jitter, and other real-time status information. It also provides multiple approaches for accurately identifying application information among ever-flowing, massive traffic. Combining these two functions, SD-WAN can dynamically adjust traffic paths based on the type of applications, increasing the flexibility and convenience of traffic scheduling.

With integrated algorithms to optimize applications, SD-WAN can further ensure high-quality experience of key applications. One example comes to the FEC algorithm that quickly recovers lost packets while minimizing the number of redundancy packets, eliminating freeze-frame or artifacts in the case of packet loss in key audio and video applications. Another example comes to the transmission optimization algorithm that minimizes the delay in file transfer and service access.

SD-WAN also ensures a high-quality experience in terms of reliability and security. On the one hand, SD-WAN achieves high reliability at the link, device, and networking levels. On the other hand, it provides comprehensive security protection measures that ensure secure and stable operation of the system architecture. All this ultimately protects enterprise services from network security threats.

2.2.2.3 High Performance: Superb Branch Devices, Ideal for Building a New Forwarding Engine

In contrast to traditional enterprise branch devices that feature packet forwarding at Layer 1–Layer 3, SD-WAN branch devices deliver application-based full-service processing at Layer 3–Layer 7. Their high-performance forwarding capabilities help enterprise branches build a new forwarding engine to ensure enterprise services operate as normal.

In SD-WAN, enterprise branch devices incorporate a wide range of functions, including VPN, QoS, application identification, monitoring, routing, security, and application optimization, to meet enterprise service requirements. When traffic bursts occur at a branch site, high-performance devices can alleviate or prevent traffic congestion.

2.2.2.4 Easy O&M: Intent-Driven, Simplified Network O&M for Branches

Inheriting the centralized control concepts and intent-driven advantages from SDN, SD-WAN enables intent-driven centralized management and control on the entire network.

In terms of centralized control, SD-WAN provides the network-wide monitoring function to obtain the status of branch links in real time, visualizing the network status. In addition, SD-WAN quickly deploys and rolls out branch services through ZTP — a plug-and-play deployment mode that significantly reduces technical barriers and saves labor costs by eliminating the need for network engineers to manually deploy services onsite.

SD-WAN also provides different methods of O&M to minimize operation permissions, and users can set different access permissions for different roles.

Furthermore, SD-WAN integrates a wide array of fault diagnosis tools to quickly demarcate complex issues. Compared with traditional enterprise WANs, SD-WAN provides automated and intelligent O&M capabilities, reducing management costs and improving O&M efficiency.

CHAPTER 3

Introduction to the SD-WAN Solution

THIS CHAPTER STARTS BY providing an overview of the SD-WAN solution, exploring the general architecture, principles, key products and components, key interaction interfaces and protocols, and key service processes.

It then describes the main characteristics of SD-WAN business models and solution deployment from the perspective of two business models: carrier resale and enterprise-built. This chapter concludes by looking at a panorama of the SD-WAN solution, laying a solid foundation for subsequent SD-WAN solution design.

3.1 SD-WAN SOLUTION

3.1.1 SD-WAN Architecture Design

The logical architecture of the SD-WAN solution consists of the network layer, control layer, and service layer. Each layer provides different functions and contains several core components, as illustrated in Figure 3.1.

3.1.1.1 Network Layer

From the perspective of services, enterprise sites are composed of enterprise branches, headquarters, DCs, and cloud-based IT infrastructure. The network layer consists of two parts: network devices used for WAN interconnection at different sites and the intermediate WANs.

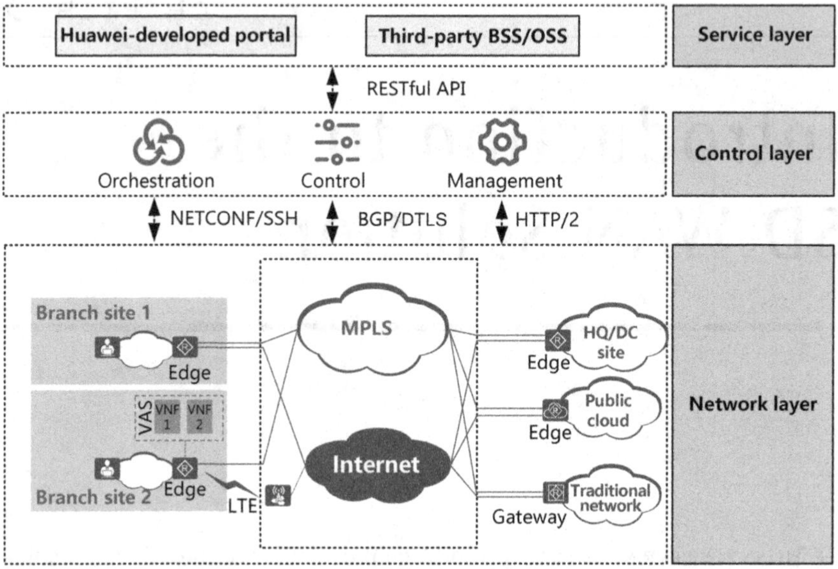

FIGURE 3.1 Hierarchical architecture of SD-WAN.

According to network function, enterprise SD-WAN can be divided into a physical (underlay) network and a virtual (overlay) network, each of which is decoupled from the other, as shown in Figure 3.2.

1. Physical network

   A physical network refers to a WAN created and maintained by carriers. Examples of such a network include a private line network, MPLS network, and the Internet. Typically, the performance of physical networks and devices continuously evolves according to Moore's Law to provide ultra-broadband access and forwarding capabilities.

2. Virtual network

   The SD-WAN solution adopts the virtualization technology — overlay — to construct one or more overlay networks over one underlay network. Service policies are deployed on the overlay networks in order to decouple complex services from complex WAN interconnection. One or more overlay networks can serve different services (e.g., from multiple departments) of one or different tenants. This is the core networking technology at the network layer of SD-WAN.

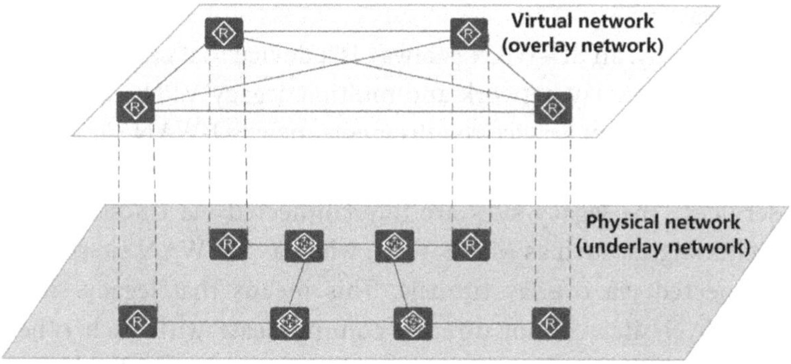

FIGURE 3.2 Decoupled physical network and virtual network.

From the perspective of network device functionality, the network layer of SD-WAN is mainly constructed using two types of network elements (NEs), namely, edge and gateway. In addition, VAS devices are deployed on demand according to service requirements.

3. Edge

SD-WAN edge is the egress CPE of an enterprise's SD-WAN site. At least one edge is deployed at each enterprise site (including cloud sites) for site interconnection. To ensure network reliability, two edges can be deployed at a site. Each edge can connect to one or more WAN links, such as traditional private lines, MPLS private lines, and Internet links.

In essence, edges are SD-WAN endpoints that are interconnected through the overlay tunneling technology. To ensure the security of enterprise services transmitted on a WAN, a data encryption technology, such as IPsec, is used on overlay tunnels. In addition, to improve the transmission quality and experience of different enterprise applications, edges need to provide a wide range of functions, such as application identification, application-based traffic steering, QoS, WAN acceleration, and security.

An edge can be a traditional hardware CPE, a universal CPE (uCPE), or a virtual CPE (vCPE) that is usually deployed at public cloud sites.

In SD-WAN, the SDN controller is used for unified management of all edges, which are considered as tenants in terms of management and are created, managed, and maintained by tenant administrators.

4. Gateway

   Typically, an SD-WAN gateway is a device that can connect to an SD-WAN overlay network and multiple legacy VPNs. For example, in order to meet service requirements, new SD-WAN sites deployed by an enterprise must interwork with its legacy sites or third-party services. The legacy sites are interconnected via traditional WAN technologies, such as MPLS VPN, whereas SD-WAN sites are interconnected via overlay tunnels. This means that legacy sites and SD-WAN sites cannot directly communicate with each other. To enable communication between a legacy network (MPLS VPN) and an SD-WAN overlay network, an SD-WAN gateway can function as an intermediate device. The role of a gateway varies depending on service scenarios. For example, a gateway connected to a legacy site is referred to as an Interworking Gateway (IWG), whereas one connected to a public cloud network is called a cloud gateway. Such gateways support function extension and can interconnect with each other to construct a Point of Presence (POP) network, in which they are called POP gateways.

   A gateway can be regarded as a special edge that can be managed by the SDN controller in a unified manner. In terms of device form, deployment mode, and software features, a gateway is similar to an edge. In addition, a gateway can be shared between multiple enterprises and tenants, and is provided by a carrier or a Managed Service Provider (MSP). As such, gateways are created, managed, and maintained by network administrators of carriers or MSPs, unlike edges that are created, managed, and maintained by tenant administrators.

5. VAS

   Both edges and gateways are equipped with basic SD-WAN interconnection functions to implement WAN interconnection between enterprise sites and applications. VASs, typified by firewall functions and WAN optimization, can be deployed in SD-WAN on demand according to customers' service requirements.

   VASs are managed as independent NEs that can be deployed using one of two methods. One method is to deploy virtual network functions (VNFs) in the uCPE, and the other is to connect physical firewalls to edges or gateways in off-path mode to perform access control. Both in-house and third-party VASs are supported. Specifically, a third-party VAS management system is provided and deployed by a VAS device provider.

3.1.1.2 Control Layer

The SDN controller — the core component at the control layer — functions like a brain of the SD-WAN solution. According to the SD-WAN architecture defined by MEF[2], the SDN controller, in a broad sense, typically provides three functions: network orchestration, control, and management. These functions are described as follows:

1. Network orchestration

   The orchestration component of the SDN controller is responsible for service-oriented network model abstraction and orchestration, as well as automated configuration provisioning. All network-related configurations in SD-WAN should center on enterprises' applications or services, and personnel not versed in network technologies should be able to use them easily. To this end, the SDN controller abstracts and defines the network model of enterprise WANs, and shields users from technical details of SD-WAN deployment and implementation through modeling. In this way, users can use the northbound service orchestration GUI of the SDN controller as well as languages and interfaces matching enterprises' applications and services to drive the SDN controller to implement simple, flexible network configuration and automated service provisioning.

   Specifically, service orchestration in SD-WAN can be classified into two categories. One category is related to enterprise WAN networking, such as SD-WAN site creation, WAN link creation, VPN creation, and VPN topology definition. The other category is related to network policies, such as application identification, application-based traffic steering, QoS, and WAN optimization policies. The correlated service models and orchestration solutions will be described in subsequent chapters.

   The network orchestration component is open to external systems through RESTful APIs in the northbound direction and delivers orchestrated network configurations to network-layer devices through Network Configuration Protocol (NETCONF) in the southbound direction.

2. Network control

   The control component of the SDN controller is responsible for centralized control of the network layer in SD-WAN and implements on-demand interconnections of enterprise WANs based on user intents. The component provides a variety of functions, such as distributing and filtering VPN routes of an SD-WAN tenant, creating and modifying VPN topologies, and driving the creation and maintenance of overlay tunnels between sites or gateways. Compared with the distributed control mode of legacy networks, the centralized control mode separates the control plane from the forwarding plane, simplifying enterprise WAN O&M, minimizing network configuration errors, and optimizing enterprise WAN O&M efficiency.

   To improve network scalability and support large-scale interconnection of branch sites across different regions, the network control component supports the distributed, independent deployment mode and smooth expansion by network scale.

   In addition to providing a wide range of logical functions, the network control component also provides diversified deployment forms. Specifically, the component can be an independent SDN controller (SDN controller in a narrow sense) or a traditional network device with enhanced functions, such as a route reflector (RR) defined in BGP.

3. Network management

   The management component of the SDN controller implements management and O&M of enterprise WANs. For example, the component collects alarms and logs of SD-WAN NEs; collects and analyzes performance statistics of links, applications, and networks; and intuitively presents a multidimensional display of O&M information such as network topologies, faults, and performance. For device management, this component typically uses NETCONF to collect O&M information, such as alarms, logs, and events, and uses HTTP/2 to collect performance data.

The SDN controller has the following core capabilities in terms of functions and architecture:

- Service-oriented network orchestration and automated provisioning

- Centralized network control, and separation of the control plane and forwarding plane
- Network management capabilities of the legacy NMS

In reality, many SD-WAN solution providers are unable to support all of these core capabilities, supporting only one or two. Strictly speaking, an SD-WAN solution that lacks one or more of these capabilities cannot be regarded as a true SD-WAN solution.

Different from the narrow sense of an SDN controller that provides only network functions, the broad sense of an SDN controller means that the controller supports all three capabilities. When constructing an SDN controller that supports all of these capabilities, an SD-WAN solution provider faces a variety of choices in products and product deployment forms. As emphasized by MEF, the three core capabilities can be integrated into one SDN controller or implemented independently through three different components. An SDN controller that supports all three requires only one product component and is easy to deploy. Such an SDN controller is used throughout the book, unless specified otherwise.

Huawei's SDN controller — iMaster NCE-WAN — is a typical example of an SDN controller that supports all three core capabilities. It is ideally suited to implementing SD-WAN orchestration, control, and management. Figure 3.3 illustrates the overall architecture of Huawei's SDN controller.

As the brain of Huawei SD-WAN solution, the SDN controller provides the following crucial functions and quality attributes:

- **High reliability**: Capitalizing on the cloud computing and distributed architecture design, the SDN controller is deployed in 2+1 cluster mode to ensure service continuity if a single server component fails. In addition, geographic redundancy ensures that services can be switched to the standby controller in a timely manner if the active controller fails, thereby providing high reliability.
- **Operability**: Carriers and MSPs need to operate the SD-WAN solution so that they can resell it to or serve their enterprise users. To meet the requirements carriers and MSPs impose on the operability and manageability of SD-WAN, the SDN controller provides relevant functions, such as multi-tenant management, deployment,

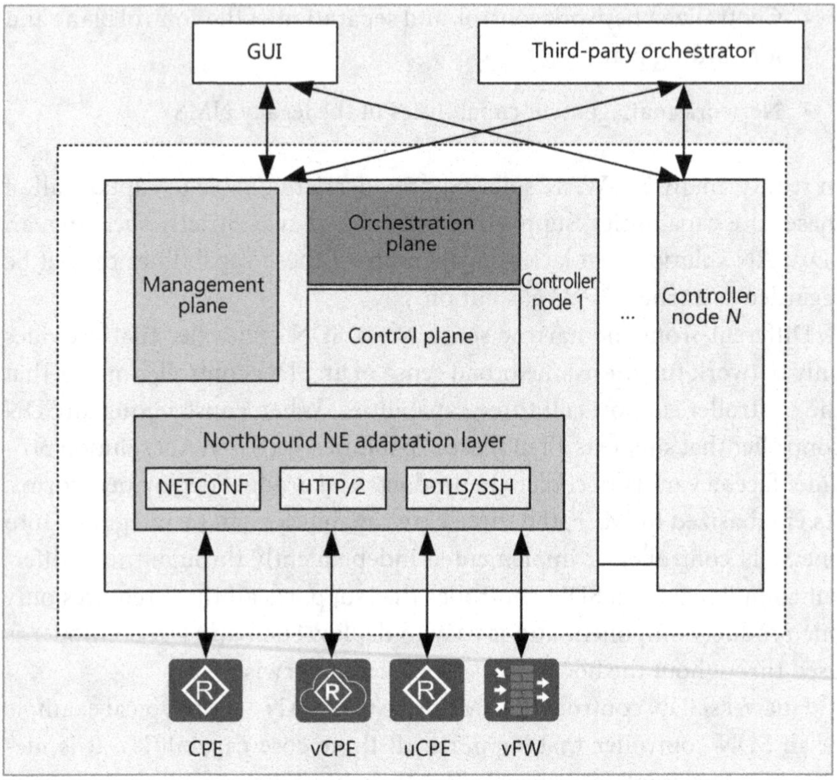

FIGURE 3.3 Logical architecture of Huawei's SDN controller.

multi-tenant gateway, and QoS. The SDN controller can also manage numerous tenants as well as large numbers of network devices belonging to a single tenant, ensuring that the SD-WAN solution is truly operable in terms of scale and scalability.

- **Cloud-based deployment**: The SD-WAN solution can be deployed in multiple modes, such as on-premises deployment and cloud-based deployment, to meet the requirements of users who want to construct and operate SD-WAN themselves. In the on-premises deployment mode, enterprises purchase and deploy the SDN controller in their own DCs or private clouds, giving enterprises exclusive use of the controller. In the cloud-based deployment mode, the SDN controller can be deployed, on demand, on various mainstream public clouds through VMs, thereby lowering requirements for enterprises' network infrastructure while also accelerating the SDN controller deployment. One cloud-based SDN controller can serve multiple MSPs concurrently and provide SD-WAN management and O&M services for their enterprise customers, implementing SD-WAN management as a service.

- **Openness**: The SDN controller provides a unique, built-in network orchestration engine to orchestrate and automatically deploy enterprise WANs and service policies. In addition, the SDN controller provides northbound, open RESTful APIs to interwork with third-party service systems of enterprises or carriers, such as BSSs and OSSs. The controller can also function as a service system of another network platform, and invoke open APIs of the platform to perform inter-network platform service orchestration. Take Amazon Web Services (AWS) — a mainstream cloud computing service platform —as an example. The SDN controller can invoke the northbound APIs of AWS to orchestrate public cloud network services, enabling vCPEs to be automatically provisioned and connecting branches to clouds.

3.1.1.3 Service Layer

The service layer of SD-WAN interconnects with the SDN controller to present and provision SD-WAN services through service interfaces. Typically, there are two types of service interfaces. One type lies in the in-house portal that contains a comprehensive, end-to-end (E2E) service configuration and provisioning process and is provided by SD-WAN solution providers. This in-house portal can be directly deployed and used by enterprise customers, and is required for solution providers to demonstrate their SD-WAN solutions. The other type comes to the northbound open API. That is, the SDN controller can be integrated with third-party service orchestration systems belonging to carriers or enterprises, such as the BSS/OSS, through such APIs. The third parties can customize the portal and service provisioning process according to their requirements for service functions and display style.

3.1.2 Key Interaction Interfaces and Protocols

Components involved in the SD-WAN solution need to collaborate with each other to provide a wealth of functions. With an SDN controller as its brain, the SD-WAN solution provides numerous interfaces and protocols.

In SD-WAN, the SDN controller uses management protocols, such as NETCONF and HTTP/2, in the southbound direction to implement unified configuration and management of the network layer, and uses control protocols such as BGP or OpenFlow to control the network layer in a unified manner. Additionally, the controller provides northbound RESTful APIs to integrate third-party service orchestration systems. As a network orchestrator, the SDN controller can integrate with third-party services through RESTful APIs provided by other service systems. The entire SD-WAN solution therefore centers on the SDN controller, which is crucial to normal operations.

These interfaces and protocols can be classified into three types of system channels, namely, management channel, control channel, and data channel, as illustrated in Figure 3.4.

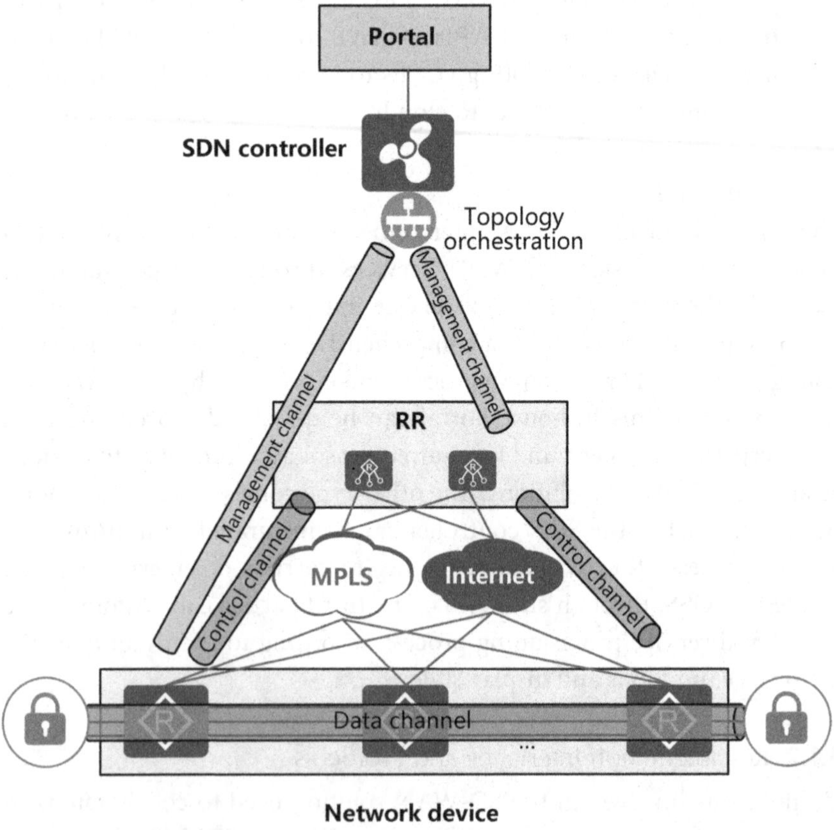

FIGURE 3.4 System channels in the SD-WAN solution.

3.1.2.1 Management Channel

Management channels are used for network configuration and O&M between the SDN controller and network devices, such as edges and gateways. After a network-layer device powers on, the device registers and establishes a management channel with the SDN controller.

Take a management channel between the SDN controller and an edge as an example. Through this channel, the controller delivers configurations to the edge (such as basic SD-WAN configurations, VPN service parameters, traffic steering, QoS, and security policies), and the edge collects and reports information (such as alarms, logs, and network traffic of network devices) required for network O&M to the controller.

The management channel is implemented as follows: The SDN controller delivers configurations to network devices through NETCONF over Secure Shell (SSH), safeguarding data transmission. Network devices use HTTP/2 to report performance data and use NETCONF to report alarms to the controller. Bidirectional certificate authentication is mandatory for network device login and registration. The system software and signature database of network devices are transmitted using Hypertext Transfer Protocol Secure (HTTPS). HTTPS and HTTP/2 packets are encrypted using Transport Layer Security (TLS) to ensure data transmission security. Tenants or MSPs log in to the SDN controller through HTTPS and establish connections with the controller, ensuring that tenants and MSPs can securely access the controller.

3.1.2.2 Control Channel

After a management channel is established, network-layer devices need to orchestrate networking and routes. As described above, the SDN controller enables centralized control, and defines and controls establishment of the topology and routes of the forwarding plane at the network layer. This means that a control channel needs to be established between network-layer devices and the controller. All policy information related to network forwarding paths, including the VPN topology, routing, and tunnel information, is delivered from the SDN controller to network-layer devices through the control channel.

In addition to enabling centralized control, the control component must also provide horizontal scale-out based on the network scale. It is recommended that MP-BGP EVPN be used to establish the control channel. As such, a BGP RR with enhanced functions, as part of the control plane in SD-WAN, is introduced in most cases to perform network control. Typically, an RR is deployed in distributed mode and interworks with the SDN controller to implement centralized network control. The RR can be an independently deployed CPE or vCPE, or it can be co-deployed with an edge or gateway at a legacy SD-WAN site.

3.1.2.3 Data Channel

After the management and control channels are set up, data channels need to be set up between edges and gateways for data transmission between enterprise sites and gateways. Data channels are established via overlay technology. To safeguard data transmission, IPsec is required for data encryption.

Among the three system channels, the management channel is the first to be established and is crucial to initial operations of the entire system. The new interfaces and protocols used by the management channel are also beyond the reach of traditional IP networks.

In summary, the SDN controller connects to service-layer software through northbound RESTful APIs and interacts with network devices via NETCONF in the southbound direction. The following describes two major protocols tailored to SD-WAN, namely, NETCONF and RESTful, and also explains the data modeling language used in NETCONF — Yet Another Next Generation (YANG).

1. NETCONF

   As networks continue to grow and new technologies, such as cloud computing and Internet of Things (IoT), gain popularity in numerous industries, conventional network management approaches typified by CLI and Simple Network Management Protocol (SNMP) can no longer meet customers' requirements for fast provisioning and innovation of network services.

   The CLI, as a man-machine interface, provides a portal for users to enter instructions. It parses the instructions and executes the corresponding commands already recorded in network devices, and its implementation may vary depending on device vendors. Against this backdrop, the CLI lacks a structured mechanism for displaying error messages and outputting command execution results, complicating network O&M.

   Different from the CLI, SNMP is a machine-machine interface that consists of a set of network management standards, such as application-layer protocols, database models, and a group of data objects. It facilitates device monitoring and management. To date, SNMP embraces the widest applications among all network management protocols on Transmission Control Protocol (TCP) or IP networks. As a protocol implemented based on User Datagram Protocol (UDP), SNMP is not oriented at configurations and lacks a secure, effective mechanism for committing configuration transactions. It is therefore used mostly for performance monitoring rather than network device configuration.

NETCONF was developed to address the shortcomings of the CLI and SNMP. As a new network management protocol implemented based on Extensible Markup Language (XML), NETCONF provides a set of programmable network device management mechanisms. This makes it ideal for meeting numerous core network management requirements, including ease of use, differentiated configuration data and status data, service- and network-oriented management, and configuration data import and export. It also meets requirements for configuration consistency check, standard data model, multiple configuration sets, and role-based access control.

NETCONF uses a client/server model in which the client and server communicate with each other using remote procedure call (RPC). The messages exchanged between the client and server are XML-encoded. NETCONF supports mature, secure transport protocols and can be extended through proprietary functions, offering high-level flexibility, reliability, scalability, and security. NETCONF can work with YANG to implement model-driven network management and automated network configuration with network programmability, simplifying network O&M and accelerating service provisioning. Furthermore, NETCONF enables configuration transactions and configurations to be imported or exported, and supports pre-deployment tests, configuration rollbacks, and free configuration switchovers, making it an ideal choice for cloud-based scenarios such as SDN and NFV.

a. NETCONF architecture:

   Figure 3.5 illustrates the fundamental architecture of NETCONF. In the architecture, at least one NMS must be deployed as the network-wide management center. The NMS

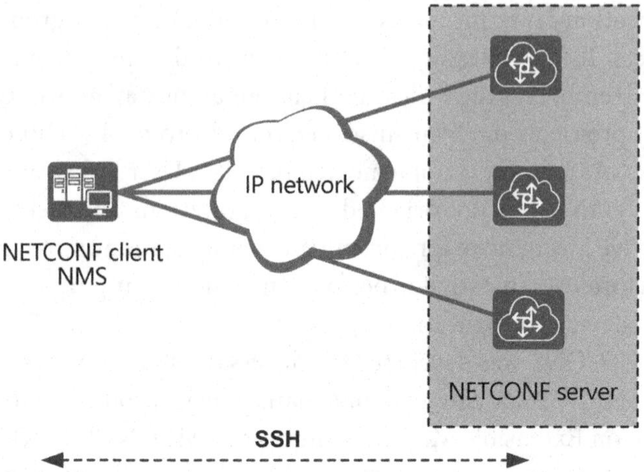

FIGURE 3.5 NETCONF architecture.

manages devices on the network. The major elements required for implementing NETCONF are the NETCONF client and the NETCONF server.

i. **NETCONF client**: systematically manages network devices using NETCONF. In most cases, the NMS functions as a NETCONF client and sends NETCONF requests enclosed by the <rpc> element to a NETCONF server to query or modify configuration data. The client can learn the status of a managed device based on the alarms and events reported by the server.

ii. **NETCONF server**: maintains information about managed devices, responds to requests from clients, and reports management data to the clients. NETCONF servers are typically network devices, for example, switches and routers. After receiving a NETCONF request from a client, a server processes the request with the assistance of the Configuration Management Framework (CMF), and then returns a response to the client. If a fault alarm is generated or an event occurs on a managed device, the NETCONF server reports the alarm or event to the client through <notification> messages. In this manner, the client can learn status changes of the managed device.

A client and server establish a connection based on a secure transmission protocol, such as SSH or TLS, and then exchange their respective capabilities through <hello> messages so that they can establish a NETCONF session. Since request exchange between a NETCONF client and server relies on NETCONF sessions, a network device that functions as a NETCONF server must support at least one NETCONF session. In this way, the client can exchange requests with the server.

A NETCONF client can typically process configurations and status data it obtains from a NETCONF server as follows:

i. The client can manipulate configurations data to make the NETCONF server operate in the user-expected status.

ii. The client cannot modify status data. Status data includes the running status of the NETCONF server and other statistics.

b. NETCONF protocol structure

NETCONF adopts a hierarchical architecture where each layer encapsulates certain NETCONF functions and provides services for its upper layer. This structure enables each layer to focus only on a single aspect of NETCONF and reduces the dependencies between different layers. In this way, the impact that internal implementation imposes on other layers can be minimized. NETCONF can be divided into four logical layers, as shown in Figure 3.6.

Table 3.1 lists each layer of NETCONF.

There are three typical scenarios for applying NETCONF in SD-WAN.

Scenario 1: The SDN controller uses NETCONF to manage network devices.

The following example focuses on how an edge registers with the SDN controller via NETCONF. After obtaining an IP address through DHCP and connecting to the network, the edge establishes a connection with the controller. Figure 3.7 illustrates the entire process.

i. The edge functions as a NETCONF server and proactively establishes a persistent TCP connection with the controller that functions as a NETCONF client.

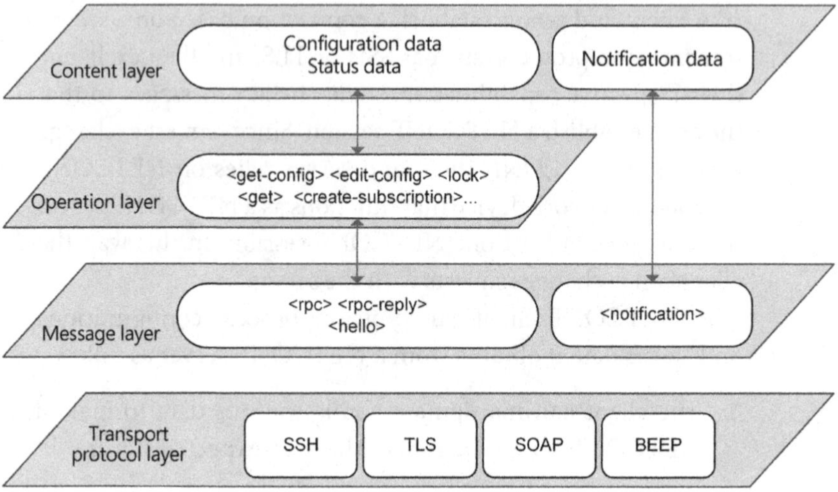

FIGURE 3.6 Layers of NETCONF.

ii. The server and client establish an SSH session, verify each other's certificates, and establish an encrypted channel.

iii. The client and server exchange <hello> messages to notify each other of the supported capability set.

iv. The client obtains the data model file supported by the server.

v. (Optional) The client initiates an event subscription. This step is required only when the server needs to report alarms and events to the server.

vi. (Optional) The client initiates full synchronization to ensure data is consistent between the client and server.

vii. The client initiates an RPC request for configuration or data query and processes the RPC response returned by the server.

Scenario 2: The SDN controller delivers configurations to devices through NETCONF.

When an edge registers with the SDN controller, the SDN controller orchestrates services and delivers configurations to the edge. Leveraging the advantages of NETCONF (model-driven

TABLE 3.1 Layers of NETCONF

Layer	Example	Description
Transport protocol layer	SSH, TLS, SOAP, and BEEP	The transport protocol layer provides a communication path for interaction between the client and server NETCONF can be layered over any transport protocol that meets the following requirements: **Connection-oriented**: NETCONF is connection-oriented, requiring a persistent connection between the NETCONF client and server. This connection must provide reliable, sequenced data delivery **Reliable**: NETCONF connections must provide authentication, data integrity, and confidentiality **Session distinguishing**: The transport protocol provides a mechanism to distinguish the session type (client or server) for NETCONF
Message layer	<rpc>, <rpc-reply>, <hello>, and <notification>	The message layer provides a simple RPC request and response mechanism that is independent of the transport protocol Typically, a NETCONF client sends RPC requests enclosed in the <rpc> element to a NETCONF server After receiving the requests, the server sends back RPC responses enclosed in the <rpc-reply> element to the client. The responses carry information at the operation and content layers The <notification> message has been added since NETCONF 1.0. Only network devices with NETCONF 1.0 or later enabled support the <notification> message
Operation layer	<get-config>, <edit-config>, <lock>, <get>, and <create-subscription>	The operation layer defines a series of basic RPC operations that constitute basic capabilities of NETCONF
Content layer	Configuration data, status data, and notification data	The content layer describes configuration data involved in network management. The configuration data heavily depends on vendors' devices So far, only the content layer has not been standardized for NETCONF. The content layer has no standard NETCONF data modeling language or data model. Common NETCONF data modeling languages include Schema and YANG. YANG, in particular, is written for NETCONF

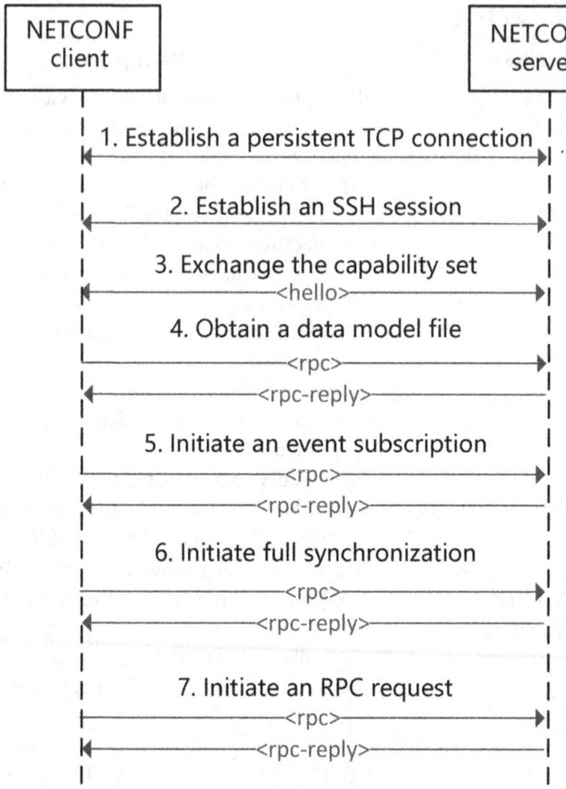

FIGURE 3.7 NETCONF interaction process.

management, programmability, and configuration transaction), the controller can automatically orchestrate service configurations based on the network model built by users and the available site and site template configurations, thus implementing software-defined networking.

Scenario 3: The SDN controller obtains the device status through NETCONF

When a device registers with the SDN controller, the controller needs to display the running status of the device, such as the CPU usage, memory usage, equipment serial number (ESN), registration status, and alarm information. The controller typically uses two methods to detect device status: proactive query and <notification> message reporting.

3.1.2.4 YANG Model

Since no operation data model is defined in NETCONF, and conventional data modeling languages cannot satisfy NETCONF requirements, we are in urgent need of an all-new data modeling language. Specifically, this language must be:

- Decoupled from the protocol mechanism.
- Easily parsed by computers.
- Easy to learn and understand.
- Compatible with the legacy Structure of Management Information (SMI).
- Capable of describing the required information and operation models.

YANG is just such a language and can be used to model the operation and content layers of NETCONF. YANG describes objects in a tree structure, which is similar to the management information base (MIB) of SNMP. MIB is described by Abstract Syntax Notation One (ASN.1) and is more flexible than SNMP. SNMP defines layers of a tree structure in a rigid manner and therefore has a limited application scope. Particularly, YANG claims to be compatible with SNMP.

The YANG model defines a hierarchical data structure dedicated to NETCONF-based operations, including configuration data, status data, RPCs, and notifications. The YANG model presents an advanced view of the data model, facilitating users' understanding about how data is encoded into NETCONF operations.

This is why YANG becomes the mainstream modeling language in the industry shortly after its emergence. YANG is also an extensible language that welcomes standard organizations, vendors, and individuals to customize and extend upon it.

3.1.2.5 RESTful API

The SDN controller interworks with the service layer through northbound RESTful APIs, including basic network APIs, network policy APIs, and network O&M APIs.

Before we go into detail about the APIs, we need to find out what RESTful is. Services that comply with the constraints and rules of

Representational State Transfer (REST) are regarded as having a RESTful architecture. REST is a software architectural style that defines a set of constraints to be used for creating web services. All things on a network can be abstracted into resources, each identified by a unique Uniform Resource Identifier (URI). Client operations on resources are stateless in that no session is stored on the server, and URIs of the resources cannot be altered. Standard methods are used to operate on resources. A RESTful architecture aims to make better use of the guidelines and constraints in legacy web standards.

RESTful APIs are designed in compliance with REST. External applications can access RESTful APIs via HTTP to implement a broad variety of functions, such as service provisioning and status monitoring. For security purposes, RESTful APIs support only HTTPS.

Standard HTTP methods for accessing managed objects include GET, PUT, POST, and DELETE, as described in Table 3.2.

3.1.3 Key Service Processes

This section focuses on the initialization of an SD-WAN system, which lays a solid foundation for smooth operations of the entire system. Figure 3.8 illustrates the process.

The SD-WAN system is initialized in five steps:

1. The network administrator defines services via the Portal of the SDN controller and invokes RESTful APIs to instruct the orchestration component of the SDN controller to orchestrate network services. Network services to be orchestrated include sites, network topologies, VPNs, and various service policies (such as routing, QoS, and security policies). To better support automated orchestration and provisioning of network topologies, the SDN controller can orchestrate and provision multiple network topologies.

TABLE 3.2 Functions of the HTTP Methods for Accessing Managed Objects

Method	Description
GET	Queries specified managed objects
PUT	Modifies specified managed objects
POST	Creates specified managed objects
DELETE	Deletes specified managed objects

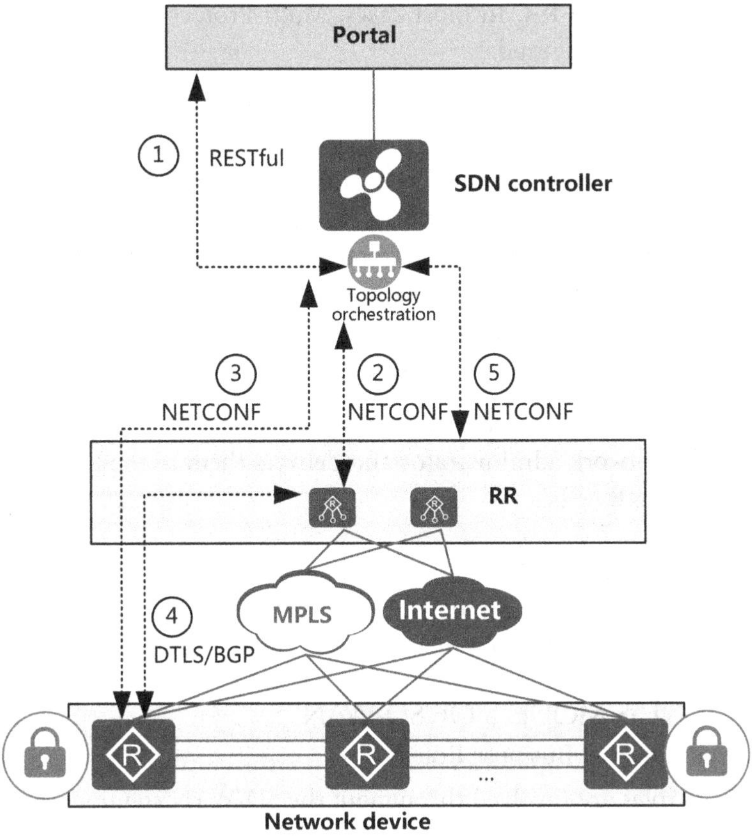

FIGURE 3.8 Widespread system initialization process.

2. The RR goes online and registers with the SDN controller. The controller allocates a globally unique management IP address (router ID) to the RR, facilitating identification and management of the RR.

3. A network device goes online and registers with the SDN controller. The SDN controller allocates a device management IP address (router ID) to the network device. The management IP address is used to identify and manage the network device. In addition, the SDN controller allocates an RR to the network device according to obtained information about the network device (such as the geographical location, WAN interface, and IP address). The SDN controller automatically configures management routes between the network device and the RR, and then configures a neighbor relationship between

the device and RR. In most cases, Multi-Protocol BGP (MP-BGP) routes are configured.

4. The network device registers with the RR. Typically, network devices use a secure communication protocol, such as DTLS, to register with the RR. Network devices notify the RR of their WAN interfaces and IPsec Security Association (IPsec SA) information. The RR also returns its own WAN interface and IPsec SA information to the network devices. Under the management of the SDN controller, the RR and network devices establish control channels with each other via a tunneling protocol, such as MP-BGP.
5. The SDN controller orchestrates the service-oriented policies defined by the network administrator and delivers them to the RR through NETCONF APIs.

The RR then delivers VPN topology, routing, and tunnel information between sites based on the policies defined by the network administrator, achieving secure, on-demand interconnection between sites.

3.2 BUSINESS MODELS OF SD-WAN

3.2.1 Definition of Business Roles

Numerous roles are involved throughout the SD-WAN solution, ranging from design, to development, and finally to actual use. The SD-WAN solution is available in multiple business models to suit the needs of different customers. These are illustrated in Figure 3.9.

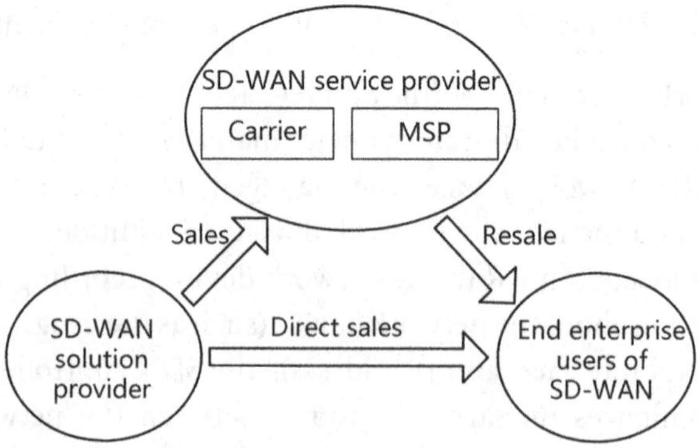

FIGURE 3.9 Roles and business models of SD-WAN.

3.2.1.1 SD-WAN Solution Provider

Typically, an SD-WAN solution provider is a device vendor that focuses on providing comprehensive SD-WAN solutions and products, including the SDN controller, network-layer edges, and gateways, in both hardware and software forms.

SD-WAN solution providers can be divided into three types:

- **Conventional network device vendors**: include vendors that provide SD-WAN solutions capitalizing on their expertise in conventional enterprise WAN products and solutions, such as Huawei, Cisco, and Juniper.
- **Conventional WAN optimization and WAN security vendors**: include vendors that reconstruct and upgrade their conventional products and claim to be capable of providing SD-WAN solutions, such as mature WAN acceleration vendors (like SilverPeak and Riverbed), and mature security vendors (like Fortinet).
- **Emerging companies focusing on providing SD-WAN solutions**: such as Velocloud and Versa.

3.2.1.2 End Users of the SD-WAN Solution

Enterprises are typical end users of SD-WAN. They can be classified by network scale, economic scale, and industry attributes. Different types of enterprises have diversified industry characteristics and service requirements. For example, differing in their network scale, economic strength, and geographical reach, enterprises can be divided into the following three types: small and midsize enterprises, large enterprises, and multinational enterprises. The following definitions are for reference:

1. Small and midsize enterprises

   Small and midsize enterprises can have as few as just one branch site. The following are some of the main features of such enterprises:

   a. There are no more than 100 employees, annual revenue is at most 10 million CNY, and there are fewer than 10 sites.

   b. Costs of WAN private lines must be kept low. There are no dedicated O&M personnel for the WAN, which is hosted by a third party.

c. Branch intercommunication is easy, mainly involving north-south traffic between branches and headquarters or DCs (if any).

d. Basic security features are required, with the built-in security features of the edge generally sufficient.

e. SaaS applications need to be accessed.

f. Basic alarm, monitoring, and visualization functions are required.

g. Plug-and-play is required at branch sites.

2. Large enterprises

Large enterprises spread their reaches across industries, such as financial services, manufacturing, and business sectors. The following are some of the main features of large enterprises:

a. There are tens to thousands of sites, with annual sales of tens of millions of CNY.

b. Carriers' private lines (such as MPLS) are used, with the desire to use the Internet to reduce the cost of private lines.

c. Single-layer and hierarchical network topologies are both used, with both north-south and east-west traffic being transmitted.

d. Public clouds and SaaS clouds need to be accessed.

e. WAN optimization is required for a better application experience.

f. Centralized, visualized O&M is required.

g. Plug-and-play is required at branch sites.

3. Multinational enterprises

Multinational enterprises are a type of large enterprises with a wide geographical span. Typically, they carry out services in multiple countries and regions. To better support their services, such enterprises have high requirements on WAN construction and service experience. The following are some of the main features of multinational enterprises:

a. There are tens to thousands of sites spread across the globe.

b. High-value traffic requires the MSP to provide private line services spanning the globe.

c. WAN optimization is required for a better application experience.

d. Centralized, visualized O&M is required.

e. Plug-and-play is required at branch sites.

Classifying enterprises facilitates summarization and understanding of enterprises' service characteristics, as well as the core requirements of enterprise services for SD-WAN.

3.2.1.3 SD-WAN Service Provider

The intermediary responsible for reselling SD-WAN solutions is known as an SD-WAN service provider, which is typically a carrier or MSP. Carriers and MSPs keep ties with numerous enterprise customers and have a large installed base in the To Business (2B) market. As such, many enterprise customers appeal to WAN lines, devices, and enterprise WAN construction and O&M services offered by carriers and MSPs. To date, it is the most popular business model. As SD-WAN picks up pace among enterprises of all sizes, this model will continue to exist and embrace further development, leveraging the operability and serviceability of SD-WAN.

3.2.2 Carrier/MSP Resale Model

In this business model, carriers attract enterprise customers by operating and providing WAN private lines and Internet services. There is no doubt that carriers' services will thrive if SD-WAN services are resold to such a huge number of enterprise customers. Similar to carriers, MSPs are responsible for providing WAN connection services. Regardless of whether MSPs have self-built WANs, they can resell SD-WAN as a brand-new hosting service. This will definitely become a new driving force for service growth.

To summarize, carriers and MSPs resell SD-WAN services to a huge number of enterprise customers and reconstruct, operate, and maintain SD-WAN for the enterprises. For these reasons, carriers and MSPs have the following requirements on the SD-WAN solution:

3.2.2.1 Multi-tenant Management

Typically, a carrier or MSP manages networks of multiple enterprises at the same time, with each enterprise as a tenant. In this manner, one SDN controller manages multiple tenants simultaneously. This is called multi-tenant management. In addition, a carrier or MSP allows multiple tenants to share a single network device, which is called device multi-tenancy.

3.2.2.2 IWG

The IWG is a type of SD-WAN gateway that is mainly used to implement interworking between new SD-WAN sites and legacy MPLS network sites.

The IWG emerges along with the carrier/MSP resale model, where IWGs are typically shared by multiple tenants, and are created, managed, and maintained by carriers or MSPs.

3.2.2.3 POP Networking

Carriers and MSPs can utilize their legacy backbone private line networks to provide differentiated SD-WAN services. That is, a POP gateway is deployed at the edge of the backbone network of a carrier or an MSP, and overlay tunnels are established between the POP gateway and the edges of enterprise branches via SD-WAN. In this manner, these edges can traverse WANs of third-party carriers and access the backbone network.

This WAN access solution is flexible and cost-effective, making it popular among carriers and MSPs. Typically, a POP gateway is also a multi-tenant device shared among enterprises. A software-based edge with flexible deployment and outstanding scalability functions as the POP gateway, and is controlled and managed by the SDN controller.

3.2.2.4 VAS

To keep up with enterprise customers' requirements for VASs such as security and WAN optimization, carriers need to provide VAS operation capabilities allowing for flexible customization and on-demand provisioning.

3.2.2.5 IPv6

As IPv4 addresses are running out, carriers are accelerating WAN evolution toward IPv6. To keep up, SD-WAN must support IPv6.

3.2.2.6 Northbound Open APIs, Easy to Integrate

Carriers generally have their own BSS or OSS, and the SDN controller needs to provide northbound open APIs to get integrated with such systems.

To summarize, new SD-WAN requirements, such as operable management, robust network connectivity, and VASs, are emerging as SD-WAN service providers resell the SD-WAN solution. SD-WAN must embrace further development to satisfy these requirements, and provide more diversified networking modes and scenarios than ever before.

3.2.3 Enterprise-Built Model

Enterprises with a certain scale and economic strength usually purchase the SD-WAN solution from SD-WAN solution providers, and then deploy and maintain the solution by themselves. For example, a bank purchases an SDN controller and network devices from an SD-WAN solution provider and installs the SDN controller in its own DC or private cloud. It also purchases WAN lines from a carrier to interconnect sites. However, the bank manages and maintains its SD-WAN independently.

As described above, the carrier/MSP resale model focuses on multi-tenant management via SD-WAN, whereas the enterprise-built model places more emphasis on adopting SD-WAN to implement hybrid networking between Internet and legacy private lines, application-based traffic steering, flexible access to the Internet, and centralized O&M.

3.3 PANORAMA OF THE SD-WAN SOLUTION

SD-WAN is a complicated solution in terms of both functionality and architecture design. As such, it is necessary to interpret the SD-WAN solution component by component to get a panoramic picture of all sub-solutions and support systems involved in the solution, as well as the SD-WAN solution as a whole.

Enterprises should comprehensively analyze an SD-WAN solution from multiple factors, such as service characteristics and core SD-WAN requirements, as well as overall solution focuses and simplicity. After such analysis, enterprises can typically divide the SD-WAN solution into four sub-solutions: networking sub-solution, application experience sub-solution, security sub-solution, and O&M sub-solution, with the SDN controller being a crucial foundation system.

Figure 3.10 shows the components of the SD-WAN solution.

The following will describe the functionality and design objectives of the four sub-solutions.

3.3.1 Networking Sub-solution

The prime mission of SD-WAN is to implement flexible, reliable interworking of enterprise WANs. Fulfilling this mission lies in the networking

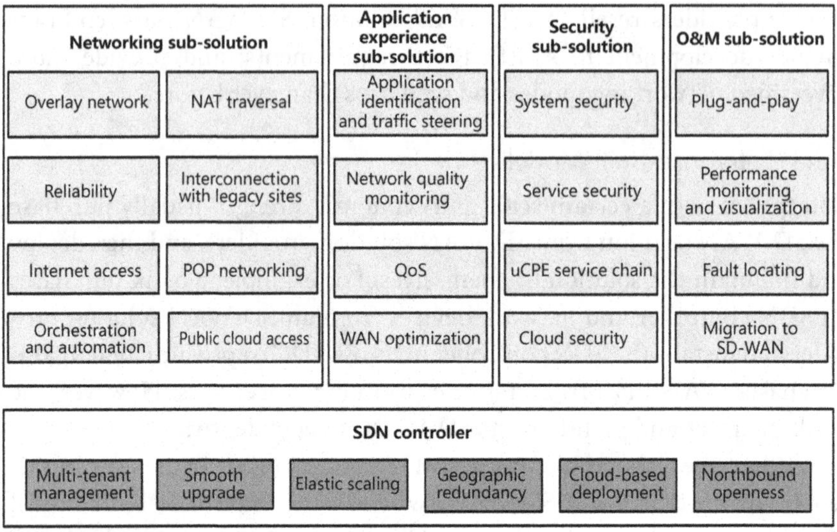

FIGURE 3.10 SD-WAN solution components.

sub-solution. Founded on hybrid WAN links and overlay networks, this sub-solution integrates conventional network technologies such as Layer 2 switching, Layer 3 routing, and VPN isolation so as to achieve on-demand, flexible, and automated connectivity between enterprise branches, DCs, and clouds under the control of the SDN controller.

The networking sub-solution can be further broken down into network infrastructure and network services. The network infrastructure consists of fundamental network technologies, such as overlay networking, underlay networking, reliability, and automated networking. With these fundamental network technologies, SD-WAN can implement more advanced network services, such as Internet access, VPN isolation, interconnection with legacy sites, public cloud access, and POP networking, as illustrated by Figure 3.11.

Network infrastructure consists of the following:

- Overlay networking

  SD-WAN is constructed via overlay technology to get various types of sites connected. Overlay technology is mainly oriented at the WAN side of enterprise sites and must be applicable to hybrid WANs with private lines, MPLS links, Internet links, and LTE links.

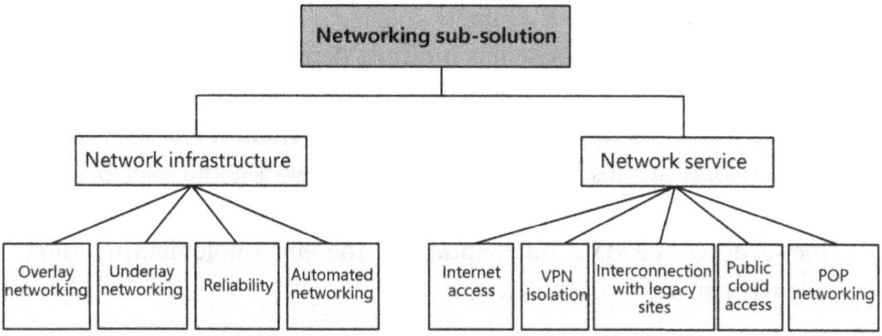

FIGURE 3.11 Networking sub-solution.

- Underlay networking

  Overlay networks interwork with each other via an underlay network that supports hybrid WAN links.

- Reliability

  Networking reliability is crucial for services. Networking reliability is used to solve network-level and site-level reliability problems. For example, when a line or device fault occurs on the network, it ensures a backup path is available and fast route convergence is performed to guarantee that the experience of the service is not affected or the impact is minimized.

- Automated networking

  Network orchestration is the core of automated networking. This technology automates the working process so that network services can be configured on multiple devices and resources can be deployed on demand, thereby making the network more flexible and responsive.

The network services include the following:

- Internet access

  Three access modes are supported: local Internet access, centralized Internet access, and hybrid Internet access (the combination of the preceding two modes). In this way, different types of enterprises can access the Internet according to their requirements.

- VPN isolation

  Typically, an enterprise has multiple departments whose services are independent of each other. SD-WAN is required for E2E VPN isolation in these departments. With this function enabled, underlay networks at sites, WAN-side overlay networks, and user-oriented LANs are logically isolated. This ensures that different departments forward service data independently, thereby implementing inter-department security isolation.

- Interconnection with legacy sites

  While SD-WAN is proving popular, many legacy non-SD-WAN sites still exist. In most cases, these sites interwork with each other via MPLS private lines from carriers. Tearing down all the legacy sites before building an SD-WAN is impossible, and new SD-WAN sites need to coexist and interwork with the legacy sites to ensure normal, continuous running of enterprise WAN services.

- Public cloud access

  Enterprises' requirements for accessing public clouds are increasing as public clouds pick up pace. In SD-WAN, a Virtual Private Cloud (VPC) is defined as a cloud site. The SDN controller implements centralized network orchestration and control to implement on-demand, automated connectivity between enterprise branch sites and cloud sites.

- POP networking

  Many MSPs have their own cross-region backbone networks. However, they cannot fully cover the "last-mile" access for enterprises. A combination of POP networking and SD-WAN is the best choice for this situation.

  In POP networking, POP gateways interwork with each other through a POP backbone network and are managed by the SDN controller. In addition, branch sites connect to POP gateways through SD-WAN tunnels. These approaches achieve high-quality interworking between enterprise branch sites and cloud sites.

3.3.2 Application Experience Sub-solution

This sub-solution focuses on application experience. It uses SD-WAN devices to monitor WAN quality and adjusts and optimizes paths for transmitting application traffic in real time based on the applications' SLA requirements on WAN links. This sub-solution employs plenty of advanced technologies, such as application identification, network quality monitoring, application-based intelligent traffic steering, and WAN optimization.

- Application-based intelligent traffic steering

  Enterprise applications are classified into high-value and common applications, which pose diversified requirements on network quality.

  In a hybrid WAN, network quality varies over different links. To meet application-specific requirements, devices adopt application-based intelligent traffic steering to map different applications to WAN links with different SLA levels. This significantly improves network link utilization and optimizes application experience for enterprise customers.

  Priorities also vary between applications. Applications with high priorities must be preferentially guaranteed upon link congestion and high-value links be fully used. Priority-based load balancing emerged to meet these requirements.

- WAN optimization

  Link transmission efficiency needs to be optimized from two aspects: maximizing utilization of legacy WAN link bandwidth and delivering guaranteed application experience when link quality deteriorates, such as with packet loss. This places great emphasis on WAN optimization.

3.3.3 Security Sub-solution

Security is fundamental to SD-WAN.

The Internet is turning itself into one of the major networking technologies of SD-WAN, and an increasing number of enterprises are moving to Internet-based services, such as network access for branch sites and enterprises' access to SaaS services via the Internet. All these bring great challenges to system security and service security.

Abiding by the zero trust principle, SD-WAN builds an all-round security protection system that delivers high-level security for the SDN controller, enterprise data transmitted on WANs, and Internet access for users.

3.3.4 O&M Sub-solution

Optimal O&M capabilities guarantee normal running of the entire SD-WAN solution. The O&M sub-solution covers all major activities involved throughout the O&M lifecycle (Day 0, 1, and 2). To go into more detail, this sub-solution monitors and visualizes key information during system running, such as alarms, link status, and application status, so as to optimize O&M for enterprise WANs.

The sub-solution also collects, monitors, analyzes, and visualizes network performance data, facilitating O&M. This enables users to conveniently and intuitively monitor SLA adherence, traffic volume of high-value applications, and bandwidth statistics.

The O&M sub-solution also provides fault locating. The SDN controller can analyze root causes of network faults and provide troubleshooting suggestions. Suggestions are based on key alarms and logs of the network as well as the execution result of mature diagnosis tools such as ping and traceroute. The SDN controller can also capitalize on big data analytics to provide preventive fault locating and warning functions, significantly improving WAN O&M efficiency.

As a fundamental support system of SD-WAN, the SDN controller firmly supports reliable, smooth operation of the four sub-solutions. To guarantee normal operations of SD-WAN, the SDN controller must be reliable, scalable, secure, and highly performing.

This chapter takes a panoramic look at the SD-WAN solution. The following chapters will go into detail about the fundamentals and implementation of SD-WAN based on the four sub-solutions.

CHAPTER 4

Starting from Sites

AFTER READING THE PRECEDING chapters, we have already got a full picture of the SD-WAN solution, including its system architecture, core components, mechanisms, and panorama.

Secure and reliable sites are critical for deploying SD-WAN, and therefore this chapter will focus on describing how to deploy SD-WAN.

4.1 SITE TYPES

"Site" is a broad concept in enterprise WANs. Typically, a site refers to an office location or a key IT infrastructure deployment point of an enterprise. For example, branch sites, headquarters sites, and DC sites are typical enterprise sites. The ultimate goal for enterprises to construct WANs is to get their sites connected and interworking with each other to facilitate seamless business and communication.

From the technical implementation perspective, sites involved in SD-WAN can be classified into SD-WAN sites and legacy sites. Mastering the concepts and characteristics of the two types of sites facilitates the understanding of inter-site interworking.

- SD-WAN site

  SD-WAN sites are sites that interwork with each other via SD-WAN technology and are managed by the SDN controller that also orchestrates services for the sites.

  Besides the branch sites, headquarters sites, and DC sites, there is another form of site. As cloud picks up pace among enterprises of all sizes, enterprise services are gradually migrated onto the cloud (such as the public cloud). In a sense, cloud is an extension of enterprise networks. Both the headquarters and branches of enterprises need to communicate with the public cloud. Considering this, the public cloud can also be considered as a special site — cloud site.

A typical cloud site comprises a large quantity of VMs that run a broad variety of applications. The network environment of a cloud site differs a lot from that of headquarters and branch sites.

In the following chapters, the unique interworking mode of cloud sites will be described in detail.

- Legacy site

  Legacy sites are non-SD-WAN sites and, unlike SD-WAN sites, are not managed by the SDN controller. Rather, they interwork with each other through conventional network technologies. For example, legacy sites access an MPLS network through a Provider Edge and interwork with each other via MPLS VPN.

  Legacy sites do not adopt SD-WAN technology. In a strict sense, they do not belong to SD-WAN, but they need to interwork with SD-WAN sites. This communication process will be described in following chapters.

4.2 CPE: "SPOKESPERSON" OF SITES

Edges need to be deployed to facilitate communication between SD-WAN sites. The edge is also known as a CPE, which represents a site.

Different SD-WAN sites have diversified service requirements, and therefore a CPE must both adapt to the site's network environment and match the site's service model.

4.2.1 Diverse CPE Forms

- Hardware form

  As a network device, the CPE was first deployed at a site as a fixed-configuration hardware device. From the hardware perspective, a CPE typically includes service routing units (SRUs), interface cards, multi-core CPUs, and various other hardware components. From the software perspective, a CPE provides both Layer 2 switching and Layer 3 routing to connect to the internal and external networks of a site. In most cases, such a CPE is regarded as a traditional CPE.

It should be noted that a so-called traditional CPE does not signify that it can only be used at a legacy site.

The boom in enterprise service types means that basic routing and forwarding functions already cannot meet enterprises' requirements for security, WAN acceleration, load balancing, and other future-oriented services. These services require dedicated hardware devices, thereby complicating service deployment. With more device types added, network O&M becomes further complex.

The rapid development of cloud computing and NFV technologies is ushering in an inevitable trend — networks evolving toward cloudification and virtualization. The functions of traditional dedicated hardware devices can be implemented via software, and industry-proven functions such as security, WAN acceleration, and load balancing are already implemented via VNFs. Deploying all these functions on the CPE can not only slash device costs and power consumption, but also achieve flexible, rapid service provisioning.

This incubates a new CPE form — Universal Customer Premises Equipment (uCPE). Although the uCPE is just a fixed-configuration hardware device, a wealth of network functions, such as firewall, WAN acceleration, and load balancing, run on the uCPE as VNFs.

Figure 4.1 illustrates the uCPE architecture.

After all the preceding network functions are offered through the uCPE, two issues need to be considered.

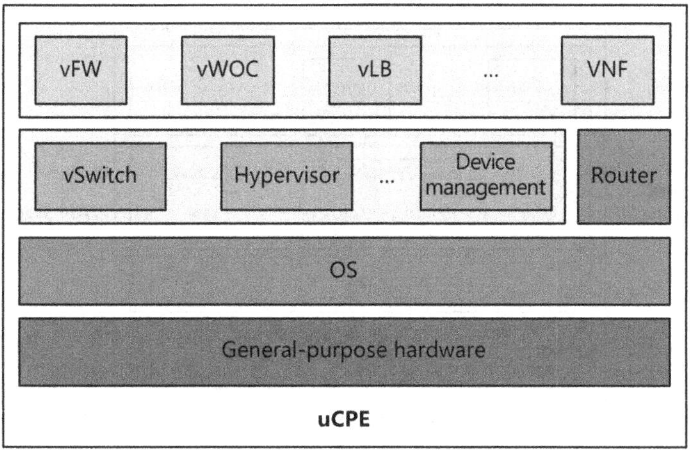

FIGURE 4.1 uCPE architecture.

The first issue concerns how to deploy these network functions. In SD-WAN, the SDN controller, as a "supervisor," can properly arrange and manage these VNF-based network functions throughout their lifecycles to implement fast service provisioning, thereby helping enterprise customers obtain network services on demand.

Now, we will look at the second issue. The network functions on the uCPE are logically independent, so how do these network functions process service traffic entering the uCPE? The key to resolving this issue also lies in the SDN controller. The controller processes service traffic by orchestrating a service function chain (SFC). An SFC is a chain of VNF functions that comply with service traffic requirements, and different service traffic may pass through diversified NEs.

As illustrated in Figure 4.2, when service traffic sent from a LAN-side interface arrives at the uCPE, the traffic is sequentially processed by the virtual firewall (vFW) and virtual WAN optimization controller (vWOC) and then sent out from the WAN-side interface of the uCPE. Similarly, the service traffic returned from the WAN-side interface is processed by the SFC and then sent out from the LAN-side interface.

- Software form

  Enterprise services are gradually migrating to the cloud, which requires network devices to be more software and virtualization oriented, in order to keep in line with this trend. Against this backdrop, traditional CPEs evolve into a new form — vCPE.

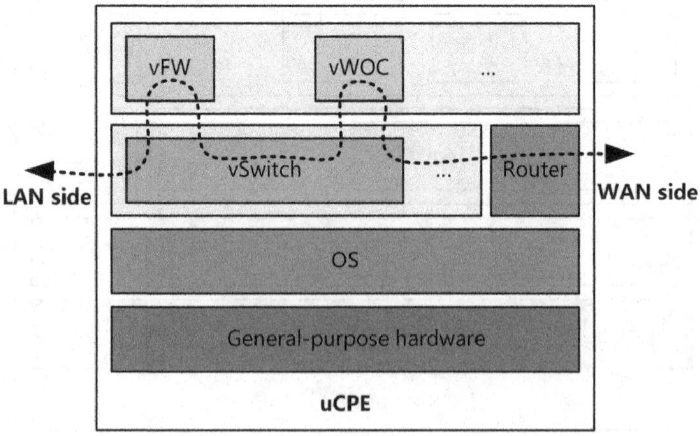

FIGURE 4.2 Service chain of the uCPE.

In this new form, the network functions of the traditional CPE are extracted from the fixed-configuration hardware device and implemented via software, completely decoupling software from hardware.

vCPEs can replace dedicated hardware devices by implementing functions specific to traditional CPEs via software. This mode facilitates convenient, fast service deployment, enhances service scalability, and slashes deployment and operation costs.

vCPEs, not limited in fixed-configuration hardware devices, can be flexibly deployed as software and run on universal servers, public clouds, and uCPEs as VNFs, as illustrated in Figure 4.3.

Overall, SD-WAN supports traditional CPEs, uCPEs in hardware form, and vCPEs in software form. Despite having similar names, these CPEs are applicable to different scenarios. Specifically, traditional CPEs and uCPEs can be deployed at the headquarters, branch, and DC sites of enterprises, while vCPEs can be deployed at public cloud sites, as shown in Figure 4.4. In actual deployment, the CPE form can be flexibly selected depending on the networking environment and service requirements specific to a site.

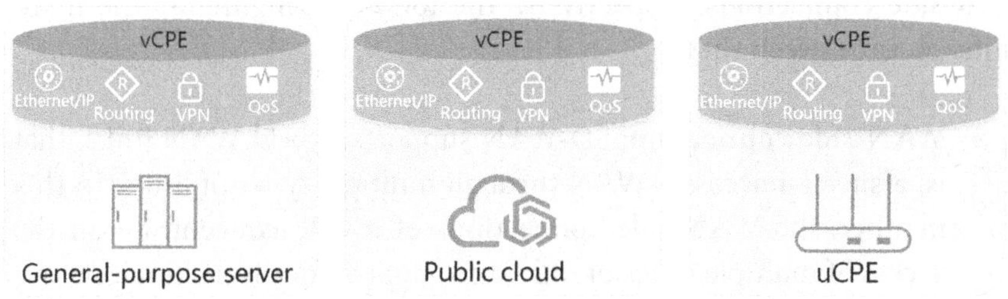

FIGURE 4.3 Operating scenario of the vCPE.

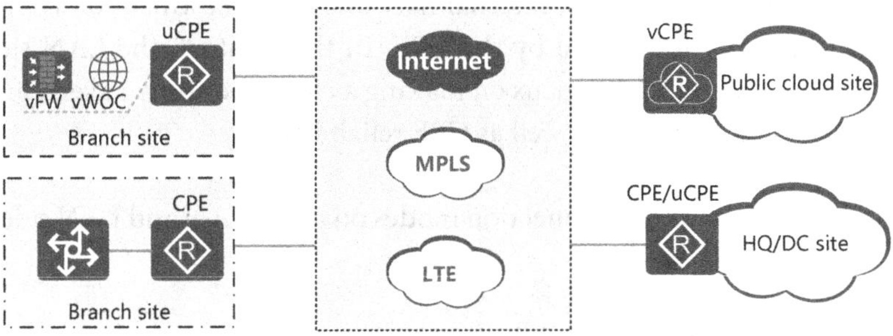

FIGURE 4.4 Usage scenarios of different CPEs.

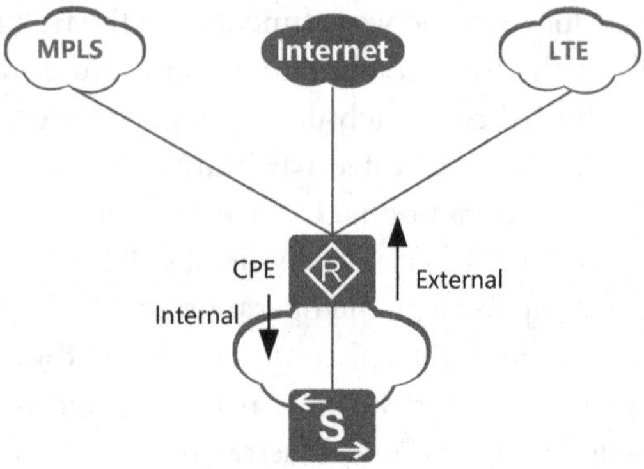

FIGURE 4.5 CPE connecting to the internal and external networks of a site.

4.2.2 CPE: Connecting Sites

As a site's traffic hub, the CPE internally connects to the internal network of the site and externally connects to the external network of the site, as shown in Figure 4.5.

External and internal connections of a CPE are called WAN-side and LAN-side connections, respectively. The following highlights the main differences between them.

- **WAN-side connection**: SD-WAN supports hybrid WAN links, that is, a site can access a WAN through multiple types of links. In this manner, the WAN-side connections of a CPE concentrate on the access of multiple types of links and support for diverse interfaces.
- **LAN-side connection**: At the LAN side, a CPE typically connects to a site's internal network, and functions as the gateway of the internal network. Therefore, internal network traffic destined for WANs is uniformly forwarded by the CPE. In this context, the LAN-side connections of a CPE focus on making a choice between Layer 2 and Layer 3 connections as well as CPE reliability.

The following details the connection modes on both WAN and LAN sides.

4.2.2.1 WAN-Side Connection

As previously described, the WAN-side connections of a CPE mainly consider how to access a WAN through multiple links, since hybrid WAN links are one of the fundamental characteristics of SD-WAN and multiple links provide greater choices for site interworking. WAN-side connections of CPEs vary depending on the actual network environment where the specific site resides, and can be classified into four types: single-site single-WAN link, single-site dual-WAN link, single-site multi-WAN link, and single-site dual-CPE.

1. Single-site single-WAN link

   Figure 4.6 illustrates an example where a CPE at a site connects to the WAN side through one link: either MPLS or Internet, featuring simple WAN-side connection.

2. Single-site dual-WAN link

   As shown in Figure 4.7, a CPE at a site connects to the WAN side through dual links: both MPLS and the Internet.

3. Single-site multi-WAN link

   Another link is added to the previous single-side dual-WAN link scenario. This provides another option as well as higher link reliability. As illustrated in Figure 4.8, an LTE link is added to the WAN side of the CPE that already holds one MPLS link and one Internet link.

   In most cases, the LTE link is charged by traffic and used as an escape link. A so-called escape link refers to a last surviving link, which has the lowest priority. This is because it forwards data only if both the primary and secondary links fail, which improves site reliability to a certain extent.

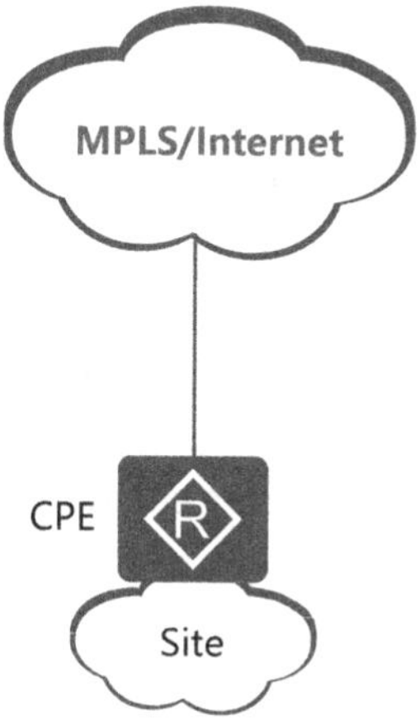

FIGURE 4.6 Single-site single-WAN link.

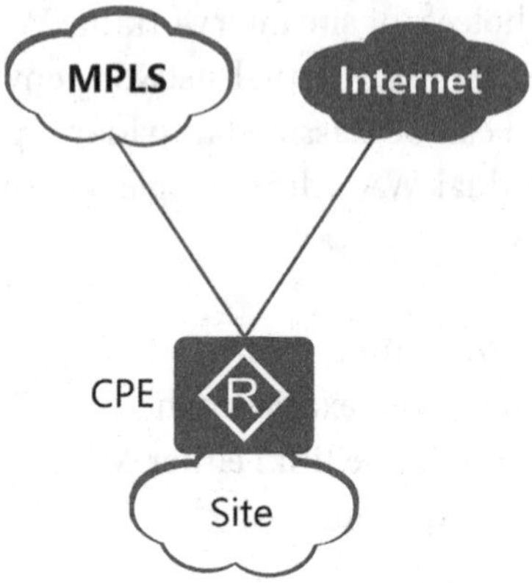

FIGURE 4.7 Single-site dual-WAN link.

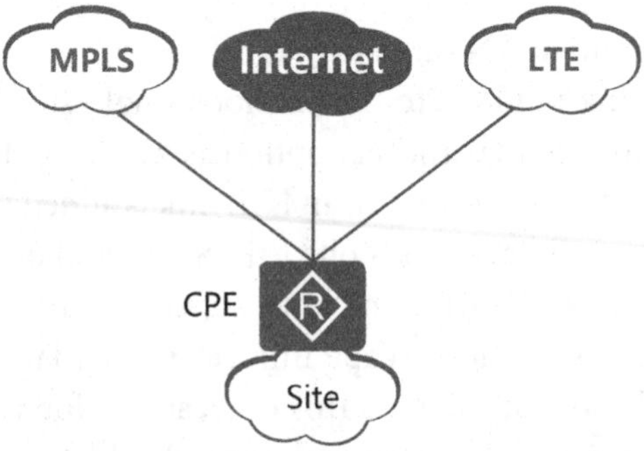

FIGURE 4.8 Single-site multi-WAN link.

4. Single-site dual-CPE

   From the perspective of device reliability, if only one CPE is deployed at a site, there is a possibility that a single point of failure occurs, posing potential risks to site stability. If the only CPE fails at a site, the entire site can no longer communicate with external networks.

In SD-WAN, two CPEs can be deployed at a site to construct a single-site dual-CPE (gateway) network. If one CPE fails on such a network, the other CPE takes over its work, guaranteeing site reliability.

As illustrated by Figure 4.9, two CPEs at a site are connected to an MPLS network, an LTE network, and the Internet, respectively. The WAN-side connections of each CPE in this scenario are the same as those of the CPE in single-site single-CPE scenarios.

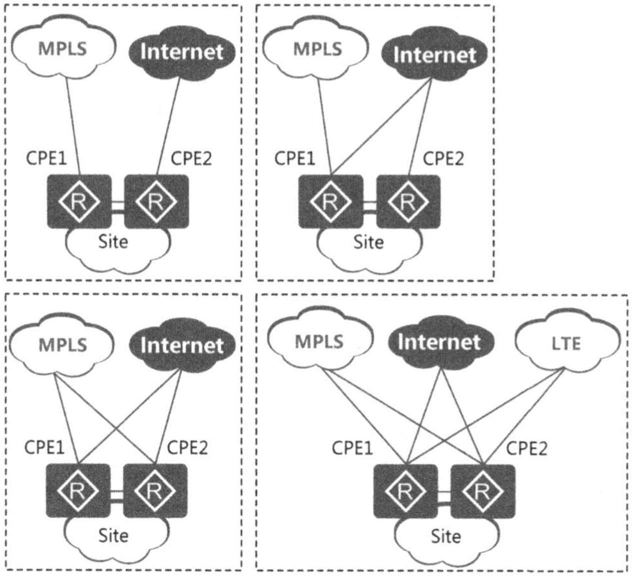

FIGURE 4.9 WAN-side connections in single-side dual-CPE scenarios.

On a single-site dual-CPE network, the two CPEs are typically interconnected. Service traffic is forwarded and service information is synchronized through the interconnection link between the two CPEs. Interconnection between the two CPEs is complicated and will be described further in the subsequent part.

4.2.2.2 LAN-Side Connection

A CPE connects to the internal network of a site at the LAN side, which is related to the actual network environment of the site. In most cases, the CPE functions as the gateway of the internal network and connects to the internal network at Layer 2 or Layer 3. The following will describe the two modes in detail:

1. Layer 2 mode

   If the scale of a site is small and the internal network has a simple structure, the CPE can connect to the site's internal network at Layer 2, as illustrated in Figure 4.10.

   A total of four methods are available to set up LAN-side connections of a CPE at Layer 2:

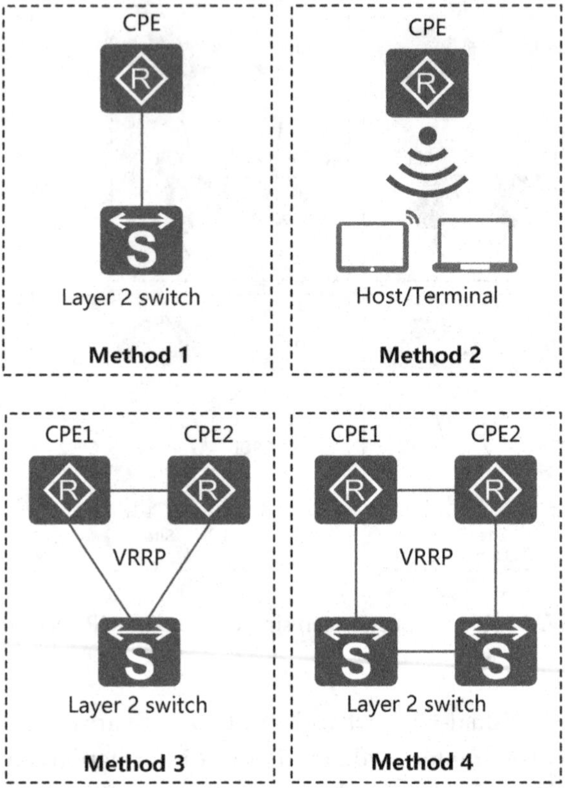

FIGURE 4.10 LAN-side Layer 3 connection of CPEs.

a. **Method 1**: Only one CPE is deployed at a site and directly connects to the Layer 2 switch at the site.

b. **Method 2**: Only one CPE is deployed at a site and directly connects to a host or a terminal at the site via Wireless Local Area Network (WLAN).

c. **Method 3**: Two CPEs are deployed at a site and are connected to the Layer 2 switch at the site. Typically, Virtual Router Redundancy Protocol (VRRP) is deployed on the two CPEs. The VRRP virtual IP address is used as the gateway address of the site's internal network, improving network reliability.

d. **Method 4**: Two CPEs are deployed at a site, and each of them is connected to a Layer 2 switch at the site. Similarly, VRRP is deployed on the two CPEs to improve network reliability.

2. Layer 3 mode

   A large-scale site has a complex network structure. For this reason, the CPEs connect to the site's internal network at Layer 3 and have specific routing protocols configured based on the actual situation of the site's internal network, in order to communicate with devices at the site, as illustrated in Figure 4.11.

   There are also four methods to build LAN-side connections of a CPE at Layer 3:

   a. **Method 1**: Only one CPE is deployed at a site. The CPE directly connects to the Layer 3 switch at the site and communicates with the site's internal network via a specific routing protocol.

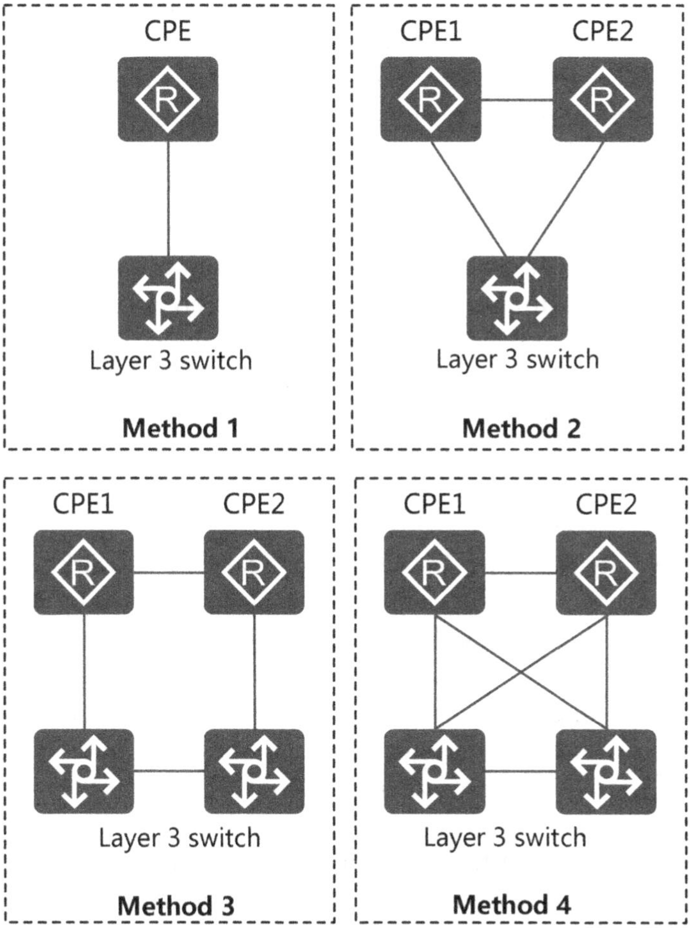

FIGURE 4.11 Challenges in single-site dual-CPE networking.

b. **Method 2**: Two CPEs are deployed at a site. The CPEs communicate with the internal network of the site through static routes or a specific routing protocol, such as Open Shortest Path First (OSPF) or BGP, by connecting to the Layer 3 switch at the site.

c. **Method 3**: Two CPEs are deployed at a site, and each of them is connected to a Layer 3 switch at the site to implement square looped networking. Additionally, to enable communication between the CPEs and the site's internal network, static routes or a proper routing protocol, such as OSPF or BGP, should be configured on the CPEs according to the actual internal network situation of the site.

d. **Method 4**: Two CPEs are deployed at a site, and each of them is connected to two Layer 3 switches at the site to implement dual-homed networking. Similarly, static routes or a suited routing protocol, such as OSPF or BGP, must be configured to enable the CPEs to communicate with the internal network at the site.

In the scenario where the internal network of a site is connected at Layer 3, routing configuration must comply with certain principles in order to avoid routing loops. One of these principles is not to advertise routes learned from the LAN side back to the LAN side. In most cases, the SDN controller automatically orchestrates routing principles with no intervention of network administrators.

4.2.3 Dual-Gateway Interconnection

To improve site reliability, the single-site dual-CPE networking, which is also known as single-site dual-gateway, is implemented by deploying two CPEs at a site. This ensures service continuity at the site even if one CPE fails.

This networking model also gives rise to numerous service challenges, such as application identification, application-based traffic steering, site-level QoS, and WAN optimization, as illustrated in Figure 4.12.

- **Application identification**: CPEs need to comprehensively analyze the uplink and downlink traffic of some applications to accurately identify them. If the uplink and downlink traffic of an application

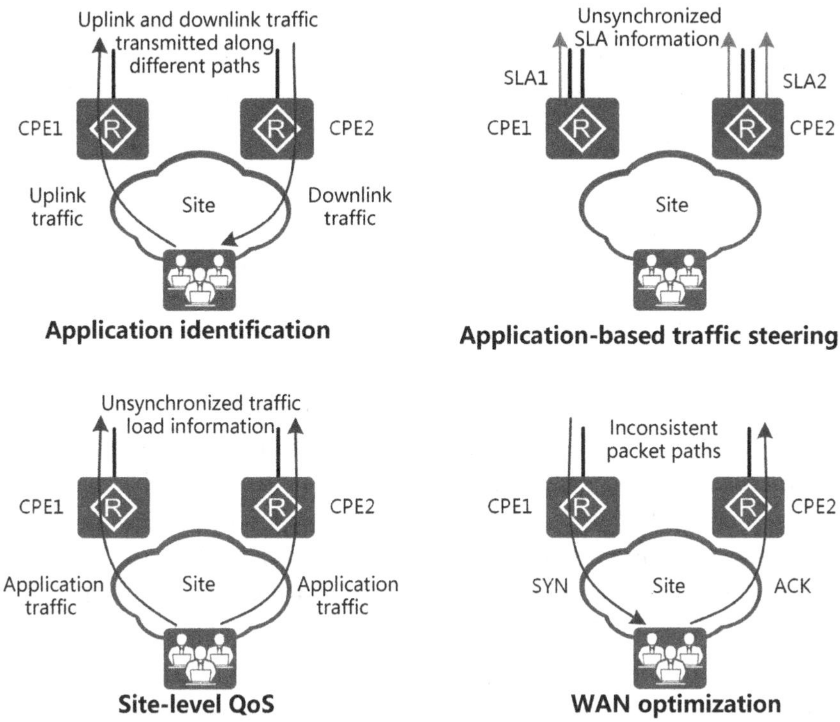

FIGURE 4.12 Interconnection link between two CPEs at a single site.

is forwarded by different CPEs at a site, no CPE can identify the application.

- **Application-based traffic steering**: Two CPEs at a site may be connected to different links. The CPEs cannot obtain the SLA information of all their peers' links. As a result, application-based traffic steering cannot be performed.
- **Site-level QoS**: When traffic limiting is performed at a site with QoS configured, two CPEs at the site cannot obtain bandwidth information of each other. As a result, the total traffic volume of the site cannot be accurately limited.
- **WAN optimization**: If the incoming and outgoing paths of TCP traffic are inconsistent when WAN optimization is performed, TCP traffic is interrupted, thereby invalidating the WAN optimization function.

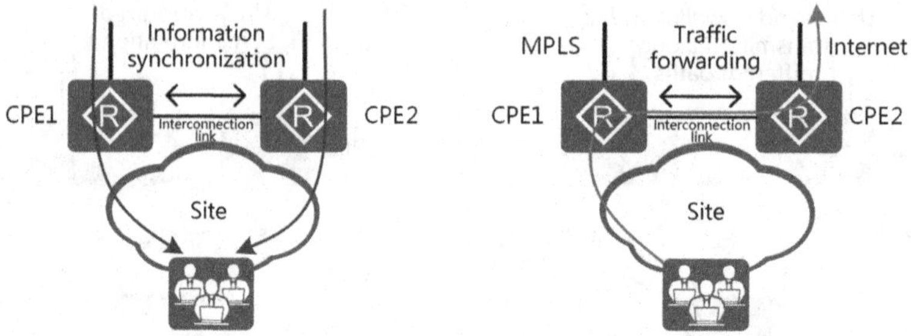

FIGURE 4.13 LAN-side Layer 2 connection of CPEs.

To address these challenges, link usage is optimized in the single-site dual-CPE networking scenario. This means the links of the two CPEs at a site are considered as a whole and shared by the two CPEs.

In response, SD-WAN considers two CPEs as one device. The two CPEs are connected through an interconnection link, allowing them to synchronize their respective key information and forward service traffic through the intermediate interconnection link, as illustrated in Figure 4.13.

- **Information synchronization**: Two CPEs can synchronize their respective key information through the intermediate interconnection link. Such key information includes routing, link, tunnel connection, tunnel connection SLA, traffic bandwidth, and application identification results. Synchronizing this information ensures the service information on the two CPEs is consistent. This means that the resources on these two CPEs can be processed in a unified manner, thereby implementing site-level QoS and cross-CPE traffic steering.
- **Traffic forwarding**: Two CPEs can also redirect their service traffic via the intermediate interconnection link to achieve inter-CPE traffic forwarding. For example, CPE1 is connected to an MPLS link, and CPE2 is connected to an Internet link. A traffic steering policy is configured for application A to forward its traffic through the Internet link. This is performed as follows: When the traffic of application A reaches CPE1, CPE1 forwards the traffic to CPE2 through the interconnection link based on a preset traffic steering policy. CPE2 then forwards the traffic through the Internet link. Furthermore, redirecting traffic through the interconnection link

can also eliminate inconsistency between incoming and outgoing paths of application traffic in application identification and WAN optimization scenarios.

4.3 KEY CAPABILITIES OF CPEs

In pursuit of more advanced functions and higher performance, CPEs have historically evolved from traditional CPEs with simple functions to convergent CPEs featuring multi-service processing, toward SD-WAN CPEs that stand out for their high-performance data forwarding. Figure 4.14 shows the evolution of CPEs.

There are three types of CPEs:

4.3.1 Traditional CPE

In most cases, a traditional CPE with basic routing capabilities functions only as a gateway to provide routing and forwarding functions, and connect the internal and external networks of a site.

4.3.2 Convergent CPE

Building on the capabilities of traditional CPEs, convergent CPEs further integrate a wealth of functions, such as Wi-Fi, QoS, VPN, and firewall, and provide comprehensive multi-service processing capabilities.

4.3.3 SD-WAN CPE

In addition to basic routing and switching capabilities, SD-WAN CPEs also need to provide service processing capabilities that are oriented to applications, such as application identification, intelligent traffic steering, multiple types of VPNs, QoS, and security. To fully leverage these capabilities,

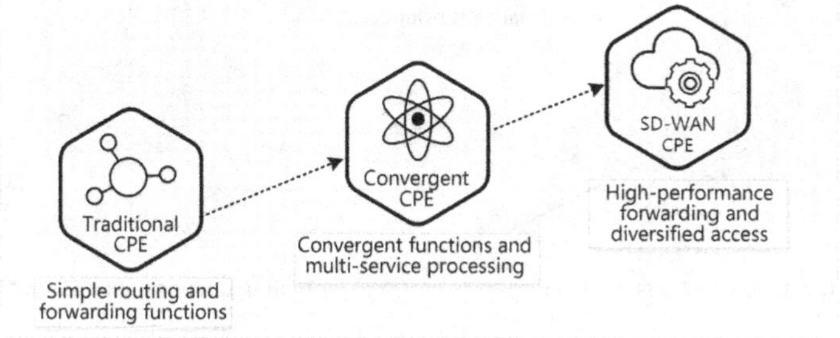

FIGURE 4.14 Evolution of CPEs.

SD-WAN CPEs must be able to forward data at high performance and provide diversified access capabilities to better adapt to different WAN access environments. The following details these high-performance forwarding and diversified access capabilities.

1. High-performance forwarding

   Alongside traditional enterprise WANs' journey of development toward SD-WAN, the performance bottlenecks of CPEs are the biggest obstacle to large-scale commercial deployment of SD-WAN. As illustrated in Figure 4.15, traditional CPEs mainly forward packets at Layer 1 to Layer 3, for example, the CPEs provide routing, QoS, and VPN functions. In contrast, SD-WAN CPEs are more oriented to applications, providing full-service processing at Layer 3–Layer 7. These future-oriented CPEs provide a variety of functions, such as application identification, multi-scenario connection, dynamic link adjustment, VPN, VAS, and simplified O&M, to ensure services are congestion-free and operate smoothly.

   SD-WAN CPEs must break through the performance bottlenecks of traditional CPEs and are capable of processing all services at Layer 3–Layer 7 at a high performance, posing higher requirements on the CPEs' system architecture than ever before. Typically, an SD-WAN CPE must provide a non-blocking switching architecture based on multi-core CPUs and NPs as well as multiple hardware acceleration

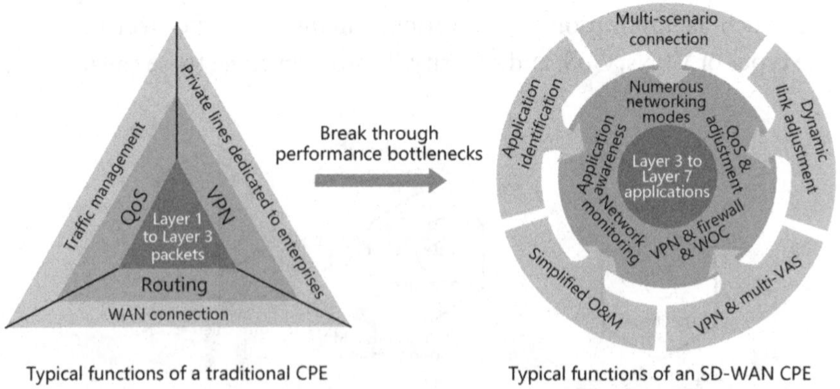

FIGURE 4.15 SD-WAN CPEs, breaking through the performance bottlenecks of traditional CPEs.

capabilities. These improvements enable these CPEs to smoothly and efficiently process various services. Figure 4.16 illustrates the typical system architecture of SD-WAN CPEs.

The system architecture of high-performance SD-WAN CPEs has the following distinct characteristics:

a. "Multi-core CPU+NP" heterogeneous architecture

A heterogeneous architecture refers to a set of highly integrated computing units. A typical chip dedicated to this kind of architecture contains a conventional universal computing unit, such as a CPU, and a high-performance dedicated computing unit. It is the heterogeneous architecture that SD-WAN CPEs appeal to.

A CPU that integrates a multitude of complete computing engines (cores) is called a multi-core CPU. Each core has an independent logical structure that consists of an abundance of logical units, such as packet buffering, execution, and instruction units, in addition to a bus interface, and interwork with each other through high-speed buses and memory sharing. These strengths

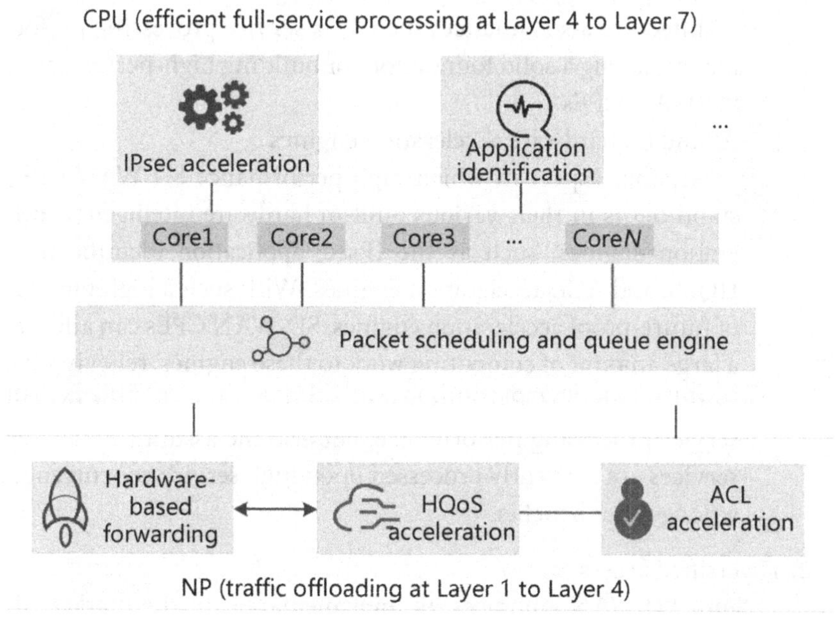

FIGURE 4.16 Typical system architecture of SD-WAN CPEs.

make multi-core CPUs ideal for concurrently processing a wide range of services, thereby boosting the overall efficiency of CPUs. This advanced CPU also significantly raises the service processing efficiency at Layer 4–Layer 7 using a dedicated packet scheduling mechanism and a queue engine, ensuring congestion-free, smooth multi-core service processing.

An NP is a programmable component that processes data packets. It innovatively integrates the flexibility of a CPU and high performance of an Application Specific Integrated Circuit (ASIC), and has a strong hardware parallel processing capability. The system architecture of SD-WAN CPEs introduces an independent NP to perform hardware-based forwarding and also supports traffic offloading at Layer 1–Layer 4 — distributing Layer 1–Layer 4 traffic from the CPU to the NP for processing. This significantly relieves the load on the CPU and improves the overall forwarding capability.

This architecture enables collaborative computing and acceleration between the CPU and NP, making a great breakthrough in enhancing the performance of CPUs, effectively reducing energy consumption and improving scalability. This architecture also enhances the CPE's service processing performance, laying a solid foundation for building high-performance SD-WAN CPEs.

b. Abundant hardware acceleration engines

Another aspect in which high-performance SD-WAN CPEs stand out is in their various built-in hardware intelligent acceleration engines, such as the IPsec, application identification, HQoS, and ACL acceleration engines. With such a high number of future-proof acceleration engines, SD-WAN CPEs can allocate a large amount of computing work to these engines, relieving the workload on the CPU. This brings not only a boost to the overall service processing performance, but also the assurance that key services are efficiently processed upon high service concurrency, among other benefits.

2. Diversified access

Since SD-WAN embraces its implementation in the market, all parties in the industry have turned their attention from Software Defined (SD) development to WAN development. As a result, the

WAN-side access capabilities of CPEs are a cornerstone for the implementation of flexible networking.

A wide range of WAN protocols are available and are selected depending on the network environment of a specific site. If a CPE supports only one interface type, it can access a WAN only in a single mode, making it impossible to flexibly adapt to the network environments of different sites. As such, CPEs must support various types of interfaces to adapt to different access environments. Table 4.1 lists the interface types supported by CPEs.

TABLE 4.1 Interface Types Supported by CPEs

Interface Type		Description
Low-speed interface	• Serial interface	• A serial interface is one of the most commonly used WAN interfaces. It can work in synchronous mode as a synchronous serial interface or in asynchronous mode as an asynchronous serial interface Link layer protocols such as Point-to-Point Protocol (PPP) can be configured on synchronous serial interfaces. Parameters such as the stop bit and data bit can be set on asynchronous serial interfaces
	• Async interface	• An async interface is a dedicated asynchronous serial interface and is one of the most commonly used WAN interfaces. A variety of parameters related to the asynchronous serial interface, such as the stop bit and data bit, as well as the link layer protocol — PPP — can be configured on the async interface
	• Channelized E1 (CE1)/Primary Rate Interface (PRI)	• A CE1/PRI is a physical interface in the E1 system that transmits voice, data, and image signals. It can work in E1 (unchannelized) or CE1/PRI (channelized) mode
	• Channelized T1 (CT1)/PRI	• A CT1/PRI is a physical interface in the T1 system that transmits voice, data, and image signals
	• E1-F/T1-F interface	• An E1-F/T1-F interface is an E1/T1 interface working in fractional channelized mode and a simplified CE1/PRI or CT1/PRI interface To reduce costs, E1-F/T1-F interfaces can be used in the E1/T1 access scenario
Integrated service digital network (ISDN) basic rate interface (BRI)		An ISDN BRI focuses on accessing an ISDN. Link layer protocols such as PPP can be configured on ISDN BRIs. The network layer protocol is IP

(Continued)

TABLE 4.1 (*Continued*) Interface Types Supported by CPEs

Interface Type		Description
x digital subscriber line (xDSL) interface	• Asymmetric Digital Subscriber Line (ADSL) interface	• An ADSL interface provides asymmetric transmit and receive rates, and rapidly transmits data over copper twisted pairs by utilizing high frequencies that are not used by regular telephone lines
	• G.Single-pair High-speed Digital Subscriber Line (G.SHDSL) interface	• A G.SHDSL interface provides symmetric transmit and receive rates, and rapidly transmits data over copper twisted pairs by utilizing high frequencies that are not used by regular telephone lines
	• Very high data rate Digital Subscriber Line (VDSL) interface	• Based on the DSL, the VDSL interface integrates various interface protocols, and multiplexes upstream and downstream channels to provide high rate transmission
Packet over SONET/SDH (POS) or channelized packet over SDH/SONET (CPOS) interface		• A POS interface uses synchronous optical network (SONET) or synchronous digital hierarchy (SDH) as the physical-layer protocol and transfers P2P IP data reliably at high speeds. A CPOS interface is a channelized POS interface that optimally uses SDH to improve devices' convergence capabilities for low-speed access
Ethernet interface		• Ethernet interfaces are the most widely used interfaces, including Fast Ethernet (FE), Gigabit Ethernet (GE), and 10GE interfaces. They can process Layer 3 protocol data and provide routing functions
Passive Optical Network (PON) interface		• A PON is a pure media network that allows access to data, voice, and video services through optical fibers PON interfaces are classified into Ethernet PON (EPON) and gigabit PON (GPON) interfaces. They transmit data at a high rate
3G/LTE interface		• A 3G interface is a physical interface that supports 3G technologies. It provides enterprise-level wireless WAN access services An LTE interface is a physical interface that supports LTE technology. Compared with 3G technology, LTE technology provides enterprises with high-bandwidth wireless WAN access services
5G interface		A 5G interface is a physical interface that supports 5G technologies, which it uses to implement high-speed, reliable wireless WAN access

4.4 CPE PLUG AND PLAY

To roll out a new enterprise site, an enterprise must deploy at least one CPE — a network device that can operate properly only after the necessary configuration and commissioning is implemented. In a conventional WAN environment, a number of issues may arise during deployment of CPEs. For example, dedicated technical personnel need to visit the site to deploy CPEs. In addition, manual CPE configurations are prone to errors and have a low efficiency. As a result, enterprise sites cannot be rolled out quickly. Resolving the issues of manual configuration has been the concern of IT management personnel at enterprises for a long time, and if not addressed, will hold enterprises back amid fierce business competition.

As such, IT management personnel are hoping for a simple, fast CPE deployment method that allows CPEs to be automatically deployed and configured, enables sites to be quickly rolled out, and instantly provisions services with several simple operations, relieving deployment personnel of performing complex operations as well as negating the need for highly professional skills. With such a deployment method, personnel can prepare the sites quickly without the need for specialized skills, significantly mitigating the workload of enterprises' IT management personnel.

The emergence of SD-WAN makes this a reality. It provides ZTP that perfectly meets all the aforementioned requirements. ZTP also features plug-and-play, enabling a CPE to go online immediately after it is connected to the network and is powered on.

So how can the SD-WAN solution achieve plug-and-play of CPEs as well as rapid site rollouts? The answer to this question will be explored in the following section.

4.4.1 Fundamentals

In conventional deployment, the IT management personnel manually configure and commission CPEs, and connect the CPEs to WANs at specific sites, which is time-consuming and cost-intensive. To make the deployment quicker and more cost-effective, the SD-WAN solution introduces a centralized network management and control system — SDN controller — to automate CPE registration and rollout as well as service configuration. Using an SDN controller, information about CPEs to be deployed and related service configurations are imported in the controller before the CPEs are delivered. In this way, the CPEs, on their arrival

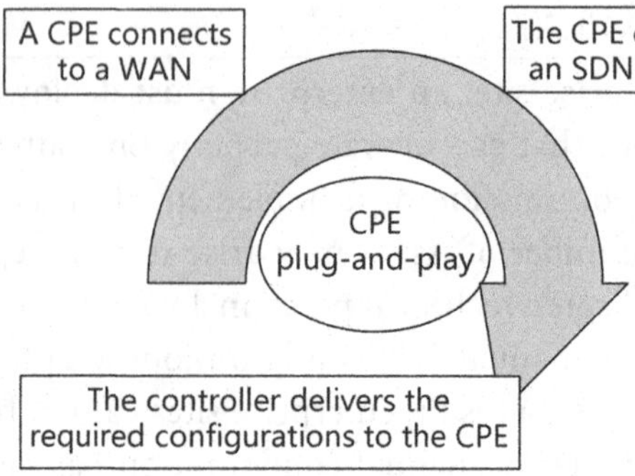

FIGURE 4.17 Process of CPE plug-and-play.

at the destined sites, can automatically connect to the controller. The controller then delivers service configurations to the CPEs. Essentially, with plug-and-play enabled, deployment personnel at sites can connect CPEs to WANs with just a few simple operations, reducing the deployment time as well as deployment costs.

As illustrated in Figure 4.17, a total of three steps must be followed before a CPE can automatically establish a connection with the SDN controller and the controller can deliver service configurations to the CPE.

4.4.1.1 A CPE Connects to a WAN

The WAN-side interface of a CPE must already be assigned an IP address before the CPE can communicate with external networks. With plug-and-play enabled, this IP address does not need to be manually configured by site personnel through the CLI. Rather, the SDN controller assigns an IP address to this interface.

As the brain of the SD-WAN solution, the controller manages all sites and CPEs in a unified manner. To this end, an administrator will set the WAN-side interface settings on a CPE to be deployed before the CPE is delivered to a site. Depending on the actual network environment of the site, the administrator can assign a static IP address to the WAN-side interface or configure the interface to dynamically obtain an IP address. The administrator also performs WAN-side gateway configurations at this stage.

The configurations set on the controller will be delivered to the CPE for loading. Typically, the controller delivers configurations to a CPE in one of two ways: with a USB flash drive or through an email. When using a USB flash drive, the site deployment personnel transfer configuration information to it and insert it into the CPE. The CPE then reads and loads this information. When delivering configurations through email, the controller sends an email containing configuration information to the site deployment personnel. The site deployment personnel then enables the CPE to read and load configuration information from the email.

4.4.1.2 The CPE Connects to the SDN Controller

Before connecting to the SDN controller, a CPE must already obtain the IP address of the controller in one of four methods: with a USB flash drive or through an email, similar to the preceding methods for delivering configurations, as well as through Dynamic Host Configuration Protocol (DHCP) or a service query server. When obtaining an IP address through DHCP, if the CPE can access the underlay network at the site through DHCP and the corresponding DHCP server is controllable, the CPE can obtain the IP address through DHCP Option 148. Through the service query server, after accessing the public network where the server is deployed, the CPE can request the controller's IP address from the server.

4.4.1.3 The SDN Controller Delivers Configurations

As described above, the offline service design can be performed on the SDN controller in advance. In another word, the services of a site are already configured before the specific CPE goes online. In this way, the SDN controller automatically delivers the configurations to the CPE after it goes online. However, this requires the controller to locate a CPE to a specific site after the CPE goes online.

Specifically, the controller determines the site to which a CPE belongs based on the identity of the CPE with either of the following methods:

- **The CPE's ESN is bound to the corresponding site**: Each CPE has a unique ESN, which can be regarded as the CPE's identity. The CPE's ESN is bound to the target site on the SDN controller in advance, so when the CPE attempts to go online, the controller can quickly determine the target site based on the ESN of the CPE. The ESN of a CPE is input into the SDN controller manually or through barcode scanning. Through barcode scanning, the SDN controller provides an API to integrate with a barcode scanning system. After a CPE arrives at the target site, the deployment personnel scans the CPE's barcode with a barcode scanning application on a smart terminal. Then the ESN is inputted into the controller and bound to the target site.

- **The CPE's ESN is decoupled from the specific site**: In addition to binding the ESN to a specific site in advance, the SDN controller can also issue a temporary identifier to a CPE for subsequent identification. The controller generates a token that functions as identification based on the relationship between the site and CPE, and then transfers the token to the CPE. Subsequently, when the CPE attempts to go online, the controller can accurately locate the site where the CPE resides based on the token. After the CPE goes online, the controller binds its ESN with the target site.

The preceding two methods both have their respective advantages and disadvantages. Binding the CPE's ESN to the site requires the ESN to be obtained in advance, which in itself requires a complex query and input process. Moreover, the CPE must also strictly correspond to a site. Decoupling the ESN from the site, however, is a relatively flexible method of determining the site, as it negates the need for inputting and binding the ESN in advance, and the CPE also does not need to strictly correspond to a site.

4.4.2 Deployment Practices

This section focuses on the roles and responsibilities of the personnel involved in typical site deployment.

1. Network administrator

   Network administrators plan and manage networks throughout the network's lifecycle. Alongside the plug-and-play deployment journey of a CPE, a network administrator prepares the SDN controller for deployment, which includes creating sites, setting WAN-side interfaces of CPEs, and configuring other services.

2. Device administrator

   Device administrators manage CPEs, including obtaining the CPEs' ESNs and sending them to network administrators, and delivering CPEs to the target sites. In specific deployment scenarios, device administrators import initial configurations into CPEs before delivering them.

3. Site deployment personnel

After receiving a CPE delivered by a device administrator, the site deployment personnel connect certain network cables and power on the CPE on site for deployment, and if necessary, the personnel also need to perform some simple operations, such as accessing a Uniform Resource Locator (URL) and clicking some buttons. In addition, site deployment personnel need to check the result after deployment is complete to ensure that the CPE goes online successfully.

Depending on the media of information transmission, site deployment can be classified into four methods: email-based, registration center-based (via a query server), USB-based, and DHCP-based deployment, as detailed below:

4.4.2.1 Email-Based Deployment

Through this method, deployment-dependent information is transferred over email to implement plug-and-play of CPEs. The email contains a URL that is already processed by the SDN controller. This URL points to the IP address of the management port on the CPE to be deployed, and contains the CPE configuration, IP address of the controller, and a token generated by the controller. In most cases, the IP address of the management port is defaulted before delivery and is fixed, enabling the deployment personnel to connect to a CPE and allowing the CPE to load initial configurations with the IP address.

Figure 4.18 illustrates email-based deployment.

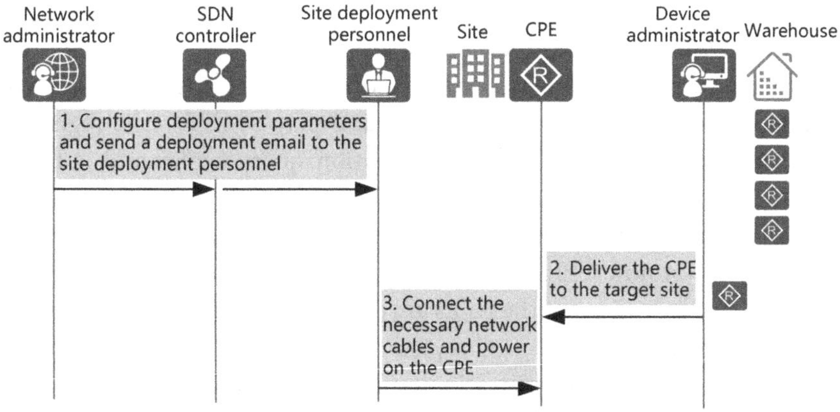

FIGURE 4.18 Email-based deployment.

Email-based deployment is performed through the following steps:

1. The network administrator performs a series of configurations on the SDN controller, including creating a site and setting the WAN-side interface parameters of a CPE. During this process, the network administrator can choose whether to enter the CPE's ESN. If choosing not to enter the ESN, the administrator only needs to specify the model of the CPE for the site. The network administrator then sends a deployment email to the site deployment personnel through the SDN controller. The URL carried in the email is typically encrypted, and the site deployment personnel needs to be notified of the password for decrypting it during site deployment.
2. The device administrator delivers a CPE of the specified model to the target site. If the CPE's ESN is not input in step 1, the device administrator does not need to strictly map the CPE to the target site at this stage.
3. The site deployment personnel use the deployment terminal — a PC or an intelligent terminal — to receive the email sent by the network administrator. The personnel then follow the email to connect the necessary network cables and power on the CPE. After the CPE connects to the SDN controller, the controller determines the site to which the CPE belongs based on the token of the CPE, binds the ESN reported by the CPE to the target site, and delivers service configurations specific to the CPE at the site.

4.4.2.2 Registration Center-Based Deployment

In this mode, deployment-dependent information is transferred through a registration center to implement plug-and-play of CPEs. The registration center is also regarded as a registration query center and is typically deployed on a public network. CPEs obtain the SDN controller's IP address from the registration center. Note that the registration center focuses only on how CPEs obtain the controller's IP address. WAN-side interface configurations of the CPEs are obtained through other methods, such as through DHCP.

Figure 4.19 illustrates registration center-based deployment.

Registration center-based deployment is performed through the following steps:

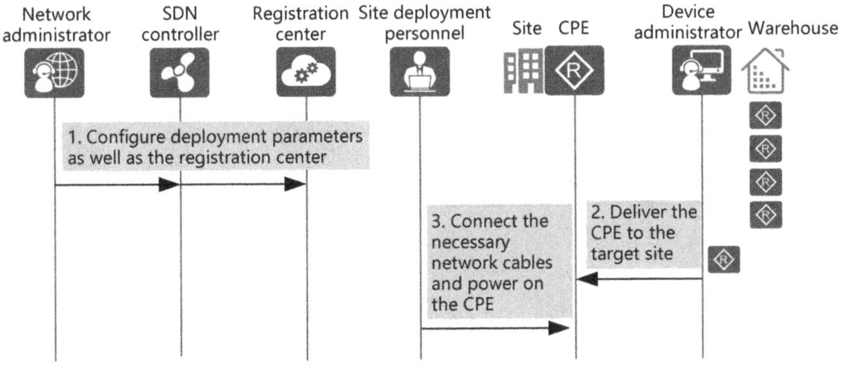

FIGURE 4.19 Registration center-based deployment.

1. The network administrator performs a series of configurations on the SDN controller, including inputting the CPE's ESN, creating a site, binding the ESN to the site, setting the WAN-side interface parameters of the CPE, and configuring the registration center. This enables the registration center to receive the CPE's request for the controller's IP address and return the IP address to the CPE in time.
2. The device administrator delivers the CPE to the target site.
3. The site deployment personnel connect the necessary network cables and power on the CPE on site. The CPE then obtains the WAN-side interface parameters from the DHCP server and uses the built-in domain name to send a request to the registration center to obtain the IP address of the SDN controller. After obtaining the IP address, the CPE connects to the SDN controller. The controller then determines the site to which the CPE belongs based on the CPE's ESN, and delivers service configurations specific to the CPE at the site.

4.4.2.3 USB-Based Deployment

Using a USB flash drive, information is transferred to implement plug-and-play of CPEs. Device administrators can also use a USB flash drive to deploy CPEs in a centralized manner, or the site deployment personnel can directly deploy CPEs using a USB flash drive on site. Typically, the USB-based deployment mode is applicable to deploying CPEs in batches. Device administrators initially configure CPEs in batches before delivery

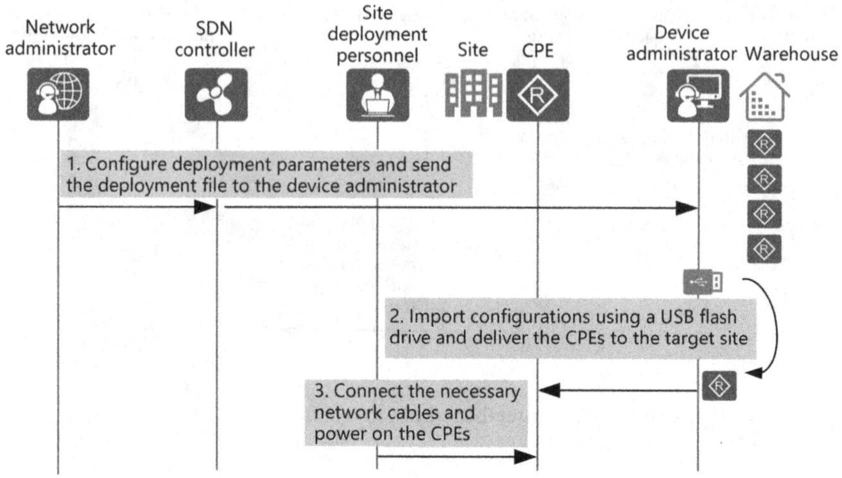

FIGURE 4.20 USB-based deployment.

and then deliver the CPEs to the target sites. The site deployment personnel then connect the necessary cables and power on the CPEs.

Figure 4.20 illustrates deploying CPEs in a batch with a USB flash drive. USB-based deployment is performed through the following steps:

1. The network administrator performs a series of configurations on the SDN controller, including entering the CPEs' ESNs, creating a site, binding the ESNs with the site, and setting the WAN-side interface parameters of the CPEs, to generate a deployment file. The administrator then sends the file to the device administrator.

2. The device administrator uses a USB flash drive to import configurations to the CPEs and deliver the CPEs to the target site.

3. The site deployment personnel connect the necessary network cables and power on the CPEs on site. After the CPEs connect to the SDN controller, the controller determines the site to which the CPEs belong based on their ESNs and delivers service configurations specific to the CPEs at the site.

When USB flash drives are used for deployment, regardless of whether the device administrators deploy CPEs in batches before delivery or the site deployment personnel deploy CPEs on site, the deployment file needs to be prepared and stored in the USB flash drives in advance, which is a complex process.

4.4.2.4 DHCP-Based Deployment

In this method, information is transferred through a DHCP server to implement plug-and-play of CPEs. The DHCP server can assign parameters, such as an IP address and gateway information, to the CPE's WAN-side interface, and transmit the controller's IP address and token information to the CPE through DHCP Option 148. In this case, the site deployment personnel only need to connect the necessary network cables and power on the CPE on site. Note that DHCP-based deployment is available only when the CPE has been connected to the underlay network through DHCP and the DHCP server can be configured by the network administrator.

Figure 4.21 illustrates DHCP-based deployment.

DHCP-based deployment is performed through the following steps:

1. The network administrator performs a series of configurations on the SDN controller, including creating a site and setting the WAN-side interface parameters of a CPE. During this process, the network administrator only needs to specify the CPE model for the site. The administrator then enables the DHCP server to transmit the controller's IP address and token information to the CPE through DHCP Option 148.

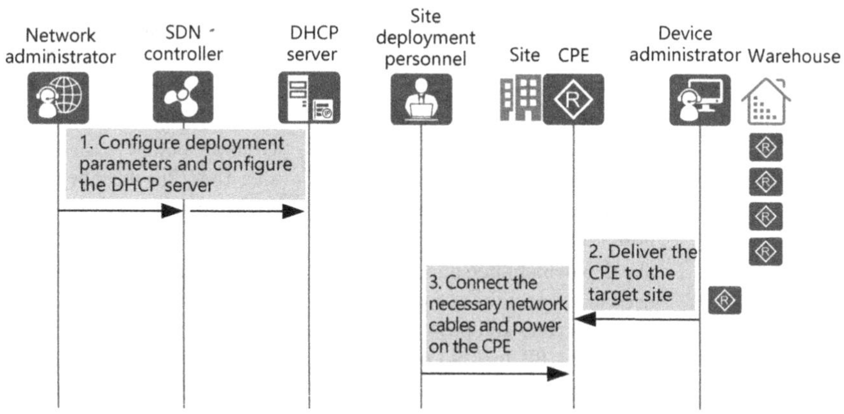

FIGURE 4.21 DHCP-based deployment.

2. The device administrator delivers a CPE of the specified model to the target site.
3. The site deployment personnel connect the necessary network cables and power on the CPE on site. The CPE obtains the WAN-side interface parameters, the controller's IP address, and token information from the DHCP server, with which the CPE can connect to the SDN controller. The controller then determines the site to which the CPE belongs based on the CPE's token and delivers service configurations specific to the CPE at the site.

Table 4.2 compares the four methods of plug-and-play deployment.

TABLE 4.2 Comparison between Four Plug-and-Play Deployment Methods

Deployment Method	Advantages	Disadvantages
Email-based deployment	• This is applicable to various modes of access. It is relatively flexible for site deployment personnel who only need to perform a few simple operations and do not need to strictly match CPEs with sites	• Site deployment personnel need to prepare deployment terminals, connect them to CPEs to be deployed, and access the URL contained in the deployment email
Registration center-based deployment	• Site deployment personnel only need to connect the necessary network cables and power on CPEs	• A registration center server needs to be deployed on the public network to provide the query service
USB-based deployment	• This method is applicable to deploying CPEs in batches. After the device administrator imports deployment-dependent configurations into CPEs before delivery, the site deployment personnel only need to connect the necessary network cables and power on the CPEs	• The device administrator prepares a USB flash drive as well as the deployment file and transfers the file into the USB flash drive. The ESNs of CPEs are bound to the target site. The CPEs must be sent to the correct site
DHCP-based deployment	• Site deployment personnel only need to connect the necessary network cables and power on the CPE. The CPE does not need to strictly match a site. As such, this mode is relatively flexible	• This method applies only to the scenario where the CPE accesses the underlay network through DHCP and the DHCP server on the underlay network must be configurable

Email-based deployment stands out for its few limitations, simple operations, and low requirements on the technical skills of deployment personnel. Therefore, email-based deployment is recommended. In addition to this method of deployment, registration center-based deployment is also commonly used, in which CPEs are automatically deployed by interacting with the registration center server. During this process, site deployment personnel do not need to perform additional operations, making this an extremely convenient method of deployment. USB-based deployment is applicable to scenarios where device administrators deploy CPEs in batches before CPEs are delivered. DHCP-based deployment has specific requirements on the underlay network and requires that a DHCP server be deployed on the underlay network and the server can be configured by the network administrator. This deployment mode applies only to carrier/MSPs resale scenarios. Compared with the email-based and registration center-based methods of deployment, the USB-based and DHCP-based deployment methods require complex operations and their application scenarios are limited. As such, these two modes are not recommended.

CHAPTER 5

Site Interconnection

As the Chinese saying goes, "If you want to get rich, build roads first." What this refers to is that we can effectively communicate with each other and create more value only when everything is connected with roads.

Similarly, SD-WAN needs to provide fast and flexible networking capabilities to facilitate communication and collaboration between an enterprise's branches, despite them being located far apart.

This chapter begins by describing the networking scenarios and design principles of SD-WAN, and then illustrates the SD-WAN networking sub-solution.

5.1 SD-WAN NETWORKING OVERVIEW

5.1.1 Networking Scenario Analysis

The SD-WAN networking sub-solution implements interconnection between enterprise branches, headquarters, DCs, cloud platforms, and legacy sites, enabling these sites to access a broad range of enterprise applications and services, such as the Internet, software as a service (SaaS) applications, and public clouds. To this end, the SDN controller needs to implement unified orchestration and control for CPEs and gateways.

Figure 5.1 shows the typical SD-WAN networking.

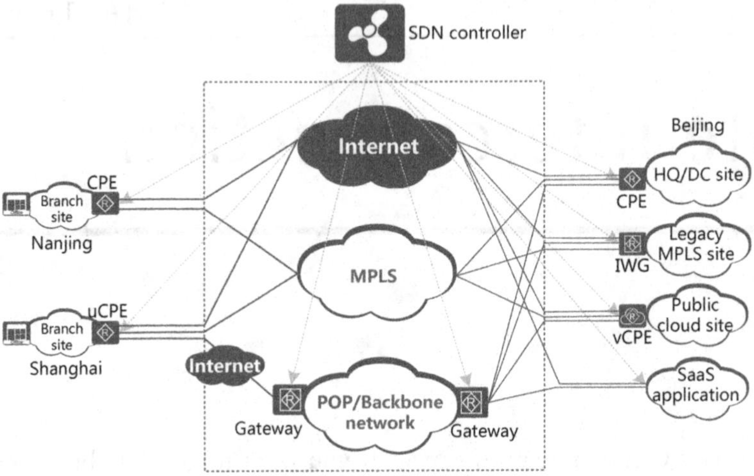

FIGURE 5.1 SD-WAN networking.

Enterprises require SD-WAN networking to support the following common service scenarios:

- Site to site

  Branches, headquarters, and DCs are three typical types of enterprise sites. Interconnection between enterprise sites is the most traditional and common service on an enterprise's WAN and can be further classified as follows: access from branch sites to the headquarters/DC, access between branch sites via the headquarters, and direct access between branch sites.

- Site to Internet

  Today's enterprises need access to the Internet anytime and anywhere. To achieve this, the SD-WAN networking sub-solution needs to support flexible Internet access modes, such as local Internet access, centralized Internet access, and hybrid Internet access.

  - **Local Internet access**: Branch sites directly access the Internet. This mode is applicable to small enterprises or scenarios where Internet traffic does not need to be centrally managed and controlled.
  - **Centralized Internet access**: Branch sites access the Internet through a centralized site, such as the headquarters. This mode applies to large and midsize enterprises, or scenarios where Internet traffic needs to be centrally managed and controlled.

- **Hybrid Internet access**: Branch sites use both local and centralized Internet access, depending on the services. This mode applies to large enterprises or scenarios where Internet traffic needs to be centrally managed and controlled, excluding traffic of specified services (e.g., Office 365).

- Site to public cloud

  A growing number of enterprises are migrating their service systems to public clouds. Against this backdrop, off-cloud sites need to quickly connect to applications deployed on public clouds on demand.

- Site to SaaS

  SaaS allows users to connect to and use cloud-based applications over the Internet, and an increasing number of enterprises are adopting this approach. As such, it is becoming increasingly important for improving access speed and reliability. To this end, the SD-WAN networking sub-solution must be able to provide optimal paths for the sites to access SaaS applications.

- SD-WAN site to legacy site

  New SD-WAN sites and legacy sites may need to communicate with each other, meaning that the SD-WAN overlay network and the traditional network must be able to communicate with each other.

  Currently, enterprise sites can interconnect with each other over the following three types of WANs:

- Private line network (guaranteed quality)

  Common private line networks use MSTP or MPLS, and offer high bandwidth and good SLA but are expensive and take a long time to deploy.

- Internet (unguaranteed quality)

  As a public network, the Internet has a wide coverage and can be quickly deployed, but it cannot guarantee SLAs are met.

- POP network

  The underlay backbone network built by a carrier or MSP is a special type of WAN, which often cannot provide the last-mile access capability for enterprise sites. To achieve cross-area interconnection among sites, an enterprise can use SD-WAN to connect its overlay network to the POP gateway of the carrier/MSP, which is then connected to the underlay backbone network.

5.1.2 Networking Design Principles

To implement on-demand, flexible, and secure SD-WAN networking, the SD-WAN networking sub-solution must be able to deliver the following capabilities:

- Interconnect sites through the IP overlay network, which is decoupled from the underlay network provided by the carrier; and deploy hybrid WAN links, such as MPLS, Internet, and LTE links.

  SD-WAN can establish overlay tunnels between each pair of sites as long as there are reachable IP routes between CPEs at the two sites, regardless of the WAN type. In this way, interconnection can be implemented between sites while the availability and flexibility of the SD-WAN networking are ensured.

- Support encryption and isolation to meet the security requirements of different service departments within an enterprise.

  Data transmitted between enterprise sites can be encrypted. In addition, an enterprise's departments can be isolated by configuring VPN isolation based on overlay technology.

- Provide a variety of network topologies to allow for interconnection between sites.

  Depending on factors such as geographical distribution, administrative management characteristics, and service needs, enterprises can choose from a wide range of topology models, including hub-spoke, full-mesh, partial-mesh, and hierarchical networking. By selecting the most suitable topology, enterprises can deliver an excellent inter-site access experience while ensuring access security.

- Support network service orchestration and automated provisioning, improving network agility.

  The enterprise WAN network model can be abstracted and defined, eliminating the need for customers to understand technical details during networking design. In addition, network service orchestration is used to implement automated network configuration, significantly improving network agility while simplifying the use of enterprise WANs.

With these principles in mind, let's delve into the detailed description of the SD-WAN networking sub-solution.

5.2 OVERLAY NETWORK DESIGN

5.2.1 Diversified Network Topologies

Based on enterprise WAN service and internal management needs, a variety of enterprise WAN topologies need to be supported. Enterprise WAN topologies are generally classified into the single-layer network topology and hierarchical network topology.

5.2.1.1 Single-Layer Network Topology

The single-layer network topology, also known as the flattened network topology, enables branch sites of an enterprise to interconnect with each other directly or through one or more hub sites, as shown in Figure 5.2.

In most cases, small and midsize enterprises and large enterprises with a small number of sites use the single-layer network topology, which can be hub-spoke, full-mesh, and partial-mesh.

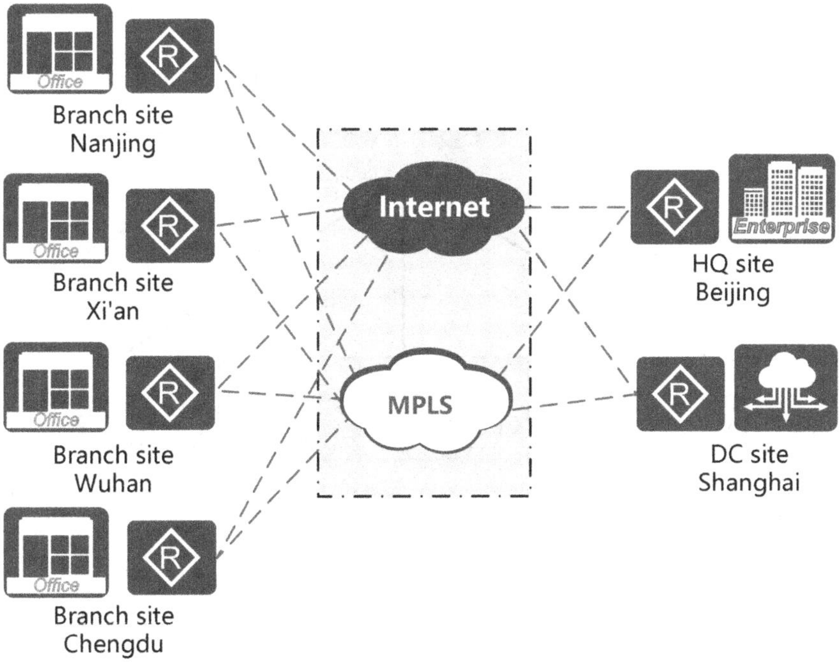

FIGURE 5.2 Single-layer network topology.

1. Hub-spoke topology

   In the hub-spoke topology, the enterprise headquarters or DC function as a hub site, while enterprise branches function as spoke sites and access servers deployed at the headquarters or DC through WANs. If branch sites of an enterprise need to communicate with each other, inter-site traffic needs to be forwarded through the hub site, as shown in Figure 5.3.

   The hub-spoke topology is applicable to enterprises with most service traffic originated from branch sites and destined for the hub site and small traffic between branch sites. Applications of such enterprises are centrally deployed on servers of the headquarters or DC, with a typical example being a chain enterprise.

   The hub-spoke topology's advantages include simplicity, scalability, and support for a relatively large number of branch sites (specifically 100–1000). However, a larger network imposes more stringent

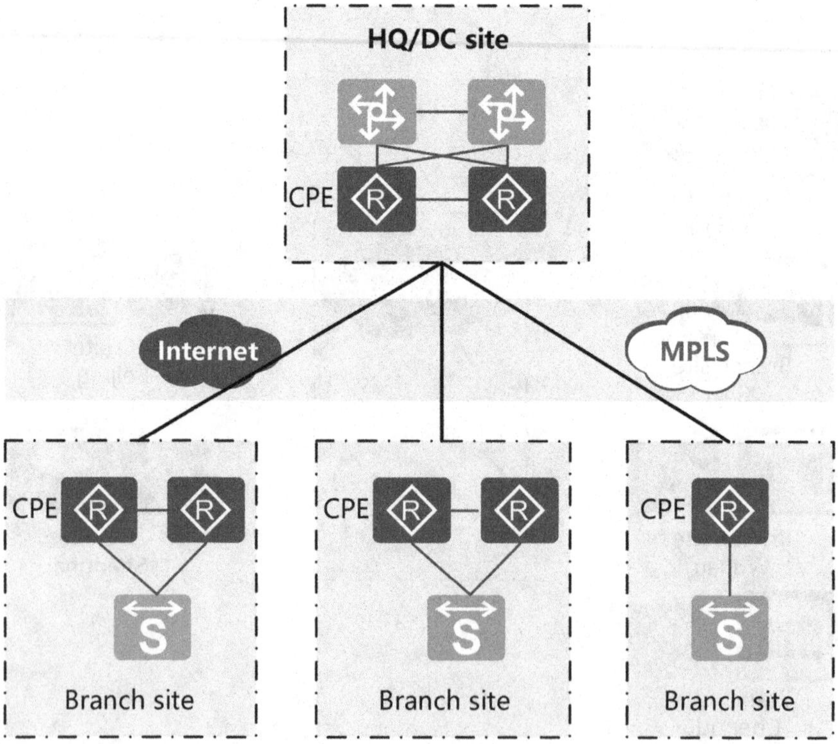

FIGURE 5.3 Hub-spoke topology.

performance requirements on devices deployed at the hub site, where CPEs that perform well on the forwarding and control planes are recommended.

2. Full-mesh topology

In the full-mesh topology, every branch of an enterprise can directly communicate with every other branch, as shown in Figure 5.4.

The full-mesh topology is applicable to small enterprises as well as large enterprises that require a high level of collaboration between branches. For example, large enterprises have a wide range of collaborative services, including high-value applications such as Voice over Internet Protocol (VoIP) and video conferencing. Such services have stringent requirements on the packet loss rate, latency, and jitter. As such, branches are recommended to directly communicate with each other.

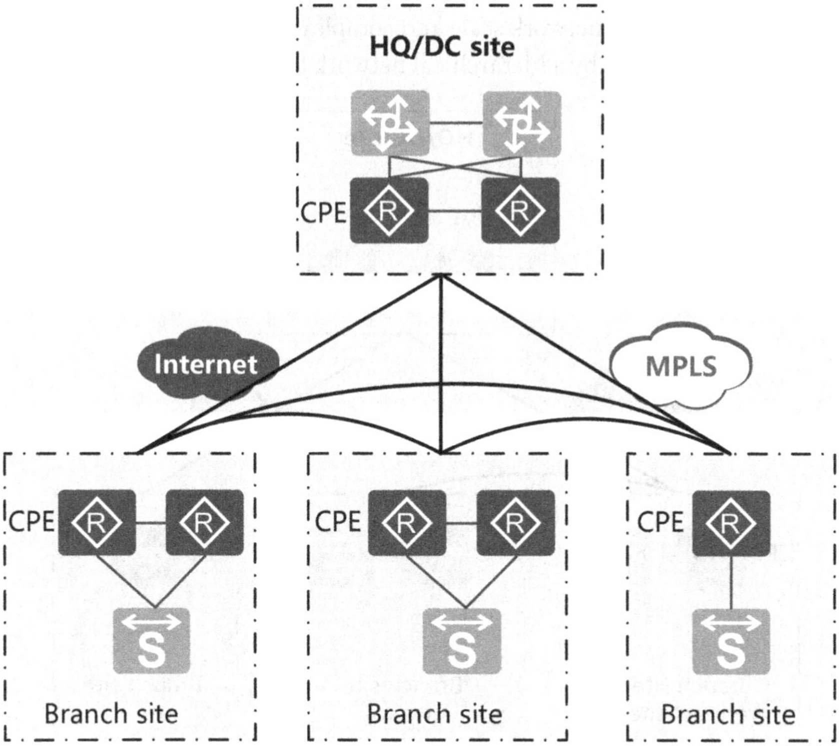

FIGURE 5.4 Full-mesh topology.

The full-mesh topology is simple and features high communication efficiency but average scalability, making it applicable to networks with only a limited number of branches (specifically 10–100).

3. Partial-mesh topology

   In reality, not all branch sites can directly connect to the underlay network. As such, the partial-mesh topology — a special topology derived from the full-mesh topology — takes shape. For enterprises that want to implement full-mesh networking, some branch sites (referred to as P sites) are unable to directly connect to the other branch sites due to insufficient WAN coverage from the local carrier. In this case, you can specify a redirect site, through which P sites can be connected to all the other sites, as shown in Figure 5.5.

5.2.1.2 Hierarchical Network Topology

If an enterprise has a large number of branch sites that span multiple provinces or even countries and continents, a flattened network topology cannot resolve issues of large network scale and complex management, which however can be addressed by a hierarchical network topology.

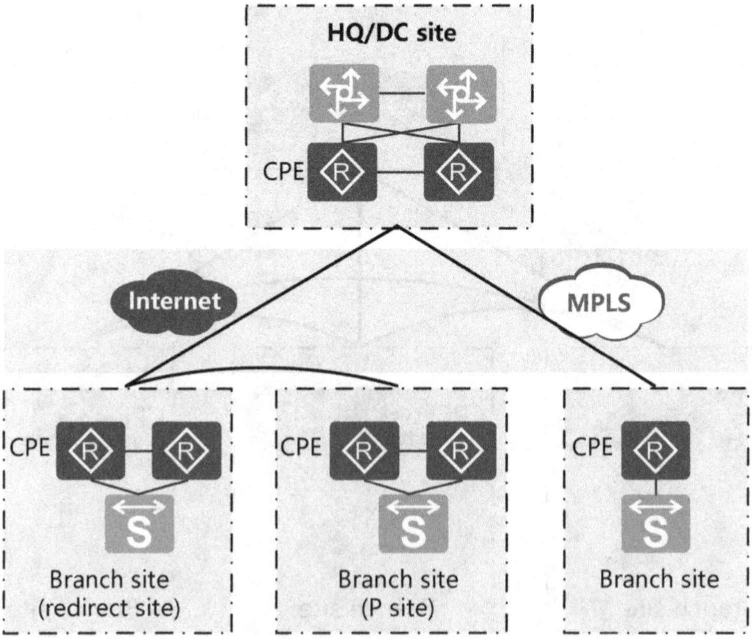

FIGURE 5.5 Partial-mesh topology.

The hierarchical network topology is a combination of multiple single-layer network topologies. The WAN that adopts the hierarchical network topology is divided into multiple areas, which are interconnected through the centralized backbone area. In this way, a large number of sites can communicate with each other across areas, as shown in Figure 5.6.

Let's take the example of a multinational enterprise that, based on the enterprise management structure, is divided into three areas: China, Europe, and America. Each area uses a single-layer network topology (hub-spoke or full-mesh) with one or more sites as border sites. The border sites constitute the backbone area (level-1 area network), which implement interconnection between areas. Border sites connect to both the level-2 and level-1 area networks. Typically, a level-2 area network is constructed based on the local public network or local MPLS private line network provided by the local carrier. Due to a wide geographical span and stringent network performance requirements, a level-1 WAN network is generally constructed using carriers' high-quality private lines, such as MPLS private lines, which are high in cost.

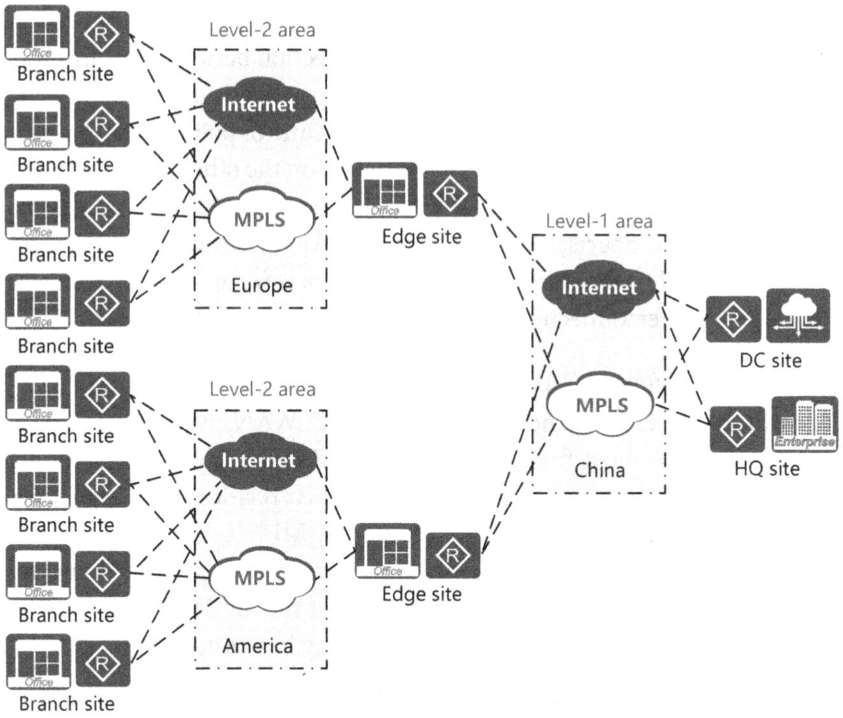

FIGURE 5.6 Hierarchical network topology.

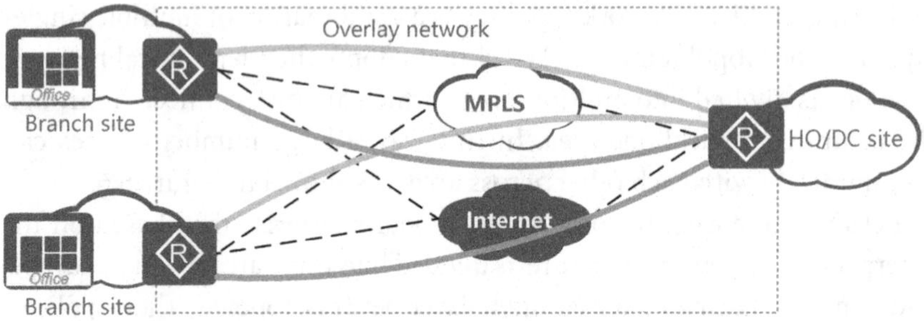

FIGURE 5.7 Overlay network.

The hierarchical network topology features a clear network structure and excellent scalability and is ideal for enterprises that have a large number of sites or multinational enterprises with sites widely distributed across countries or regions.

5.2.2 SD-WAN Network Model

After analyzing the service types and WAN networking models of enterprises, we need to consider how to build a network for interconnection between sites. As mentioned already, interconnection between sites involves the underlay and overlay networks. The underlay network, also known as the WAN, is constructed based on the WAN private lines or public networks provided by different carriers. The overlay network, on the other hand, connects different sites and ensures IP reachability based on the underlay network. The overlay network is not related to the specific WAN interconnection technology. In SD-WAN networking, an overlay network is set up between different sites to enable interconnection between them, as shown in Figure 5.7.

5.2.2.1 MEF-Defined SD-WAN Network Model

For your better understanding of how the SD-WAN overlay network is implemented, we must first define the SD-WAN overlay network model and key elements on the overlay network and their relationships. Figure 5.8 shows the SD-WAN network model defined by MEF[3].

According to MEF, the SD-WAN service is an application-aware, policy-driven connectivity service. In terms of the SD-WAN service, the SD-WAN network model involves the following key concepts:

- **SD-WAN virtual connection (SWVC)**: IP overlay connection service.
- SD-WAN virtual connection end point (SWVC EP).

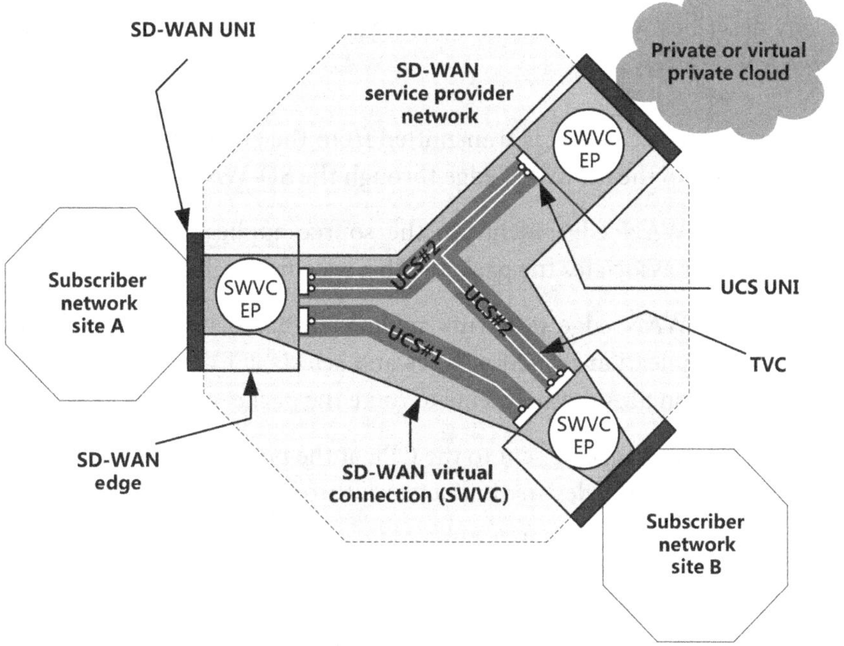

FIGURE 5.8 MEF's definition of the SD-WAN network model.

- SD-WAN service provider network.
- **SD-WAN user network interface (UNI)**: This interface is the demarcation point between the subscriber network and service provider network, and is a LAN-side service interface.
- **Subscriber network**: local service network of an enterprise site.
- **SD-WAN edge**: CPE of an enterprise site.
- **Underlay connectivity service (UCS)**: network service offerings that provide connectivity between the sites.
- **UCS UNI**: a WAN-side service interface.
- **Tunnel virtual connection (TVC)**: a virtual connection for connecting edge WAN interfaces across the underlay network. Two CPEs may be interconnected through WAN private lines of multiple underlay networks. Given this, one SWVC can be established across multiple TVCs.

To offer a better understanding of the SD-WAN network model, MEF briefly describes what it believes is the typical E2E operation process of SD-WAN services.

1. An IP service packet is transmitted from the enterprise's subscriber network to the SD-WAN edge through the SD-WAN UNI.
2. The SD-WAN edge identifies the source application type of the packet or associates the packet with a specific application.
3. The SD-WAN edge performs application-based traffic steering or other application-related policies and selects a TVC that meets the application SLA requirements to route the packet out.
4. The service packet is sent to the CPE at the peer site of the TVC and then reaches the destination network through the SD-WAN UNI of the peer site.

5.2.2.2 Huawei's Definition of the SD-WAN Network Model

Figure 5.9 shows the SD-WAN network model defined by Huawei, which is enhanced and optimized based on MEF's definition.

The SD-WAN network model defined by Huawei involves the following basic concepts:

1. Site ID

   A site ID is a globally unique identifier of an enterprise site in SD-WAN and is usually a string of digits or an IP address that is automatically allocated by the SDN controller. This site corresponds to the subscriber network defined by MEF.

2. CPE ID

   A site usually has one or two CPEs. Each CPE is uniquely identified using a CPE ID, also known as the router ID of a CPE. The CPE ID, which is often the IP address of Loopback0 on the CPE, is centrally and automatically allocated by the SDN controller. The CPE corresponds to the edge defined by MEF.

3. WAN

   A WAN, also called a transport network (TN), is provided by a carrier and usually includes the carrier's private network and public network. Different WANs have different SLA levels, service provisioning processes, and charging policies. If TNs provided by different carriers are reachable to each other, they are considered to be in the same routing domain (RD). For example, in Figure 5.9, the Internet networks provided by Internet service provider (ISP) B and ISP C can communicate with each other and therefore belong to the same RD.

The SD-WAN overlay network is constructed based on TNs. The TN corresponds to the underlay connectivity service defined by MEF.

4. WAN interface

   A WAN interface, also called a transport network port (TNP), connects a CPE to the transport network. Essentially, two enterprise branch sites are interconnected through WAN interfaces of CPEs at both sites. Key information about the TNP of a CPE includes: site ID, CPE ID, TN, public IP address and port, private IP address and port, and tunnel encapsulation type.

   TNPs are endpoints of tunnels interconnecting CPEs at two sites. The TNP corresponds to the UCS UNI defined by MEF.

5. RD

   WANs provided by different carriers are usually constructed independently and cannot communicate with each other. If they can communicate with each other, they belong to the same RD.

6. SD-WAN tunnel

   An SD-WAN tunnel is a kind of overlay tunnel. To enable two sites to communicate with each other, we need to establish a tunnel between them, with the TNPs of CPEs at the two sites acting as the endpoints of the tunnel. The TNPs can belong to the same TN or different TNs in the same RD. In this way, interworking can be implemented on the underlay network, and an SD-WAN tunnel can be directly established between the two TNPs.

   The SD-WAN tunnel corresponds to the SWVC defined by MEF.

5.2.3 VPN Design

SD-WAN implements communication between sites by establishing an overlay network on top of the underlay network. For security purposes,

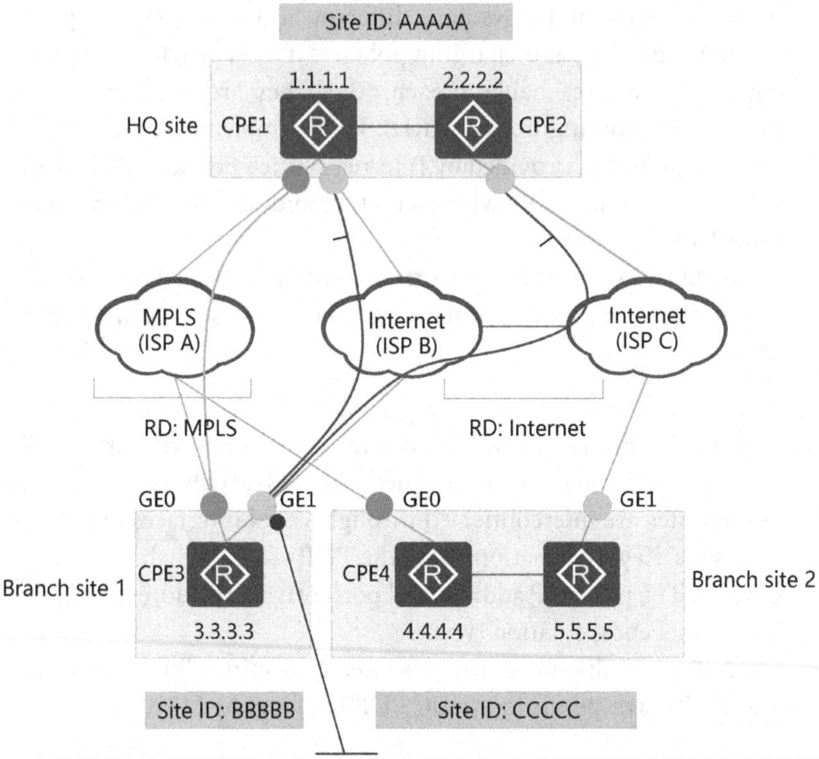

WAN Interface	Site ID	CPE ID	TN	RD	Other
GE1	BBBBB	3.3.3.3	Internet (ISP B)	Internet	Public IP address, private IP address, etc.

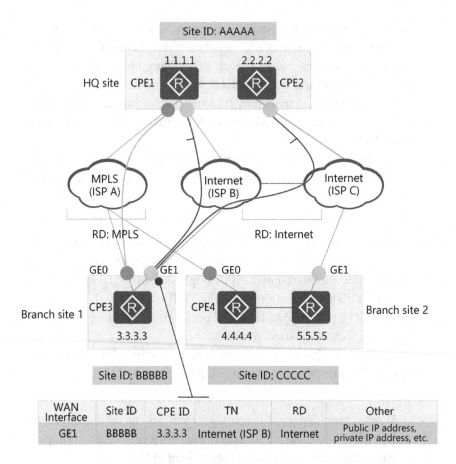

FIGURE 5.9 Huawei's definition of the SD-WAN network model.

enterprises want the overlay network to be separate from the underlay network and services of different departments to be isolated from each other. SD-WAN uses VPN technology to isolate services, with each VPN being an independent private network and logically isolated from all other VPNs, ensuring service security.

E2E VPN isolation is implemented at sites and in tunnels.

5.2.3.1 VPN Isolation at Sites

Different Virtual Routing and Forwarding (VRF) instances have different forwarding tables and cannot access each other, implementing isolation between VPNs at a site, as shown in Figure 5.10.

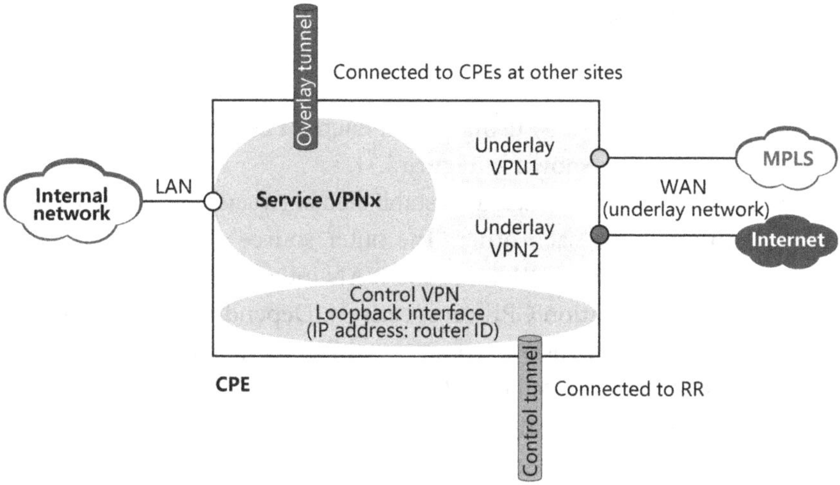

FIGURE 5.10 VPN isolation at a site.

1. Underlay VPN

   All WAN interfaces of a CPE are deployed in an independent VPN. This ensures that the underlay network is isolated from the overlay network on the CPE, and that there are no conflicts between IP addresses allocated by different carriers. In this mode, a CPE registers with the SDN controller through a VPN on the underlay network.

2. Control VPN

   CPEs at a site need to establish a connection with a remote RR. To this end, an independent control VPN with a loopback interface is planned on each CPE. The IP address of the loopback interface is used as the router ID of the CPE to establish a BGP control channel with the RR.

3. Service VPN

   Depending on the number of services to be isolated for an enterprise, one or more service VPNs can be configured to carry LAN-side services of each site and implement WAN interconnection. Service VPNs can run on static routes and numerous routing protocols, such as OSPF and BGP. What's more, VRRP can be deployed on the LAN side, and VASs such as QoS, application-based traffic steering, security, and WAN optimization controller (WOC) policies can be deployed for VPNs.

5.2.3.2 VPN Isolation in Tunnels

Through VPN isolation in an overlay tunnel, we can isolate traffic of different VPNs in the tunnel. Generally, services of an enterprise's different departments are isolated by using VPNs. Each VPN corresponds to a virtual network (VN), as shown in Figure 5.11.

One or more overlay tunnels are established between two sites that need to communicate with each other. The outer source and destination IP addresses of the tunnels are those of the WAN interfaces corresponding to the source and destination CPEs, respectively. Depending on the requirements, enterprises can choose from several tunneling technologies, such as Generic Routing Encapsulation (GRE), IPsec, and VXLAN.

To distinguish service traffic of each VPN in an overlay tunnel, an identifier needs to be added for traffic of each VPN in the tunnel. Generally, a VPN-identifying ID is added to packets that are encapsulated based on the specific tunneling technology. Typical examples include VXLAN Network Identifiers (VNIs) on VXLAN tunnels and the key field in GRE packet headers in GRE tunnels.

5.2.4 Tunnel Design

SD-WAN leverages IP overlay tunnels to carry and transmit service data between sites, making IP overlay tunneling technologies crucial. Let's have a look at the technical features of the SD-WAN tunneling technologies.

5.2.4.1 SD-WAN Tunnel Design Principles

In principle, we can use all SD-WAN tunneling technologies — such as VXLAN, GRE, and IPsec — regardless of the IP overlay technology.

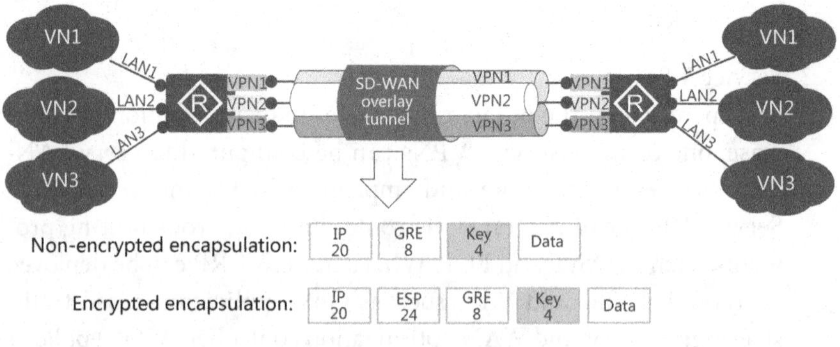

FIGURE 5.11 VPN isolation in tunnels.

In order to meet the SD-WAN networking needs, the following design requirements should be met:

1. Achieve decoupling from the underlay network to eliminate the dependency on a specific carrier's WAN networking technologies

   As long as there are reachable IP routes, SD-WAN networking uses IP overlay for WAN interconnection between enterprise sites and achieves decoupling from the specific WAN physical networking technology. Note that SD-WAN tunnels must be established based on IP tunneling technologies, including commonly used IPsec, GRE, and VXLAN. If necessary, additional protocols can be used.

2. Enable encryption technologies such as IPsec

   SD-WAN supports hybrid WAN links, including the less secure Internet links. For security purposes, if tunnels are established over Internet links, encryption technologies, such as the most commonly used IPsec technology, must be used to encrypt data. For secure WAN links, encryption is optional.

3. Isolate VPNs

   Enterprises usually need to implement isolation between their tenants and between VPNs under a single tenant. To implement this, SD-WAN needs to use a tunneling technology that supports logical VPN isolation based on VPN IDs.

4. Support network address translation (NAT) traversal

   Due to limited IPv4 addresses, many sites are without public IP addresses and can therefore only communicate using a post-NAT public IP address. SD-WAN needs to provide a mechanism similar to Session Traversal Utilities for NAT (STUN) for NAT traversal on WANs. In addition, encapsulated overlay tunnel packets between CPEs also need to support NAT traversal. Be aware that standard VXLAN does not support NAT traversal and needs to be reconstructed.

5. Support multiple IP overlay encapsulation formats to maintain good scalability

   WANs use various IP packet technologies. The mainstream Internet Protocol version 4 (IPv4) is gradually being replaced by Internet Protocol version 6 (IPv6), which will be able to support much larger scale networks. This is leading to the emergence of new IPv6-based technologies such as Segment Routing over IPv6 (SRv6). SD-WAN needs to maintain good scalability in tunnel encapsulation formats and select the desired tunnel encapsulation format based on the actual WAN IP technology that is used for interconnection between sites.

6. Maximize encapsulation efficiency

   The tunnel encapsulation format used for IP overlay must be simple and efficient, without occupying too many network bandwidth resources.

5.2.4.2 IP Overlay Tunnel Technology

VXLAN, GRE, and IPsec are common IP overlay tunneling technologies. Next, we will look at their fundamental principles before moving on to suggestions for SD-WAN tunnel design.

1. VXLAN

   VXLAN, essentially a tunneling technology, is one of the Network Virtualization over Layer 3 (NVO3) technologies defined by the Internet Engineering Task Force (IETF). It uses the MAC-in-UDP packet encapsulation technique to encapsulate Layer 2 packets using a Layer 3 protocol, extending Layer 2 networks at Layer 3. Figure 5.12 shows the VXLAN network model.

   A VN is established over VXLAN tunnels on a Layer 3 network as follows:

   a. A VXLAN tunnel endpoint (VTEP) is an edge device on a VXLAN network and is also the start or end point of a VXLAN tunnel. It encapsulates and decapsulates VXLAN packets, in which the

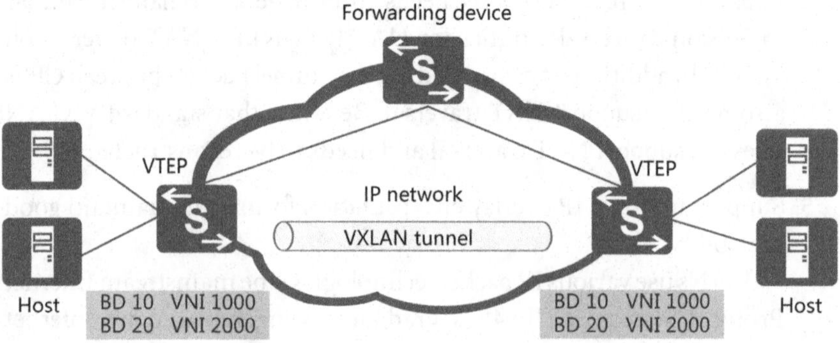

FIGURE 5.12 VXLAN network model.

source IP address is the source VTEP's IP address, and the destination IP address is the destination VTEP's IP address. A VXLAN tunnel is defined by a pair of VTEPs.

b. A VNI is similar to a traditional network's Virtual Local Area Network (VLAN) ID and uniquely identifies a subnet on a VN. Users with different VNIs cannot directly communicate with each other at Layer 2. Consisting of 24 bits, a VNI can identify up to 16 million subnets.

c. Bridge domain (BD): On a VXLAN network, every BD has a one-to-one mapping with the corresponding VNI, enabling users in the same BD to communicate with each other at Layer 2.

VXLAN is a network virtualization technology. With this technology, the source VTEP encapsulates a data packet into a UDP packet whose outer header contains the IP address and Media Access Control (MAC) address of the physical network, before sending it to the destination VTEP through the VXLAN tunnel. After receiving the packet, the destination VTEP decapsulates it and sends the original data packet to the destination device. Figure 5.13 shows the VXLAN packet format.

Table 5.1 describes the fields in a VXLAN packet

From the VXLAN network model and packet format, we can find the following characteristics of VXLAN:

a. Compared with Layer 2 isolation based on 12-bit VLAN IDs, isolation based on 24-bit VNIs has a much greater number of subnets, reaching 16 million. This enables SD-WAN to support a substantial number of tenants.

b. On a VXLAN network, a VNI can be flexibly associated with other services. A typical example is the association of VNIs with VPN instances to support complex services such as Layer 2 virtual private network (L2VPN) and Layer 3 virtual private network (L3VPN).

c. No devices on the network (apart from VXLAN edge devices) need to identify the MAC addresses of hosts.

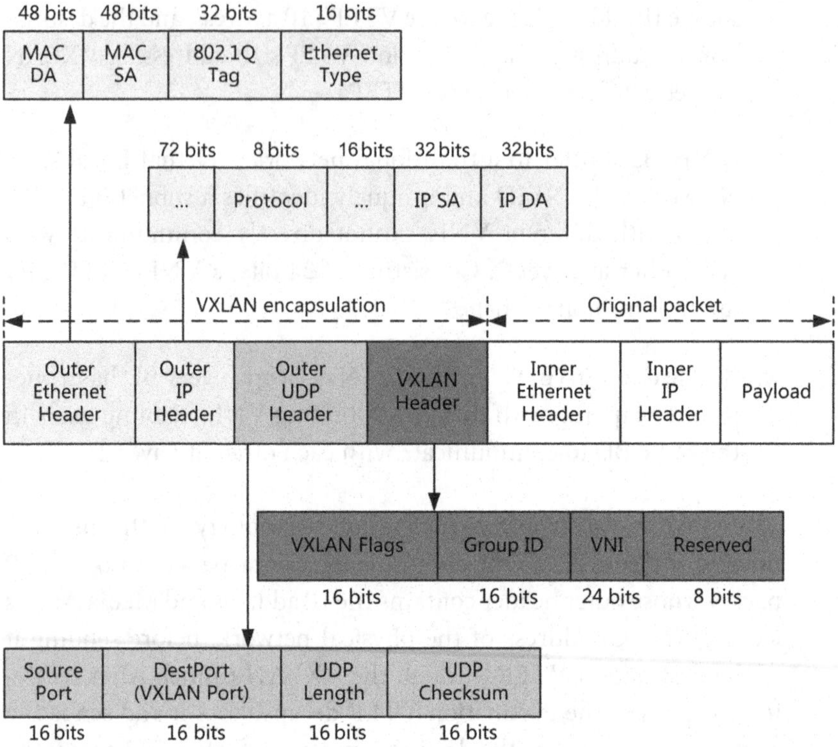

FIGURE 5.13 VXLAN packet format.

d. MAC-in-UDP encapsulation extends Layer 2 networks and decouples physical networks from VNs. Tenants can plan their own VNs, without being restricted by physical network IP addresses or BDs. This greatly simplifies network management.

e. VXLAN-encapsulated UDP source port information is obtained by performing hash calculation on inner flow information. The underlay network can perform load balancing without parsing inner packets, improving network throughput.

To sum up, VXLAN has good scalability and supports both Layer 2 and Layer 3 networks. However, it has inherent disadvantages in data encryption and NAT traversal. To compensate for this, when used as an SD-WAN tunneling protocol, VXLAN needs to be used in combination with traditional encryption protocols such as IPsec.

TABLE 5.1 Description of Fields in a VXLAN Packet

Field	Description
VXLAN header	VXLAN Flags: 16 bits Group ID: 16 bits. If the first bit of the **VXLAN Flags** field is **1**, the value of this field is a group ID; if the first bit of the **VXLAN Flags** field is **0**, the value of this field is all 0s VNI: VXLAN identifier Reserved: 8 bits. This field is all 0s
Outer UDP header	Source Port: source UDP port number, which is 16 bits long and is calculated using the hash algorithm based on the inner Ethernet packet header DestPort: destination UDP port number, which is 16 bits long. The value is 4789 UDP Length: UDP packet length, 16 bits (total length of the UDP header and UDP data) UDP Checksum: checksum of a UDP packet, which is 16 bits long. This field is used to check whether an error occurs during UDP packet transmission
Outer IP header	IP SA: 32-bit IP address of the source VTEP on a VXLAN tunnel IP DA: 32-bit IP address of the destination VTEP on a VXLAN tunnel Protocol: protocol used to transmit the data packet, 8 bits long
Outer Ethernet header	MAC DA: 48-bit destination MAC address of the next-hop device on the path to the destination VTEP MAC SA: 48-bit source MAC address of the source VTEP that sends the packet 802.1Q Tag: 32-bit VLAN tag carried in a packet. This field is optional Ethernet Type: 16-bit Ethernet frame type. The value of this field is 0x0800 when an IP packet is transmitted

2. GRE/NVGRE

GRE is a traditional IP over IP tunneling technology. Although GRE first emerged a long time ago, it is still widely used in the industry. Simply put, GRE is a protocol encapsulation format. It stipulates how to encapsulate another network protocol with one network protocol. Figure 5.14 shows the format of a GRE-encapsulated packet.

a. **Delivery header**: external protocol packet header (such as the IP packet header) that is encapsulated.

b. **GRE header**: data added to a data packet during encapsulation, including the GRE protocol and carrier protocol information.

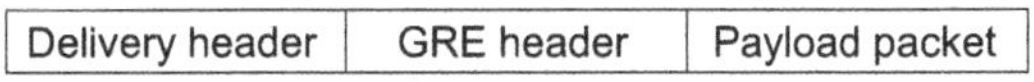

FIGURE 5.14 GRE-encapsulated packet.

c. **Payload packet**: data packet to be encapsulated and transmitted, which is referred to as a payload. When receiving a payload, the system uses the encapsulation protocol to perform GRE encapsulation on the payload and adds a GRE header to the payload to form a GRE packet. The original packet and the GRE header are then encapsulated into an IP packet, which is forwarded at the IP layer.

In contrast to VXLAN that uses a standard transport protocol, such as TCP or UDP, NVGRE uses GRE to encapsulate Layer 2 packets. NVGRE uses the first 24 bits of the **Key** field in the GRE packet header as the tenant network identifier and therefore supports about 16 million VNs, the same as VXLAN. Another difference between NVGRE and VXLAN lies in how they implement load balancing. NVGRE uses GRE for encapsulation and implements load balancing based on the extended GRE field **FlowID**. To this end, physical network must be able to identify extended information of GRE tunnels.

Instead of using flooding and IP multicast for learning, NVGRE uses a more flexible method for broadcasting, which requires support from hardware or vendors. In terms of packet fragmentation, NVGRE differs itself from VXLAN by reducing the maximum transmission unit of data packets. This reduces the size of data packets on the internal VN and eliminates the need for the transport network to support large frames. Figure 5.15 shows the NVGRE packet format.

a. **Outer Ethernet header**: In the outer frame, the source Ethernet address is set to the MAC address of the associated NVGRE terminal, and the destination Ethernet address is set to the MAC address of the destination Network Virtualization Edge (NVE).

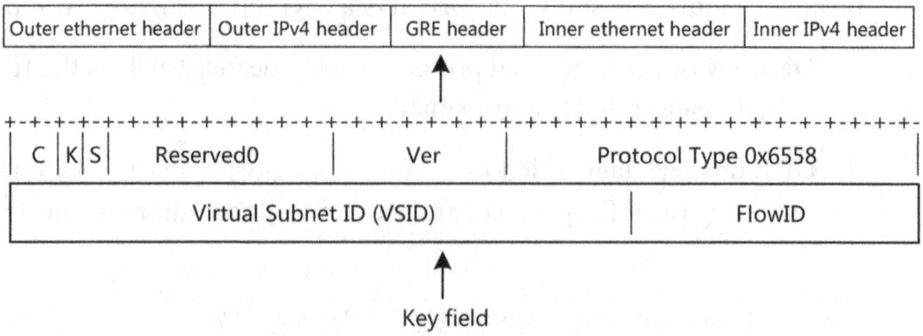

FIGURE 5.15 NVGRE packet format.

The destination terminal may be in the same physical subnet or in another subnet, and the outer VLAN tag is optional.

b. **Outer IPv4 header**: Both IPv4 and IPv6 can be used as the transport protocol of GRE tunnels. The IP address in the outer frame is called a provider address (PA).

c. **GRE header**: The key fields are described as follows:

    i. The Checksum (C) and Sequence Number (S) fields must be set to 0.

    ii. The value of the K field must be 1. The 32-bit **Key** field carries the Virtual Subnet ID (VSID) and the optional FlowID.

    iii. Protocol Type: Set this field to 0x6558 (transparent Ethernet bridge).

    iv. VSID: first 24 bits of the **Key** field.

    v. FlowID (optional): last eight bits of the **Key** field. If FlowID is not specified, it must be set to 0.

Considering the preceding GRE and NVGRE fundamental principles, we can conclude that GRE has good availability and strong protocol extension capabilities. With these advantages, it can be used in combination with traditional encryption protocols such as IPsec to meet all function requirements of SD-WAN tunnels after proper extension.

5.2.5 Routing Design

Routing design is one of the core functions of the SD-WAN networking solution. An ideal routing solution can not only ensure smooth services but also enable quick recovery from network faults based on route convergence, delivering guaranteed service experience.

SD-WAN uses BGP EVPN to design the routing solution, with local enhancements made for special scenarios such as NAT traversal on WANs. Now, let's cover the BGP EVPN fundamentals.

5.2.5.1 EVPN Fundamentals

EVPN is defined by RFC 7432 (BGP MPLS-Based Ethernet VPN), and it is a VPN technology designed for Layer 2 network interconnection. EVPN,

BGP, and MPLS IP VPN use similar mechanisms. By extending BGP Update packets, EVPN enables the control plane to learn and advertise MAC addresses between the Layer 2 networks of different sites.

EVPN uses MP-BGP to advertise the reachability of MAC and IP addresses, and it uses policy control similar to that of the traditional MPLS VPN, making EVPN distinctive with its high flexibility. MP-BGP defines a universal control plane for EVPN, introduces a new sub-address family — EVPN address family, and adds the Network Layer Reachability Information (NLRI), which is referred to as the EVPN NLRI.

The EVPN NLRI defines multiple types of BGP EVPN routes. Initially, Type 2 routes are used to carry Layer 2 MAC/IP route information. With the development of EVPN technologies, EVPN is also used to transmit Layer 3 routing information. Type 5 routes are IP prefix routes used to advertise imported external routes, and they play an important role in SD-WAN characterized by Layer 3 interconnection.

1. EVPN concepts

    a. **EVPN instance (EVI)**: Instances with the same VNI belong to the same Layer 2 BD or Layer 3 VPN.

    b. **VPN-Target**: It controls the advertising and receiving of EVPN routes, with each EVI being associated with one or more VPN targets.

    c. **VTEP**: It is the gateway at the edge of an EVPN, which is an endpoint of an IP overlay tunnel in the EVPN. In SD-WAN, CPEs and gateways can serve as the VTEPs of an EVPN.

2. Working mechanism of EVPN on the control plane

    The EVPN sub-address family has been added to BGP to negotiate BGP EVPN peer relationships. If Internal Border Gateway Protocol (IBGP) is configured, an RR can be introduced to simplify full-mesh configuration. All sites establish BGP peer relationships only with the RR, which discovers and receives BGP connections initiated by VTEPs and generates a client list to reflect the routes received from a VTEP to all the other VTEPs. The RR can be deployed independently or with an existing CPE (CPE deployed at the headquarters), as shown in Figure 5.16.

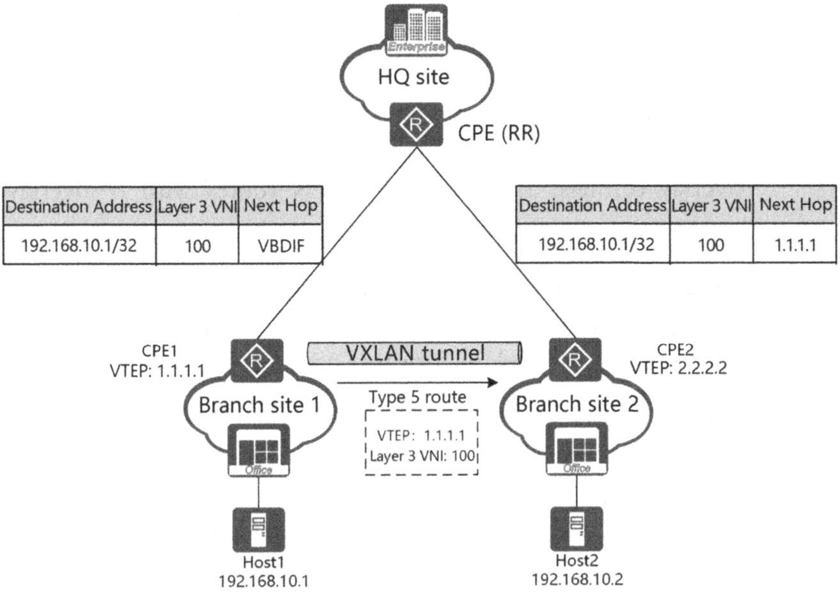

FIGURE 5.16 Working mechanism of EVPN on the control plane.

The advertised EVPN routes contain both Route Distinguisher (RD) and VPN target (also referred to as route target) information. An RD is used to identify an EVPN route, whereas a VPN target is a BGP extended community attribute used to control the advertising and receiving of EVPN routes. That is, a VPN target defines the peers that can receive EVPN routes from the local end, as well as whether the local end can receive EVPN routes from peers.

If the devices in an EVPN belong to the same autonomous system (AS), you can configure a core device as the RR to prevent IBGP peer relationships from being established between all VTEPs. In this case, the RR advertises and receives only EVPN routes, without the need to encapsulate and decapsulate VXLAN packets. The use of RR significantly simplifies network deployment.

EVPN is an IP-in-IP technology and provides an overlay architecture established based on the IP network. When a site receives a Type 5 route advertised by a remote site, and the route passes the route target check of the VRF, EVPN will attempt to establish an IP overlay tunnel with the next-hop IP address of the peer. This tunnel is used for outer encapsulation during Layer 3 forwarding.

Combining the benefits of both MP-BGP and IP overlay, EVPN offers the following advantages:

a. **Simplified network configuration and deployment**: MP-BGP is used to automatically discover edge gateways, establish IP overlay tunnels, and associate EVPN routes with IP overlay tunnels.

b. **Separated control and forwarding planes for easier management**: The control plane advertises routing information, and the forwarding plane forwards packets.

c. **Good network interoperability and scalability**: The EVPN is constructed based on the pure IP overlay technology.

With the preceding advantages, EVPN is widely used in data center networks. The technical features and advantages of EVPN are also required by SD-WAN networking.

5.2.5.2 EVPN-Based Enhanced SD-WAN Routing Solution

Enterprise WANs are more flexible and variable than both data center networks and campus networks, and they entail issues such as multiple WAN links for interconnection and WAN NAT traversal. These issues cannot be resolved through encapsulation using BGP EVPN.

Specifically, if the CPE at an enterprise site is located behind the NAT device, the CPE will not have a public IP address, and sites cannot directly communicate with each other. Additionally, due to the presence of hybrid WAN links, each site can have multiple WAN interfaces for interconnection. In terms of directly using BGP EVPN, all these factors pose new technical challenges to SD-WAN.

As such, SD-WAN enhances key information such as the next hop, with the SD-WAN routing mechanism still using BGP EVPN's basic mechanism to advertise network segment prefix routes based on Type 5 packets. This enables SD-WAN routes to carry the IP address of one WAN interface, IP addresses of multiple WAN interfaces, and IP addresses before and after NAT as the next hop. Figure 5.17 shows the design principles of the routing solution.

After the enhancements, the BGP Update packets between the CPE and RR carry the following information:

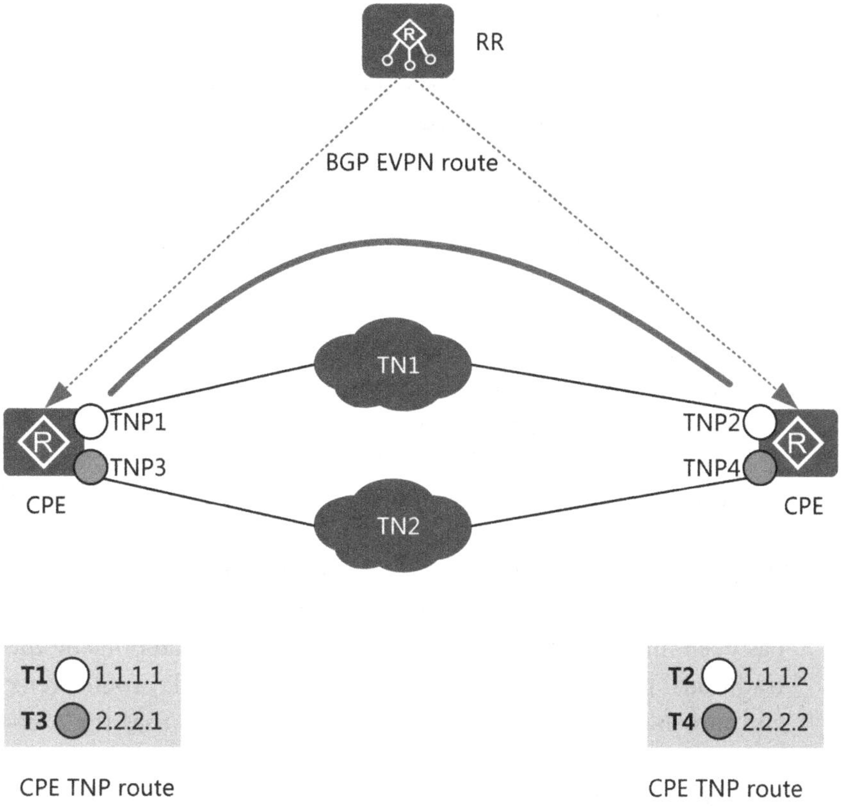

FIGURE 5.17 Design principles of the routing solution.

1. VPN route

   The VPN route information from branch sites is mainly from the LAN-side routes of related VPNs on the CPE. This information carries the next-hop TNP information and common BGP parameters such as Origin, Originator, Preference, Site ID Tag, and VPN identifier.

2. Next-hop TNP route

   The next-hop TNP route describes reachability information about the WAN interface of the CPE and TNP, including the TNP public IP address, private IP address, site ID, and route preference. A VPN route is imported to the CPE's forwarding table only when the TNP related to the VPN route is reachable. TNPs are the end-points of SD-WAN tunnels, and their key information includes the site ID, CPE ID, transport network ID, and tunnel encapsulation information (such as GRE or IPsec information). In addition, TNP routes also contain TNP information, including the TNP private IP address, public IP address, carrier, priority, site ID, and weight.

In general, an SD-WAN network may be large, consisting of hundreds or even thousands of branch sites. To prevent difficulties in network expansion due to too many route neighbors between sites, you can deploy an RR on the network. The RR exchanges routes between sites across WANs.

BGP is generally deployed between the EVPN RR and CPE to transmit overlay VPN routes of SD-WAN, including the site VPN route prefix, next-hop TNP route information, and IPsec key for data encryption on data channels between CPEs.

Sites usually have multiple uplinks, with multiple dynamic tunnels connecting to other sites. Traditional routes are established based on tunnels and next hops. Therefore, changes in the underlay link status and inter-site tunnels will cause route flapping on the entire network. To solve this problem, the indirect next hop mechanism has been introduced to the BGP control plane. Instead of using the traditional route advertisement mode where the overlay route prefix and next-hop IP address are transmitted, SD-WAN uses the overlay route prefix and next-hop site and independently advertises the next-hop site routes. This mechanism maintains stability of routes, regardless of link or tunnel changes.

Figure 5.18 shows the recommended route deployment solution for a single CPE.

- **LAN-side route**: On the LAN-side interfaces of a CPE, you can configure traditional static routing protocols, OSPF, or External Border Gateway Protocol (EBGP). The specific routing protocol to be used is the same as the one configured by the network administrator on the SDN controller for the network device that the LAN-side interface of the CPE connects to.
- **Interworking link route**: Different VPNs, for example, an underlay VPN and a service VPN, on the same CPE may need to communicate with each other. To implement this, a tunnel functioning as an interworking link needs to be established. During orchestration, the system uses a fixed routing protocol, for example, OSPF, which does not need to be configured by a network administrator.

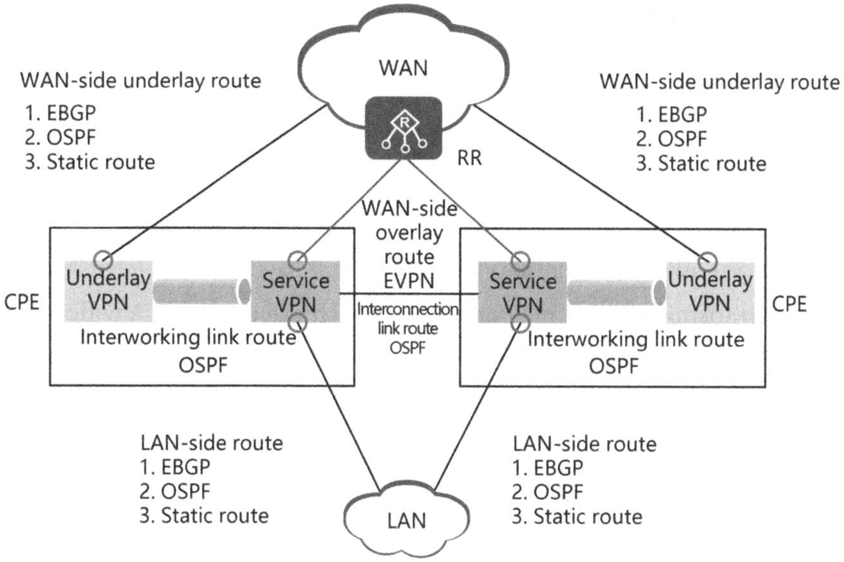

FIGURE 5.18 Routing deployment solution.

- **Interconnection link route**: If two CPEs are deployed at a site, an interconnection link needs to be established between them for interworking between the CPEs over the same VPN. By default, a specific routing protocol (e.g., OSPF) is used, which is automatically orchestrated by the SDN controller.
- **WAN-side overlay route**: EVPN is usually used to establish BGP peer relationships with the RR to advertise overlay network routes. This configuration is automatically orchestrated by the SDN controller.
- **WAN-side underlay route**: On the WAN-side interfaces of a CPE, you can configure traditional static routing protocols, OSPF, or EBGP similarly to the LAN-side route deployment mode. The specific routing protocol to be used is the same as the one configured by the network administrator on the SDN controller for the network device that the WAN-side underlay network connects to.

5.3 NETWORK RELIABILITY DESIGN

With the rapid development and popularization of information technologies, enterprises are increasingly becoming dependent on information, as well as posing increasingly higher requirements on the reliability of networks over which information is transmitted. However, building a reliable network system is a complicated and arduous process.

Network reliability design is a key and complex part of network planning and design. This section describes the reliability design in SD-WAN, including network link reliability design, network device reliability design, and core site reliability design.

5.3.1 Network Link Reliability Design

Data is transmitted over network links, which are crucial to a wide variety of sectors (such as financial services, securities, airlines, and railways) as disconnections will interrupt data transmission and cause significant economic losses.

At the physical layer, a network consists of network devices and cables and may encounter faults. For example, transport or core device failures or cable disconnections will interrupt network links. To mitigate these risks, deploy one primary and one secondary link. In most cases, traffic is not transmitted on the secondary link. This simple design enhances reliability with the need for costly private lines.

In SD-WAN, multiple uplinks are all in active state and traffic is load balanced among them based on the preconfigured traffic scheduling policies. If a link fails, SD-WAN can quickly detect deterioration in link quality and adjust the traffic policy accordingly to switch traffic to a functioning link, improving link reliability. In this way, an enterprise's private lines can be fully utilized, increasing enterprise access bandwidth and enhancing the interconnection reliability between enterprise sites.

5.3.1.1 Full Interconnection between CPEs and the Underlay Network

Enterprise sites can be interconnected using a wide range of underlay networks, including MPLS networks and the Internet.

In Figure 5.19, each CPE has one link connecting to the MPLS network, and one to the Internet, and the CPEs separately maintain information about their connections with both underlay networks. If an underlay network fails (e.g., the MPLS network), the CPEs proactively switch service traffic to the link connecting to the other underlay network (the Internet in this example), ensuring that branch sites properly communicate with each other.

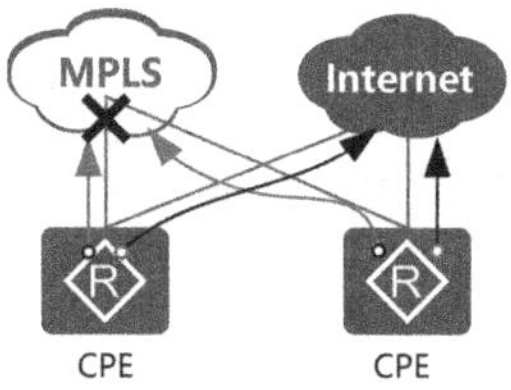

FIGURE 5.19 Full interconnection between CPEs and underlay networks.

5.3.1.2 Each CPE Connects to Only One Underlay Network

As shown in Figure 5.20, two CPEs functioning as egress gateways are deployed at a site and interconnected through an interconnection link. Each CPE connects to a different underlay network (MPLS or Internet) in the upstream direction, and only detects the status of its upstream underlay network. The two CPEs notify each other of the detection result. When one CPE detects a fault on the connected underlay network, it notifies the other CPE and adjusts the packet forwarding policy, enabling it to forward packets to the other CPE through the interconnection link.

5.3.2 Network Device Reliability Design

In addition to link faults, networks may encounter network device faults, severely compromising network reliability. Specifically, if the egress gateway of a site is faulty, sites may fail to communicate with each other; therefore, two CPEs are usually deployed at the site egress to enhance site reliability. Instead of using the traditional active/standby mode, the two CPEs work in active-active mode where if one CPE fails, the other one takes over all traffic.

Service information (including service sessions and application identification results), link statistics, and packet scheduling policies need to be synchronized between the two CPEs to enable them to work like one device. The two CPEs can detect each other's status through detection protocols.

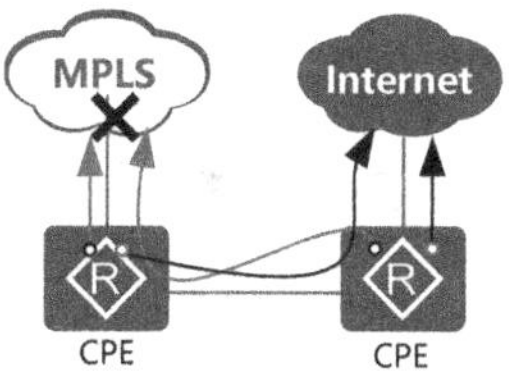

FIGURE 5.20 CPEs connected using an interconnection link, with each having a single uplink.

When one CPE detects that the other is faulty, it takes over the services of all sites, ensuring network connectivity. In addition, the packet scheduling mechanism is adjusted in real time to deliver optimal service experience.

Backup can be implemented for devices at a site in either of the following ways:

1. **LAN-side Layer 2 networking**: VRRP is used for backup, and multiple VRRP instances can be configured to implement load balancing.

   Figure 5.21 shows the LAN-side Layer 2 networking.

2. **LAN-side Layer 3 networking**: Equal-Cost Multi-Path (ECMP) is used for backup. Traditional routers learn ECMP routes from SD-WAN devices and load balance traffic based on these routes. If a CPE is faulty, the corresponding neighbor relationship is deleted and the corresponding route is withdrawn.

   Figure 5.22 shows the LAN-side Layer 3 networking.

5.3.3 Core Site Reliability Design

If a core site fails (e.g., the hub site) due to an earthquake, fire, or other force majeure event, the other sites become "information silos" and are cut off from external networks, severely affecting how the enterprise network runs. Therefore, it is imperative that core sites are reliable.

5.3.3.1 Hub Site Redundancy Design

In the hub-spoke topology, the data of all branch sites is transmitted through the hub site. If the hub site fails, the entire network will break down. To prevent this, redundancy must be implemented for the device at the hub site,

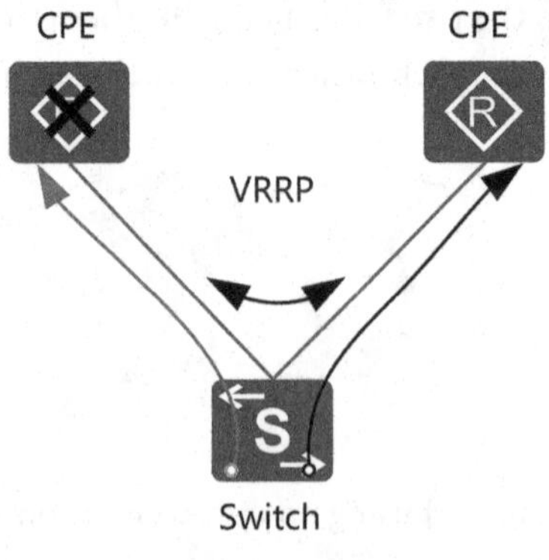

FIGURE 5.21 LAN-side Layer 2 networking.

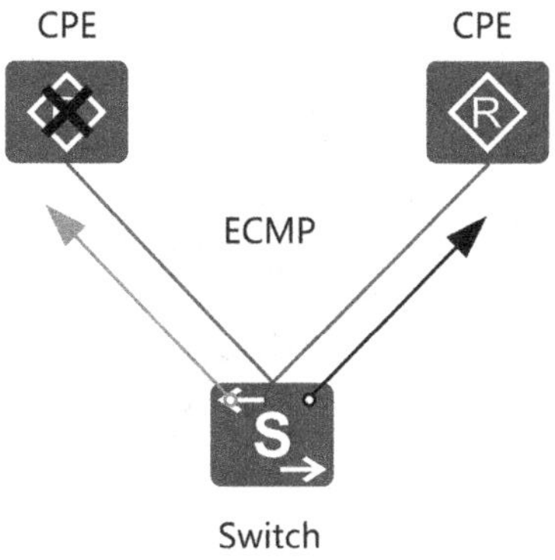

FIGURE 5.22 LAN-side Layer 3 networking.

as well as the hub site itself. SD-WAN supports the dual-hub redundancy solution. If the active hub site is faulty, all traffic is quickly switched to the standby hub site, without manual intervention.

Figure 5.23 shows the hub site redundancy design.

5.3.3.2 Redirect Site Redundancy Design

As mentioned above, in a full-mesh network, sites communicate with each other through tunnels. If the underlay network between sites is unavailable, direct tunnels cannot be established between them. To resolve this problem, SD-WAN introduces redirect sites. To enhance redirect site reliability,

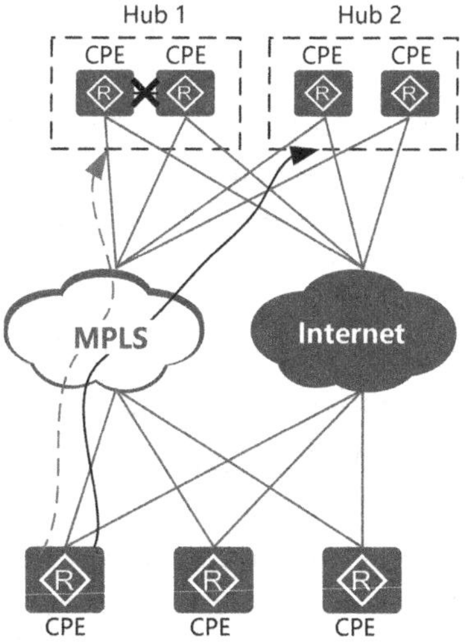

FIGURE 5.23 Hub site redundancy design.

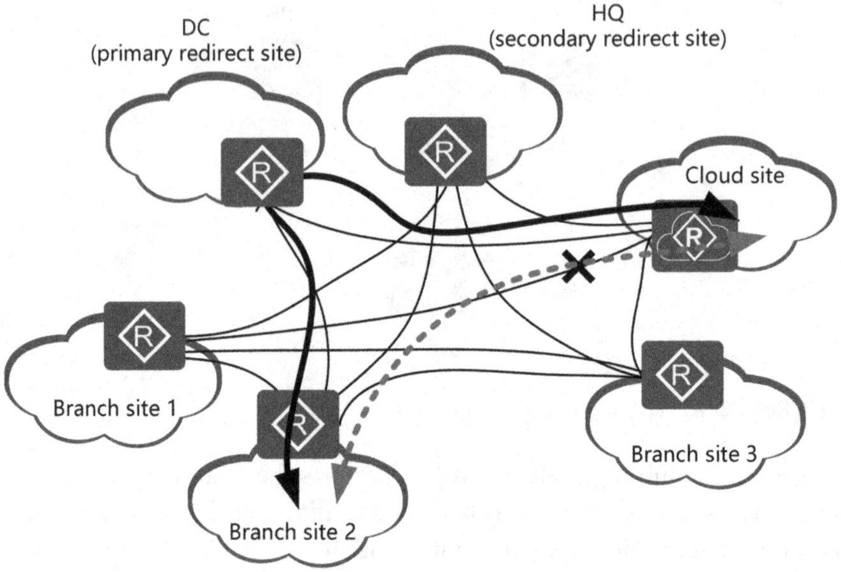

FIGURE 5.24 Redirect site redundancy design.

deploy dual redirect sites in primary/secondary mode where if the primary redirect site fails, traffic is quickly switched to the backup redirect site.

A redirect site forwards traffic transmitted between other sites and therefore poses high requirements on throughput. Generally, the redirect site is deployed at the headquarters, DC, or midsize/large branches.

In Figure 5.24, a direct tunnel cannot be established between branch site 2 and the cloud site. Therefore, to enable communication between them, two redirect sites are deployed in primary/secondary mode (one at the DC and the other at the headquarters). In this way, traffic between branch 2 and the cloud site is forwarded through a redirect site.

5.4 NETWORK ORCHESTRATION AND AUTOMATION

5.4.1 Fundamentals of Network Orchestration

On traditional enterprise WANs, deployment and adjustment are performed manually and involve switching, routing, security, WAN optimization, and other complex network technologies. This leads to an extended operation time, low efficiency, and high error rate. Therefore, high expectations are placed on network automation, which can be implemented through network orchestration.

Orchestration is a concept that originally relates to the arts field, and it involves the selection of different instruments for different parts of a musical piece.

In the network field, orchestration is a policy-driven method for implementing network automation by coordinating the hardware and software required for running applications or services.

Network orchestration is a method of automatically executing network service deployment requirements, minimizing the level of manual intervention required for services or service deployment. For example, a network administrator plans to create the same service department at two sites and implement interworking between them. To achieve this, the network administrator simply needs to create sites and a VPN on the SDN controller, which then automatically delivers the related network configurations to CPEs at the two sites.

Through abstracting network behaviors and capabilities, SDN centrally orchestrates and controls network behaviors. Simply put, network orchestration is the organizing and scheduling of abstract network objects, and it includes allocating and scheduling network resources. The abstraction of network services can be designed at three layers: point, line, and plane.

5.4.1.1 Point: Abstraction of Single-Point NE Functions

Devices of different models from various vendors may be deployed on a WAN. To minimize the impact of these differences on the network and enable the network administrator to focus on services, abstract the service functions provided by each network device.

For example, when a network administrator adds a WAN interface of a CPE to a VPN on the SDN controller, the SDN controller automatically adds the interface to the VRF corresponding to the VPN and modifies network resource information (such as VLANs, IP addresses, and QoS queues) as well as routing protocol parameters related to the interface. During this entire process, the network administrator does not need to learn or manually enter the VRF and interface information of the VPN, or manually plan interface parameters such as IP addresses. In another example, a network administrator needs to create a site with dual CPEs. To do this, the network administrator only needs to abstract the services that are to be created for the site, create services for the site, and specify two CPEs. The SDN controller then automatically converts the services into configurations and delivers the configurations to the two CPEs.

5.4.1.2 Line: Abstraction of Network Connectivity Functions

After abstracting single-point NE functions, you need to abstract network connectivity functions. For example, to interconnect two sites, you simply need to abstract a site interconnection service, create the service, and specify the two sites. The SDN controller then automatically delivers the related configurations, such as BGP peers, to the two sites.

Why don't we directly use the abstracted single-point NE functions for orchestration? Of course, this is a feasible method; however, the network administrator must have extensive professional knowledge and experience, as well as be very familiar with the network functions and technical details of each CPE. This makes service creation skill-dependent and complicates service orchestration. As such, this method is more suitable for professionals.

5.4.1.3 Plane: Abstraction of Networking Functions

Orchestration through abstracting networking functions provides network administrators with design capabilities that are closer to the service language. The difference between orchestration using this method and orchestration using abstracted single-point NE functions is like the difference between an object-oriented advanced programming language and an assembly language. During orchestration, service objects can be associated with different physical network devices in order to reuse network services.

Network service units are organized and configured based on the network model, and network model complexity determines service orchestration complexity.

Leveraging network orchestration, SD-WAN significantly simplifies the multibranch networking of enterprises.

Figure 5.25 shows the orchestration process.

The orchestration process is described as follows:

1. Create an enterprise site, which can be a branch site, headquarters site, DC site, or even a cloud site on the public cloud. Enter the basic connection information about the site, such as the type and number of CPEs at the site, number and type of WAN interfaces, and information for interconnection with the controller to implement ZTP.

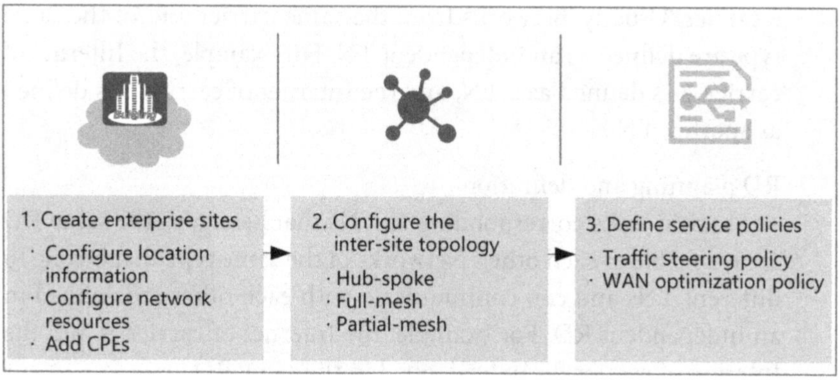

FIGURE 5.25 Orchestration process.

2. Configure the inter-site topology, which can be a combination of multiple topology modes such as hub-spoke, full-mesh, and partial-mesh. Then plan the VPN and topology between sites.

3. Define the service policies required during SD-WAN running, such as traffic steering and WAN optimization policies.

5.4.2 Tunnel Orchestration

An SD-WAN site usually has multiple physical uplinks for connecting to various types of networks provided by different carriers. Therefore, proper planning and design are required to establish overlay tunnels over the physical links.

5.4.2.1 Overlay Tunnel Establishment

After a CPE learns the TNP information of the peer CPE, the CPE checks the connectivity and determines whether to establish an overlay tunnel with the TNP of the peer CPE. Before establishing an overlay tunnel, you need to make necessary preparations, which include determining the number of WAN-side links at a site, types of networks to which CPEs are connected, and whether the networks can communicate with each other. Typically, the preparations involve: TN planning and definition as well as RD planning and definition.

1. TN planning and definition

   A TN defines the type of physical link on the WAN side of a site and is determined by the type of WAN access network provided by a carrier. Usually, networks from the same carrier and of the same type are defined as an independent TN. For example, the Internet of carrier A is defined as a TN, and the Internet of carrier B is defined as another TN.

2. RD planning and definition

Physical links corresponding to TNs that belong to the same RD are reachable to each other. Networks of the same type that belong to different TNs and can communicate with each other are defined in an independent RD. For example, the Internet of carrier A and the Internet of carrier B can be defined in the same RD.

After the TN and RD are determined, you can establish overlay tunnels based on the actual conditions. For example, if there are multiple WAN interfaces that belong to different TNs between sites, and the TNs belong to different RDs and cannot communicate with each other, create an overlay tunnel for each TN.

In Figure 5.26, an overlay tunnel can be established between the hub site and spoke site 1, and between the hub site and spoke site 2 through the MPLS or the Internet TN.

If a site has multiple WAN interfaces that belong to different TNs, and the TNs can communicate with each other (in the same RD), multiple overlay tunnels can be established between the sites, as shown in Figure 5.27.

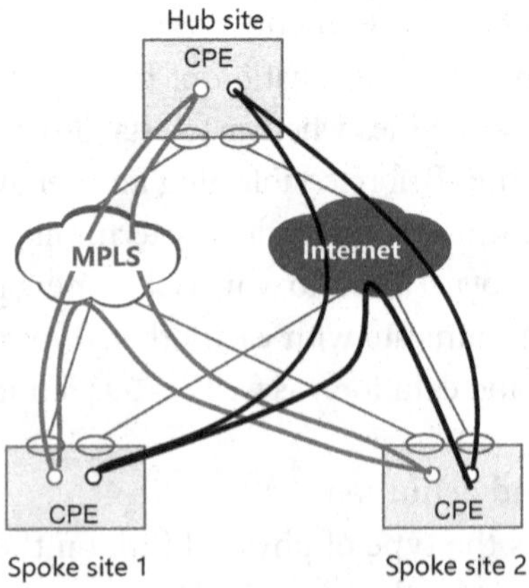

FIGURE 5.26 Overlay tunnels established between sites in different RDs.

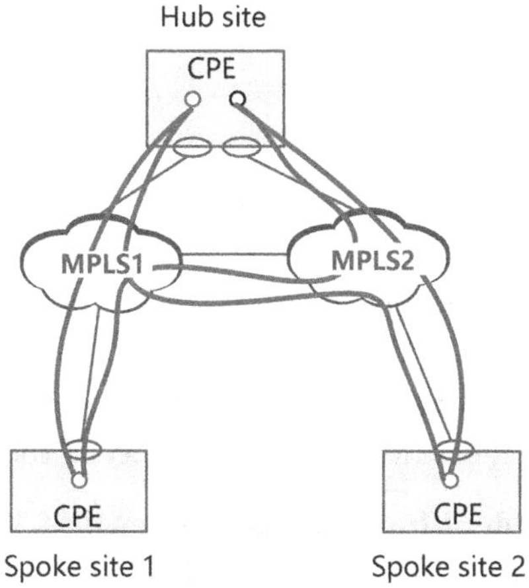

FIGURE 5.27 Overlay tunnels established between sites in the same RD.

5.4.2.2 Examples of Overlay Tunnel Orchestration

After the preceding introduction of overlay tunnel establishment principles, let's cover specific examples of overlay tunnel orchestration.

In Figure 5.28, site 1 and site 2 are both configured with dual gateways and each has two WAN links: The MPLS link connects to CPE 1, and the Internet link connects to CPE 2. Site 3 is a single-gateway site with two WAN links: an MPLS link and an Internet link.

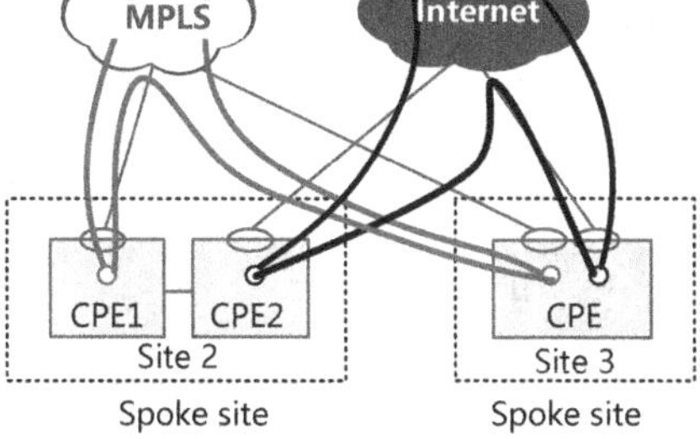

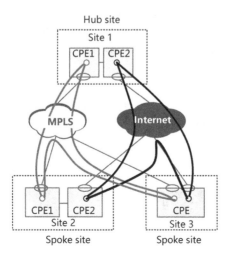

FIGURE 5.28 Overlay tunnel orchestration in dual-gateway networking.

To implement interworking between sites, design overlay tunnels as follows:

- Establish overlay tunnels among site 1, site 2, and site 3 through MPLS links.
- Establish overlay tunnels among site 1, site 2, and site 3 through Internet links.

  To orchestrate overlay tunnels on the preceding networking, define two TN types (MPLS and Internet) on the SDN controller:

- **Dual-gateway dual-link template**: The MPLS link belongs to CPE 1, and the Internet link belongs to CPE 2. This template is applied to site 1 and site 2.
- **Single-gateway dual-link template**: The MPLS and Internet links connect to the same CPE and are bound to site 3.

5.4.3 Topology Orchestration

Based on an enterprise's service requirements, SD-WAN supports the following typical topology models for inter-site interconnection:

5.4.3.1 Hub-Spoke

The hub-spoke topology model is applicable to an enterprise that needs all branch sites to communicate with each other through the headquarters for centralized security control. In this model, dual hub sites can be deployed to enhance reliability, as shown in Figure 5.29.

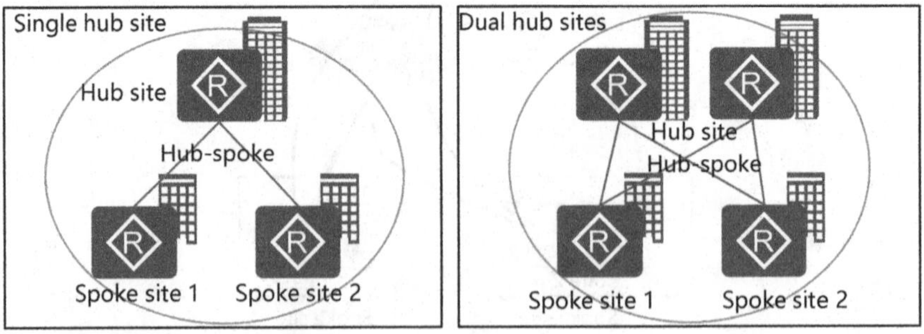

FIGURE 5.29 Hub-spoke topology model.

5.4.3.2 Full-Mesh

The full-mesh topology model is applicable to an enterprise that needs direct communication between all sites in order to eliminate the delay in diverting traffic through the headquarters, as shown in Figure 5.30.

5.4.3.3 Partial-Mesh

The partial-mesh topology model is applicable to an enterprise that requires direct communication between most sites, and the remaining sites communicate with others through a third site (also known as the redirect site). For example, in Figure 5.31, spoke site 2 and spoke site 3 cannot directly access spoke site 4. Instead, they access spoke site 4 through spoke site 1, which is the redirect site.

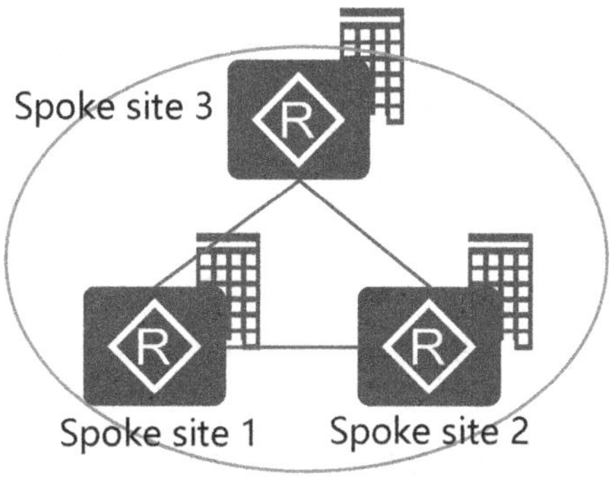

FIGURE 5.30 Full-mesh topology model.

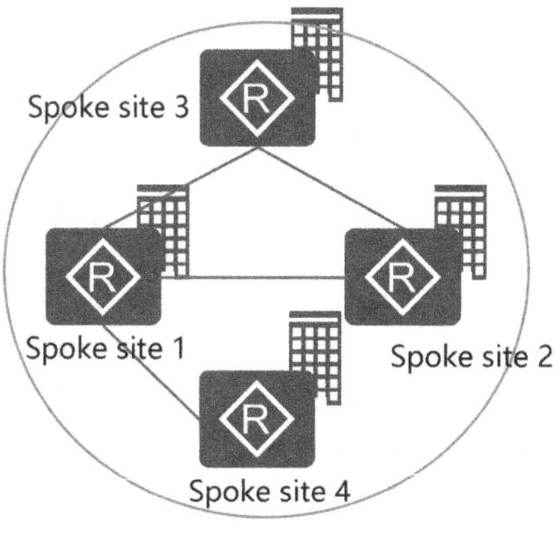

FIGURE 5.31 Partial-mesh topology model.

5.4.3.4 Hierarchical

The hierarchical topology model is applicable to a large-scale cross-area enterprise, where sites are deployed by area, sites in the same area communicate with each other directly or through the hub site, and sites in different areas communicate with each other through the hub site. Sites in an area communicate with sites in other areas through an edge site, which is also known as a border site. To enhance reliability, deploy two border sites in each area, as shown in Figure 5.32.

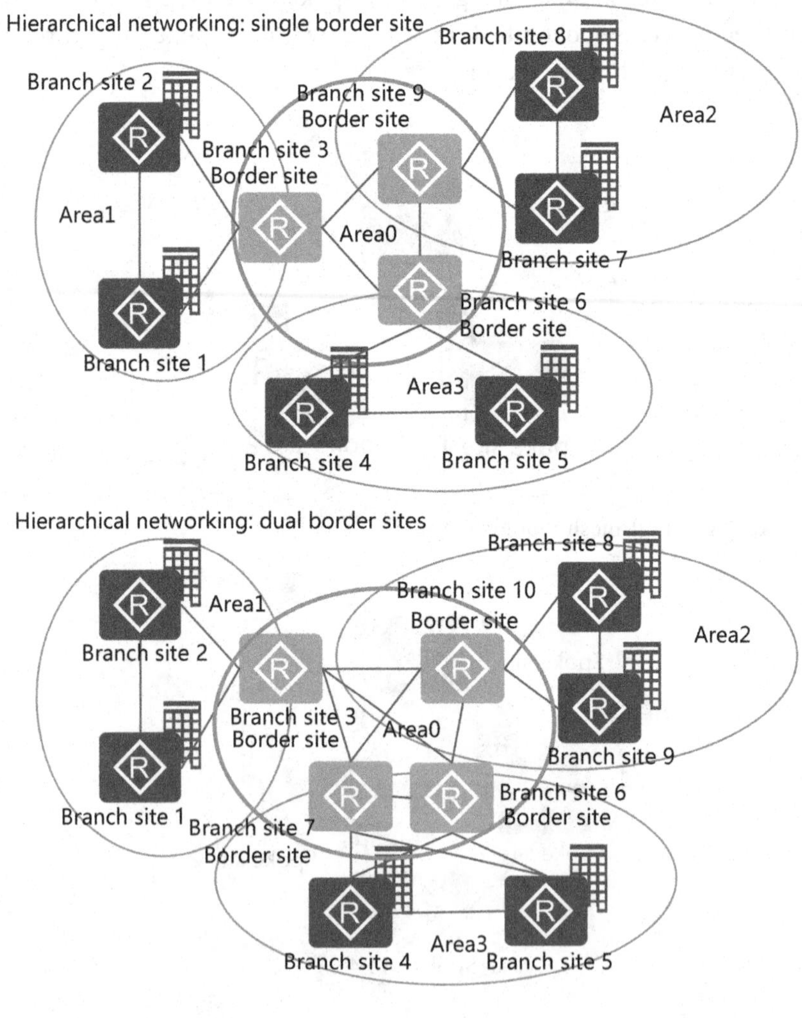

FIGURE 5.32 Hierarchical topology model.

The SDN controller implements orchestration for the hub-spoke, full-mesh, partial-mesh, and hierarchical topology models. In most cases, a network administrator specifies a topology model between the connected sites on the SDN controller. The SDN controller then generates the corresponding network model based on the topology model, converts the network model into routing policies, and delivers the routing policies to an RR. The RR controls the advertising and receiving of routes for different sites based on the routing policies delivered by the SDN controller. In this way, inter-site access is implemented based on the topology model planned by the network administrator.

Figure 5.33 illustrates the topology orchestration process.

The topology orchestration process is described as follows:

1. On the SDN controller, a network administrator specifies a topology model for each service VPN.

2. Routing policies are configured on the RR to control route learning at sites. These routing policies are automatically orchestrated and generated by the SDN controller based on the topology model configured by the network administrator.

3. The RR matches the routing policy based on the site ID to filter routes or modify the next-hop site ID.

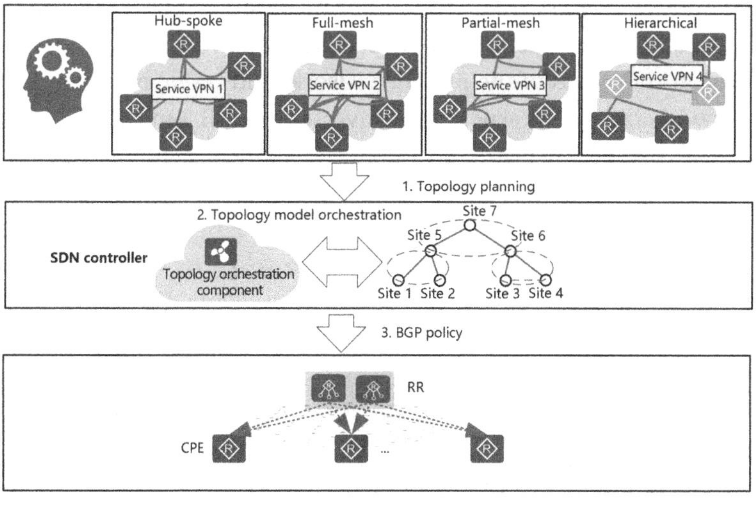

FIGURE 5.33 Topology orchestration.

Sites in different areas can use different networking modes, including:

- **Hub-spoke networking**: When a spoke site learns a route from another spoke site, the next hop of this route needs to be changed to the hub site.
- **Full-mesh/Partial-mesh networking**: A site can receive routes from all the other sites. If there are any redirect sites, add them as the backup next hop for all routes.
- **Hierarchical networking**: When a non-border site in an area receives routes from sites in other areas, the next-hop site needs to be changed to the local area's border site. When a border site in an area receives routes from sites in another area, the next-hop site needs to be changed to the other area's border site or the hub site for interconnection between areas.

5.5 INTERNET ACCESS

The globalization, digitization, and cloudification of enterprise services means that enterprise sites demand greater Internet access, leading to a wider range of available Internet access services.

In most cases, personal users use private IP addresses, and CPEs use the carrier-assigned public IP addresses to connect to the Internet. Before a user can access the Internet, their private IP addresses must be translated into a public IP address. To do this, CPEs must support the NAT function.

Figure 5.34 shows the typical NAT process. The CPE (NAT device) is deployed at the egress of the private network, and traffic between the private network and the Internet must traverse the CPE. On the CPE, a NAT table saves mapped private and public IP addresses, implementing the NAT function.

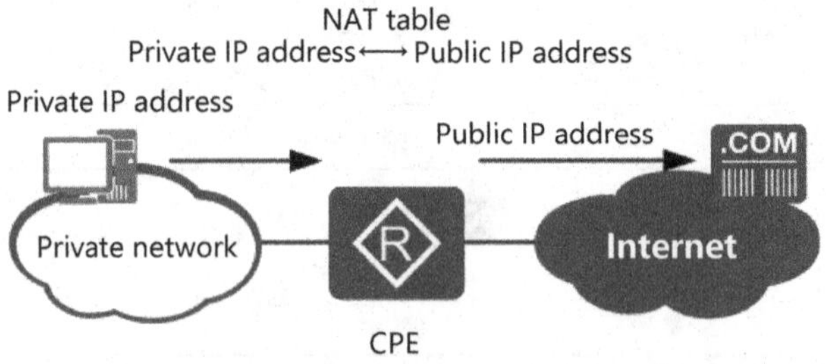

FIGURE 5.34 NAT.

In SD-WAN, the CPE provides the source IP address translation function to translate private IP addresses into public IP addresses so that users on a private network at a site can access the Internet. This function is implemented mainly through Network Address and Port Translation (NAPT) and Easy IP.

1. NAPT

   NAPT translates both IP addresses and port numbers. It does this by translating multiple combinations of IP addresses and port numbers into a public IP address, through which multiple users on the private network access the Internet.

   Figure 5.35 shows the NAPT process.

   The NAPT process is described as follows:

   a. Host A, on a private network, attempts to access a server on the Internet by sending a packet to the CPE, with the source IP address being 10.1.1.100 and source port number being 1025.

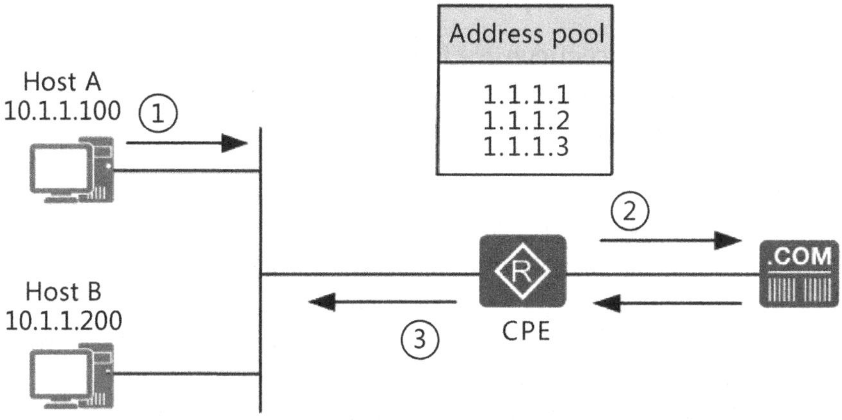

NAT table	
Source Address:Port	Destination Address:Port
10.1.1.100:1025	1.1.1.1:10025
10.1.1.200:1025	1.1.1.1:10026

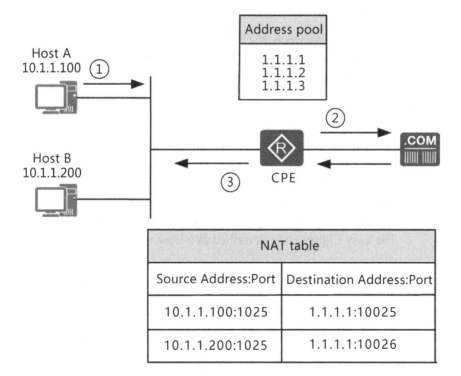

FIGURE 5.35 NAPT process.

b. After receiving the packet from host A, the CPE selects an idle public IP address (such as 1.1.1.1) from its IP address pool, replaces the source IP address of the packet with the selected public IP address, and then replaces the source port number of the packet with the corresponding port number (e.g., 10025). The CPE then creates a NAT entry and sends the packet to the Internet. After the translation, the source IP address is 1.1.1.1 and the source port number is 10025.

c. After receiving a response packet from the server, the CPE searches the NAT table for a matching entry and finds the entry created in step 2, replaces the destination address in the packet with the address of host A and the destination port number with the port number of host A, and then sends the packet to host A. If host B on the private network also accesses a server on the Internet, the CPE assigns the same public IP address 1.1.1.1 but a different port number (e.g., 10026) to host B, and creates another NAT entry.

d. NAPT is a common NAT mode that allows a large number of private network users to access the Internet by using a limited number of public IP addresses.

2. Easy IP

Easy IP uses the public IP address of an outbound interface on the CPE as the post-NAT address and translates both the IP address and port number. Easy IP can be regarded as a special type of NAPT and allows a public IP address to be used by multiple private network users for Internet access.

Figure 5.36 shows the Easy IP process.

The Easy IP process is described as follows:

a. Host A, on a private network, attempts to access a server on the Internet by sending a packet to the CPE, with the source IP address being 10.1.1.100 and source port number being 1025.

b. After receiving the packet from host A, the CPE replaces the source IP address of the packet with the public IP address (e.g., 1.1.1.1) of the outbound interface for connecting to the Internet, replaces the source port number of the packet with

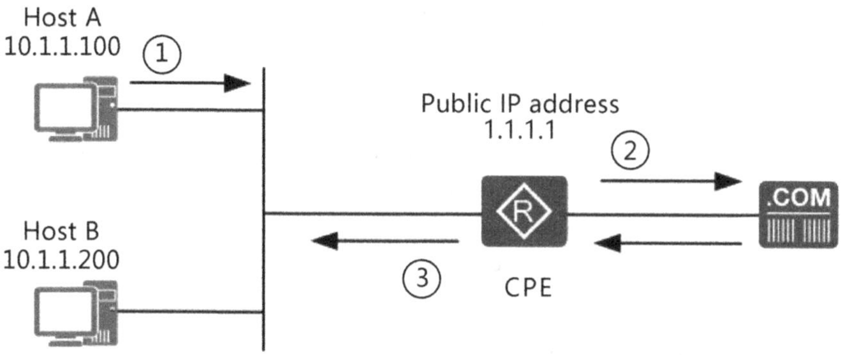

NAT table	
Source Address:Port	Destination Address:Port
10.1.1.100:1025	1.1.1.1:10025
10.1.1.200:1025	1.1.1.1:10026

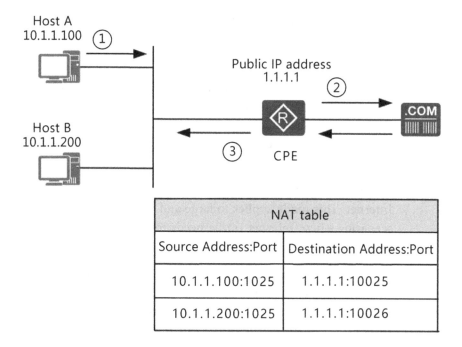

FIGURE 5.36 Easy IP process.

the corresponding port number (for example, 10025), creates a NAT entry, and sends the packet to the Internet. After the translation, the source IP address is 1.1.1.1 and the source port number is 10025.

c. After receiving the response packet from the server, the CPE searches the NAT table for a matching entry and finds the entry created in step 2, replaces the destination address and port number with those of host A, and sends the packet to host A. If host B on the private network also accesses a server on the Internet, the CPE assigns the same public IP address 1.1.1.1 but a different port number (e.g., 10026) to host B, and then creates another NAT entry.

Easy IP is especially suitable for small sites that require Internet access. This is because a small site has limited hosts on the private network, and the outbound interface of the CPE usually obtains a dynamic public IP address through dial-up. In such cases, Easy IP enables private network users to access the Internet by using the IP address of the CPE's outbound interface.

Enterprises can determine their Internet access modes based on the enterprise's scale as well as IT management and control policies. SD-WAN provides the following Internet access modes:

- **Local Internet access**: Internet traffic of all sites in an enterprise is routed from the local Internet link to the Internet. This mode is applicable to small enterprises or scenarios where Internet traffic does not need to be centrally managed and controlled.
- **Centralized Internet access**: Internet traffic of all sites in an enterprise is routed from the centralized Internet gateway to the Internet. This mode applies to large and midsize enterprises, or scenarios where Internet traffic needs to be centrally managed and controlled.
- **Hybrid Internet access**: By default, most of an enterprise's Internet traffic is routed through the centralized Internet gateway, while some service-specific Internet traffic is routed through the local internet link. This mode is suitable for large enterprises or scenarios that primarily require Internet traffic to be centrally managed and controlled, with some application-specific traffic (e.g., Office 365) being directly routed to the Internet. This minimizes the delay for application-specific traffic, improving efficiency.

5.5.1 Local Internet Access

Local Internet access, also known as local breakout, is applicable to small enterprises, where each site has an independent Internet link and Internet traffic does not need to be centrally managed and controlled. Local Internet access policies can be configured on a per-department and per-site basis.

In local Internet access mode, SD-WAN enables a site to use multiple outbound interfaces to access the Internet. Internet access policies can be formulated for each department so that they can select outbound interfaces from multiple outbound interfaces based on the outbound interface priorities, implementing Internet link backup.

In local Internet access, the NAT function (e.g., Easy IP) can be configured on an outbound interface. The IP address of the outbound interface is then used as the post-NAT public IP address. Figure 5.37 shows the local Internet access scenario.

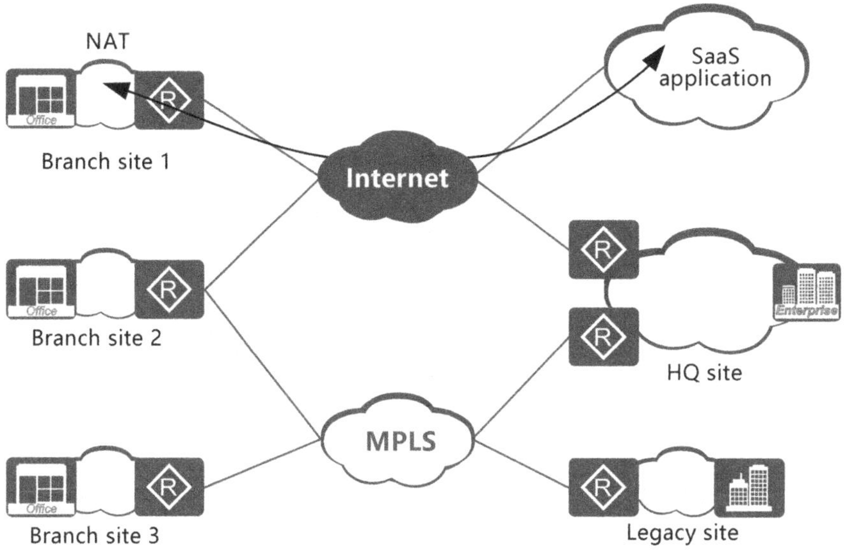

FIGURE 5.37 Local Internet access.

5.5.2 Centralized Internet Access

Centralized Internet access is applicable to midsize and large enterprises whose sites do not have links to the Internet and need to access the Internet through a centralized gateway, implementing centralized management and control of Internet traffic.

Based on the enterprise network topology plan, centralized Internet access can be further classified into intra-area centralized Internet access and global centralized Internet access. An enterprise can use both modes to enhance Internet access reliability.

5.5.2.1 Intra-area Centralized Internet Access

A centralized Internet gateway is configured for each interconnection area.

The default route advertised by the centralized Internet gateway in an interconnection area is automatically filtered out by sites in other areas. In this way, the default route is advertised only in the local area.

Figure 5.38 shows the intra-area centralized Internet access scenario.

5.5.2.2 Global Centralized Internet Access

A globally selected centralized Internet gateway gives all interconnected sites access to the Internet.

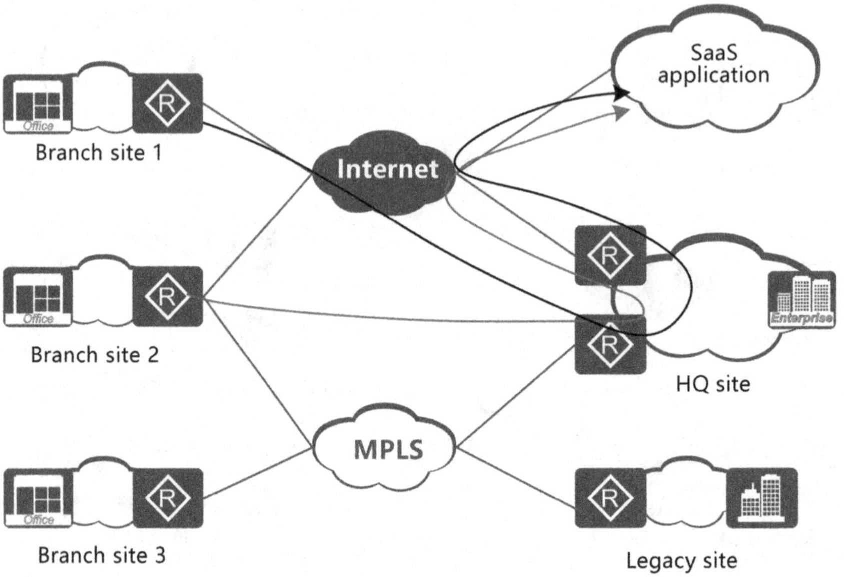

FIGURE 5.38 Intra-area centralized Internet access.

Figure 5.39 shows the global centralized Internet access scenario.

To enhance reliability, two sites can be configured as the centralized Internet gateways working in active/standby mode.

A centralized Internet gateway site can access the Internet using either of the following modes:

- **Using the LAN-side Internet egress**: All Internet traffic is routed through the LAN-side egress of the centralized gateway site. In this mode, configure a default route or a dynamic routing protocol on the LAN side to learn the default route from the LAN-side firewall.
- **Using a WAN-side interface**: All Internet traffic is routed through the WAN-side egress of the centralized Internet gateway.

5.5.3 Hybrid Internet Access

Hybrid Internet access is applicable to scenarios where most of an enterprise's Internet traffic needs to be centrally managed and controlled and some application-specific traffic (e.g., Office 365) is directly routed to the Internet to achieve the optimal application experience.

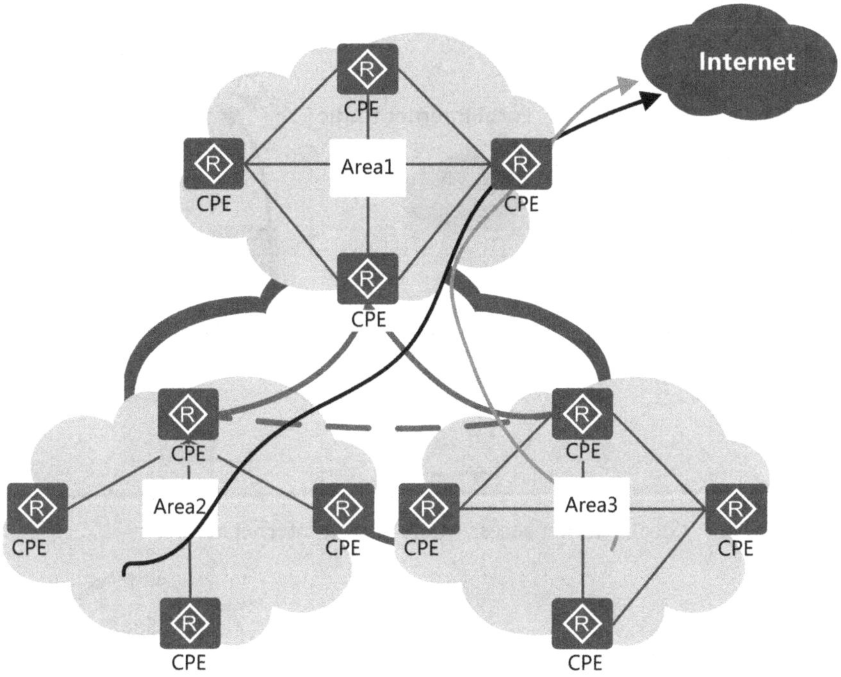

FIGURE 5.39 Global centralized Internet access.

The following hybrid Internet access modes are available:

- Local Internet access + centralized Internet access

  In this mode, all Internet traffic is routed through the local gateway by default. If the local Internet egress is faulty, Internet traffic is routed through the centralized Internet gateway, as shown in Figure 5.40.

- Centralized Internet access + Local Internet access for specific service traffic

  In this mode, all Internet traffic is routed through the centralized Internet gateway by default, and some service-specific traffic (such as Office 365) is directly routed to the Internet through the local WAN interface. When the local Internet egress is faulty, Internet traffic can still be routed through the centralized Internet gateway, as shown in Figure 5.41.

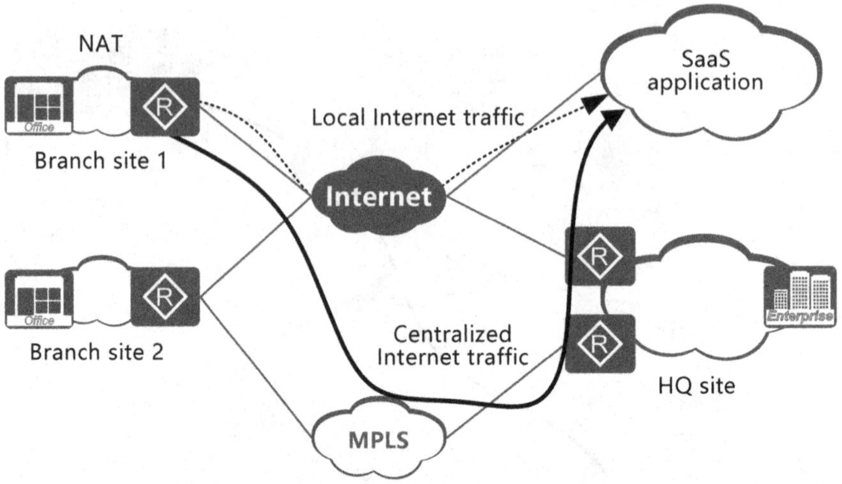

FIGURE 5.40 Local Internet access+Centralized Internet access.

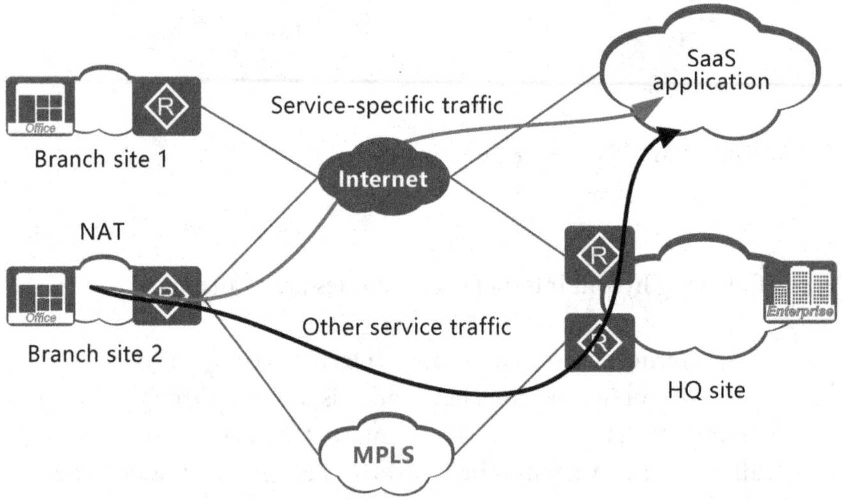

FIGURE 5.41 Centralized Internet access+Local Internet access for specific service traffic.

5.6 WAN NAT TRAVERSAL

Although NAT technology solves the issue of insufficient public IPv4 addresses, it brings its own issues. A typical example is that public network entities cannot proactively establish connections with private network entities deployed behind a NAT device. This is because the NAT device creates a NAT entry only when a packet is sent from a private network to a public network. In order for public network entities to proactively establish connections with private network entities, NAT traversal is hence introduced.

It is common to deploy the CPE behind the NAT device on an underlay network built over Internet links. In addition to the NAT function that the CPE provides, which allows private network users at a site access the Internet, SD-WAN needs to implement NAT traversal for interconnection between sites.

If CPEs are deployed behind the NAT device, SD-WAN needs to ensure that the SDN controller and CPEs at other sites establish network connections with these CPEs behind the NAT device. In SD-WAN, traffic between different components may traverse the NAT device, as shown in Figure 5.42.

SD-WAN involves the following NAT traversal scenarios:

- Connection between the CPE and the SDN controller

  A management channel is established between the CPE and the SDN controller. The SDN controller is assigned a public IP address, and the CPE proactively sends a connection request to the public IP address of the SDN controller. In this case, we can assume that NAT traversal has been implemented between the CPE and the SDN controller.

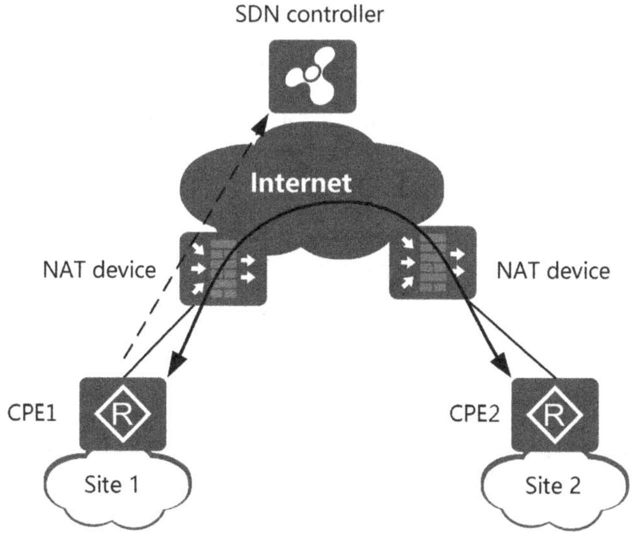

FIGURE 5.42 NAT traversal scenario.

- Connection between CPEs

  Inter-site traffic is transmitted over the data channel established between CPEs. In this case, a peer-to-peer (P2P) connection is established between CPEs. If a P2P connection needs to be established between a CPE deployed on the private network behind the NAT device and another CPE, especially between CPEs at two sites deployed on the private network behind the NAT device, NAT traversal is required.

  STUN can be used to implement NAT traversal in such scenarios. In related standards, STUN is positioned as a tool for implementing NAT traversal and is used by a terminal to check the IP address and port allocated by a NAT device to the terminal itself, as well as connectivity between two terminals.

  STUN is a client-server network protocol. The server is deployed on a public network and has a public IP address, while the client is deployed on a private network behind the NAT device. In SD-WAN, the SDN controller or RR functions as the STUN server, and a CPE functions as the STUN client. Figure 5.43 shows the STUN connection established between a CPE (STUN client) and the SDN controller (STUN server).

  By leveraging STUN, two CPEs that establish a data channel connection can obtain each other's post-NAT public IP address, based on which the TNPs of the two CPEs establish a connection, thereby implementing NAT traversal.

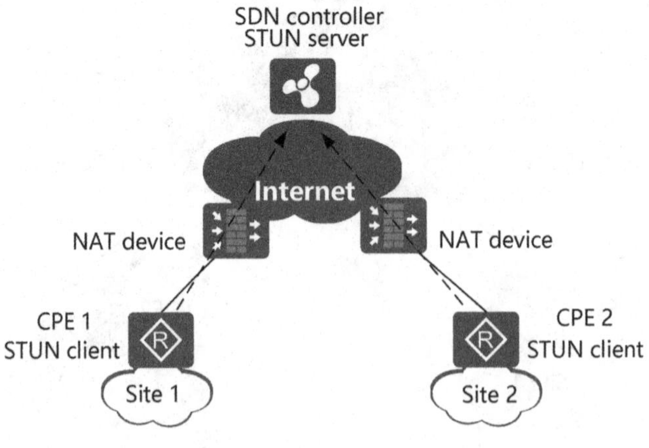

FIGURE 5.43 STUN connection.

The following describes the implementation of STUN-based NAT traversal. For details about STUN and other NAT traversal technologies, see the corresponding standards.

5.6.1 NAT Mapping and Filtering

First, we will look at the concepts of NAT mapping and filtering in STUN. NAT mapping refers to the process in which the NAT device maps the IP addresses of a group of hosts on a private network to the same public IP address so that the specific group of hosts can share a unique public IP address. In this way, all different information flows appear to come from the same IP address. NAT mapping can be achieved in the following ways:

- **Address- and port-independent NAT mapping**: A private IP address and its corresponding port are mapped to a fixed public IP address and port. In this mode, the NAT device uses the same mapping for subsequent packets that are sent from the same private IP address and port to any public IP address and port.
- **Address-dependent NAT mapping**: The NAT device uses the same mapping for packets that are sent from the same private IP address and port to the same public IP address corresponding to any port.
- **Address- and port-dependent NAT mapping**: The NAT device uses the same mapping for packets that are sent from the same private IP address and port to the same public IP address and port.

NAT filtering refers to the process in which the NAT device filters packets sent from a public network to a private network and can be achieved in the following ways:

- **Address- and port-independent NAT filtering**: Packets from any public IP address can traverse the NAT device and reach hosts on a private network.
- **Address-dependent NAT filtering**: Packets from only specific public IP addresses can traverse the NAT device and reach hosts on a private network.
- **Address- and port-dependent NAT filtering**: Packets from only specific public IP addresses and ports can traverse the NAT device and reach hosts on a private network.

STUN defines the following NAT modes for UDP packets: cone NAT and symmetric NAT. Cone NAT can be further classified into full cone NAT, restricted cone NAT, and port-restricted cone NAT. NAT mapping and filtering vary according to NAT modes.

5.6.1.1 Full Cone NAT

In full cone NAT, the NAT device maps an internal address and port to an external address and port. Furthermore, any external host can send a packet to the internal host, by sending a packet to the mapped public address. In short, as long as mappings are established between the internal host and external host on the NAT device, the external host can send data to the internal host.

As shown in Figure 5.44, host A on a private network uses IP address A and port A to send packets to host B with IP address B and port B on a public network. The NAT device maps the source IP address and port of packets from host A to IP address X (a public IP address) and port X, respectively. Through the NAT device, any host, for example, host C, on a public network can initiate a connection to host A through IP address X and port X.

Full cone NAT performs mapping for packets sent from an internal address (private IP address+port) to any destination address using the

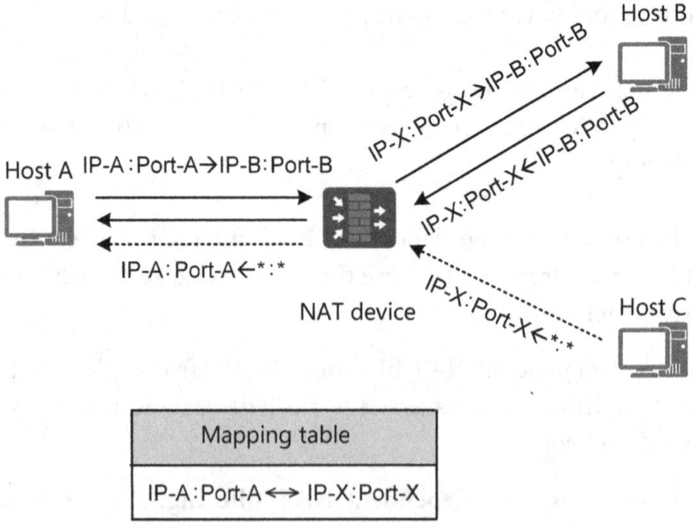

FIGURE 5.44 Full cone NAT.

same external address (public IP address+port). Packets sent from any external host can traverse the NAT device and reach an internal host based on full cone NAT mapping. Therefore, full cone NAT is independent of the address and port in both mapping and filtering.

5.6.1.2 Restricted Cone NAT

In restricted cone NAT (also known as address-restricted cone NAT), the NAT device also maps an internal address and port to an external address and port. Unlike full cone NAT, an external host can send a packet to an internal host only if the external address (and any port at that address) has received a packet from the internal address and port.

As shown in Figure 5.45, host A on a private network uses IP address A and port A to send packets to host B with IP address B and port B on a public network. The NAT device maps the source IP address and port of packets from host A to IP address X (a public IP address) and port X, respectively. Only host B can use any port to initiate a connection to host A. Host C cannot use IP address X and port X to initiate a connection to host A.

Restricted cone NAT performs mapping for packets sent from an internal address (private IP address+port) to any destination address using the same external address (public IP address+port). Based on restricted cone NAT mapping, packets sent from an external host can traverse the

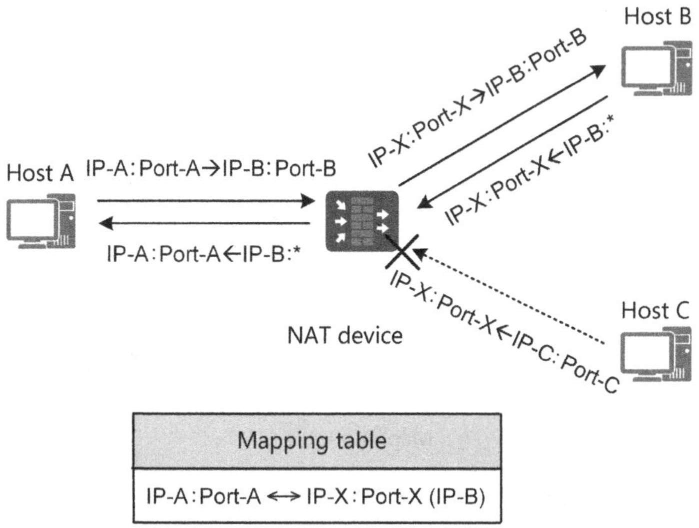

FIGURE 5.45 Restricted cone NAT.

NAT device and reach an internal host only if the internal host had previously sent a packet to the external host. Therefore, restricted cone NAT is address-independent in mapping but address-dependent in filtering.

5.6.1.3 Port-Restricted Cone NAT

In port-restricted cone NAT, all packets from the same internal IP address and port are mapped to the same external IP address and port. Port-restricted cone NAT has stricter requirements on external access than restricted cone NAT. An external host with a specific external IP address and port can send a packet to an internal host only if the internal host had previously sent a packet to the external host with the specific external IP address and port.

As shown in Figure 5.46, host A on a private network uses IP address A and port A to send packets to host B with IP address B and port B on a public network. The NAT device maps the source IP address and port of packets from host A to IP address X (a public IP address) and port X, respectively. Only host B can initiate a connection to host A from port B. If host B attempts to initiate a connection to host A from other ports or other hosts attempt to initiate a connection to host A through IP address X and port X, the connection fails.

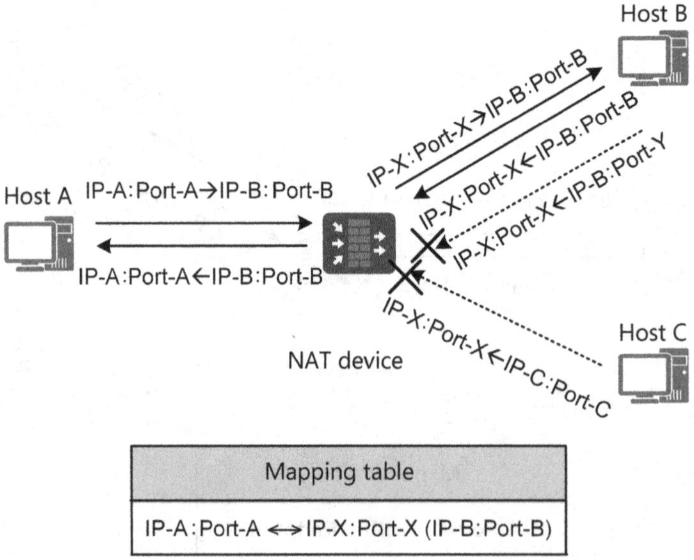

FIGURE 5.46 Port-restricted cone NAT.

Port-restricted cone NAT performs mapping for packets sent from an internal address (private IP address+port) to any destination address using the same external address (public IP address+port). Based on the port-restricted cone NAT mapping, packets sent from an external host with a specific external IP address and port can traverse the NAT device and reach an internal host only if the internal host had previously sent a packet to the external host with the specific external IP address and port. Therefore, port-restricted cone NAT is address-independent in mapping and address- and port-dependent in filtering.

5.6.1.4 Symmetric NAT

In symmetric NAT, the NAT device maps all packets sent from the same private IP address and port to a specific destination IP address and port to the same public IP address and port. Packets sent from the same internal host to different destination IP addresses and ports are mapped to different ports with the same public IP address.

As shown in Figure 5.47, host A resides on a private network, whereas host B and host C are deployed on a public network. Host A uses IP address A and port A, host B uses IP address B and port B, and host C uses IP address C

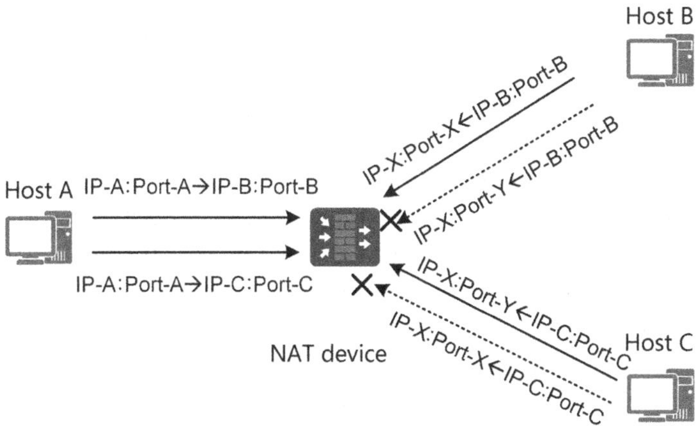

Mapping table
IP-A:Port-A ↔ IP-X:Port-X (IP-B:Port-B)
IP-A:Port-A ↔ IP-X:Port-Y (IP-C:Port-C)

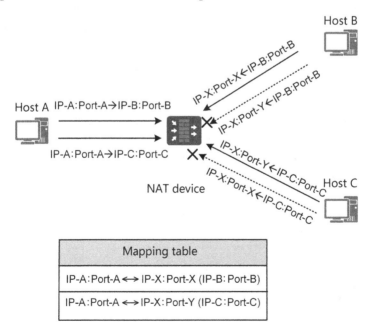

FIGURE 5.47 Symmetric NAT.

and port C. When host A sends packets to host B, the NAT device maps the source IP address and port of packets from host A to IP address X (a public IP address) and port X, respectively. When host A sends packets to host C, the NAT device maps the source IP address and port of packets from host A to IP address X (a public IP address) and port Y, respectively. Host B can initiate a connection to host A only through IP address X and port X, and host C can initiate a connection to host A only through IP address X and port Y.

Symmetric NAT performs mapping for packets from an internal host to different external hosts through different ports. Packets sent from an external host can traverse the NAT device and reach an internal host only after the external host had previously received packets from the internal host and determined the port used to send packets to the internal host. Therefore, symmetric NAT is dependent on the address and port in both mapping and filtering.

5.6.2 NAT Detection and Hole Punching

Full cone NAT, restricted cone NAT, port-restricted cone NAT, and symmetric NAT take on different characteristics in mapping and filtering. In the majority of cases, an external host cannot directly connect to an internal host. Configure hole punching on the NAT device so that packets from the external host can traverse the NAT device. This enables communication between an external host and an internal host.

In SD-WAN, a CPE behind a NAT device is required to obtain its NAT type and post-NAT IP address and port, as well as those of another CPE to be connected. Based on the NAT types, the two CPEs can then determine whether hole punching is feasible and how to perform it. After mapping, the CPEs obtain each other's NAT type and public IP address using the NAT detection function provided by STUN to establish a connection. For details about the detection and hole punching processes, see the STUN protocol standard documentation.

In SD-WAN, NAT traversal is supported only between devices of the following NAT types:

- Static NAT/Full cone NAT <-> Any NAT type
- Restricted cone NAT <-> Restricted cone NAT, port-restricted cone NAT, or symmetric NAT
- Port-restricted cone NAT <-> Port-restricted cone NAT

NAT traversal is not supported between devices of the following NAT types:

- Port-restricted cone NAT <-> Symmetric NAT
- Symmetric NAT <-> Symmetric NAT

5.7 INTERCONNECTION WITH LEGACY SITES

Before deploying SD-WAN, enterprises may have numerous sites that use conventional network technologies (such as MPLS VPN). These sites include legacy headquarters, DC, and branch sites, and even cloud networks that use heterogeneous VPN technologies. These sites are referred to as legacy sites.

After SD-WAN sites are deployed or some legacy sites are reconstructed into SD-WAN sites, an enterprise may have both SD-WAN and legacy sites, between which communication is required. The SD-WAN network domain uses overlay VPN technologies such as EVPN, while the traditional network domain uses traditional underlay VPN technologies such as MPLS VPN. For this reason, communication is unavailable between them on both the control plane and forwarding plane. A solution enabling interconnection between SD-WAN sites and legacy sites is required.

The following types of solutions are available for interconnection with legacy sites, as shown in Figure 5.48:

- **Interconnection through an SD-WAN site**: In this solution, a connection is established between an SD-WAN site and the legacy network. This solution can be further classified into back-to-back site interconnection and local interconnection.
- **Interconnection through a gateway**: In this solution, an independent IWG is deployed to implement interconnection with legacy

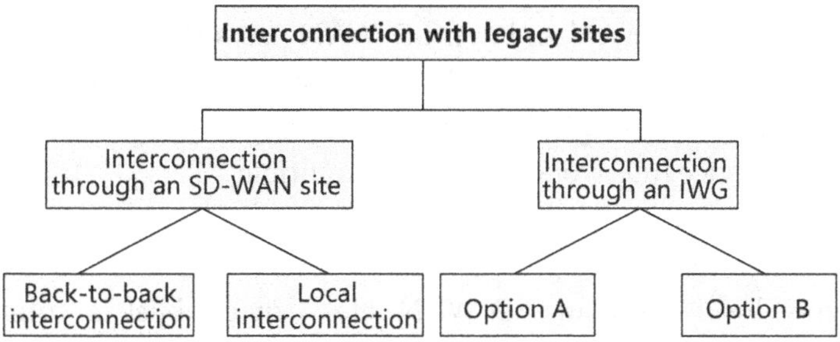

FIGURE 5.48 Interconnection with legacy sites.

sites. The IWG connects to both SD-WAN and legacy networks and therefore can function as a gateway to connect to the SD-WAN network domain and legacy network domain. This solution can be further classified into interconnection in Option A mode and interconnection in Option B mode.

In summary, interconnection through an SD-WAN site is ideal for a single enterprise that builds and operates its own network; interconnection through an IWG is ideal for the carrier/MSP resale scenario where a carrier or an MSP deploys the IWG and leases it to tenants.

Here, we use interconnection between SD-WAN sites and legacy MPLS VPN sites as an example to illustrate the implementation and characteristics of the two solutions for interconnection with legacy sites.

5.7.1 Interconnection through an SD-WAN Site

5.7.1.1 Back-to-Back Interconnection

If the WANs of SD-WAN sites cannot directly communicate with the MPLS network of legacy sites, configure an SD-WAN site and a legacy site. Make sure that each site logically functions as a gateway site to implement back-to-back site interconnection.

Specifically, use a dedicated physical link to connect a LAN interface at a legacy site and a LAN interface at an SD-WAN site. This dedicated link runs a protocol such as BGP or OSPF as required to exchange routes between the legacy MPLS network domain and SD-WAN network domain. It is also required to advertise the routes to other sites in their own domains. All the other SD-WAN sites also communicate with legacy sites through the dedicated link. This enables interconnection between SD-WAN sites and legacy sites, as shown in Figure 5.49.

If an enterprise has multiple VPNs and users within a VPN are distributed in both SD-WAN and non-SD-WAN network domains, multiple logical links can be created over the dedicated link for back-to-back site interconnection. Each logical link is added to the corresponding VPN to implement service interworking between multiple VPNs.

5.7.1.2 Local Interconnection

Leveraging local breakout, SD-WAN sites can establish a connection between the SD-WAN overlay network domain and underlay network domain and exchange routes between the two domains. This is done

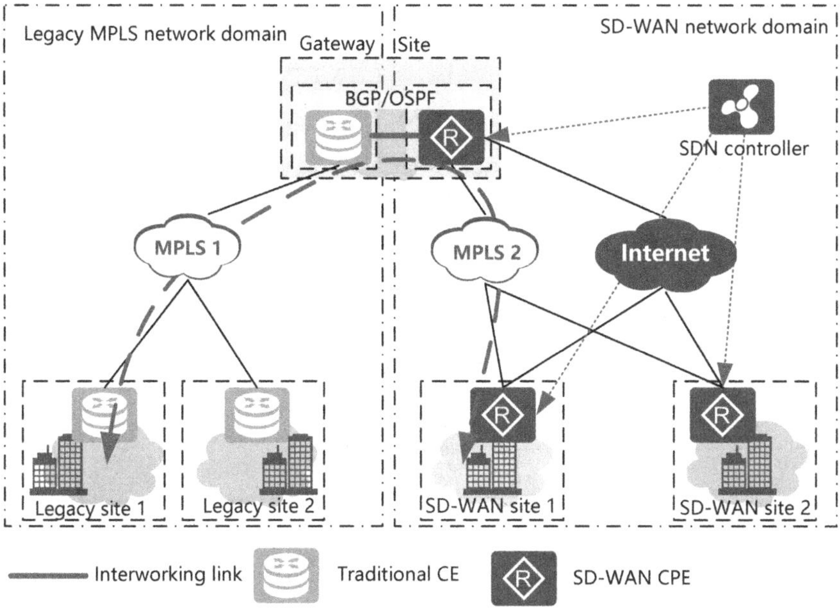

FIGURE 5.49 Interconnection through a dedicated link.

when SD-WAN sites connect to the same WAN, or when they need to communicate with the legacy MPLS network. This implements local interconnection with legacy sites.

SD-WAN enables flexible deployment and supports the following local site interconnection scenarios based on the actual service needs and networking characteristics:

- Distributed interconnection

  If the WANs of all SD-WAN sites can directly communicate with the MPLS network of legacy sites over the local underlay network, local breakout can be deployed for all SD-WAN sites to implement distributed interconnection between SD-WAN sites and legacy sites, as shown in Figure 5.50. In this mode, traffic of each site is directly forwarded, without being diverted to other SD-WAN sites through the overlay network domain, featuring high forwarding efficiency.

- Centralized interconnection

  If some SD-WAN sites cannot access the legacy network through local breakout, a site that supports local breakout functions as a centralized access site to interconnect with legacy sites and advertises

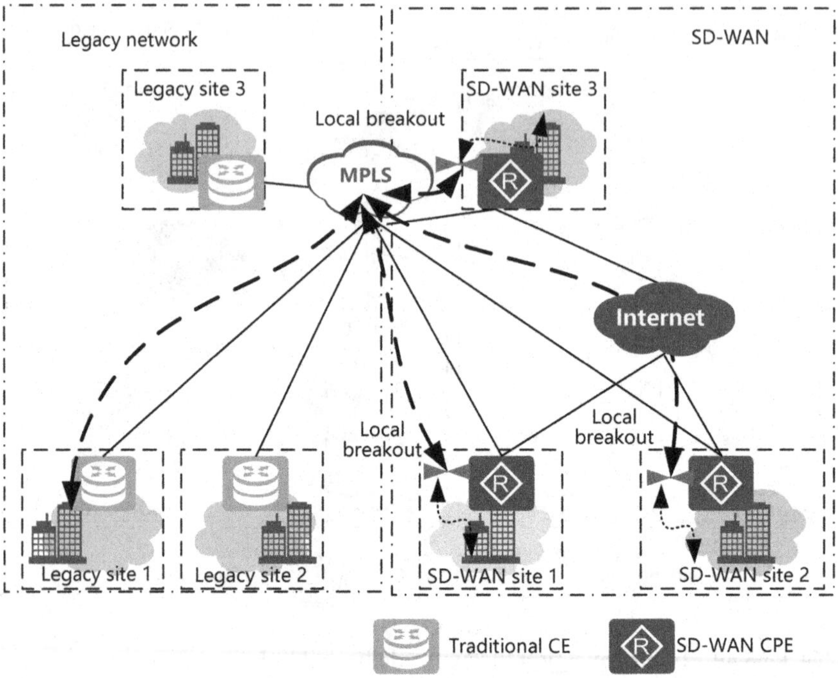

FIGURE 5.50 Distributed interconnection.

routes through the overlay network domain. Other SD-WAN sites learn the routes and forward traffic to the centralized access site through overlay tunnels. After this, the centralized access site forwards the traffic to legacy sites through local breakout, as shown in Figure 5.51. This mode features simple deployment, centralized traffic, and easy management and control.

- Hybrid interconnection

  The hybrid interconnection mode is a combination of the distributed interconnection and centralized interconnection modes. This is where SD-WAN sites communicate with legacy networks either directly or through a centralized access site, depending on if SD-WAN sites can directly communicate with legacy network domains.

  In hybrid interconnection mode, when the MPLS link for local interconnection is faulty at an SD-WAN site, traffic is automatically switched to the overlay network domain. It is then forwarded to the centralized access site through the overlay tunnel. This implements

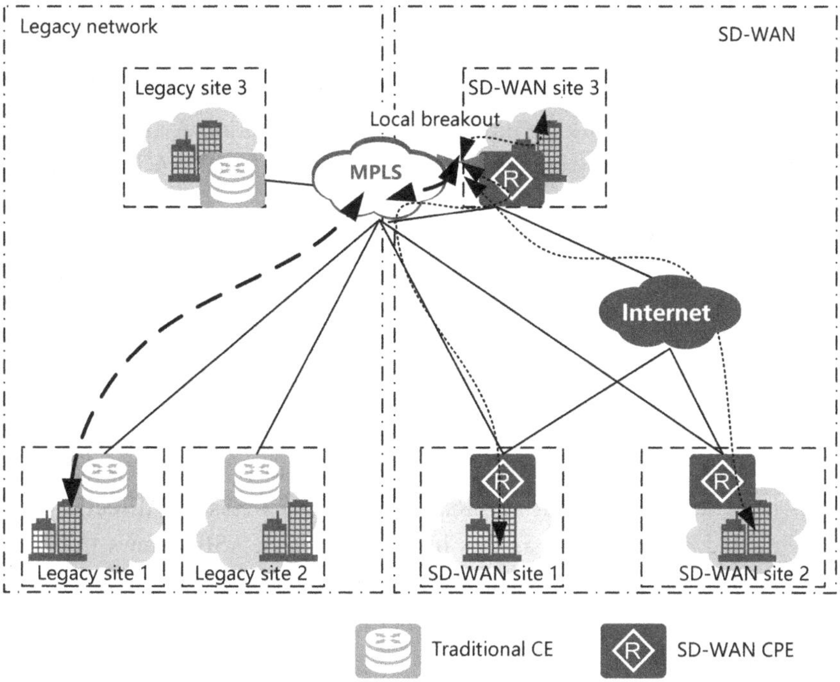

FIGURE 5.51 Centralized interconnection.

service backup between SD-WAN sites and legacy sites and offers higher reliability.

5.7.2 Interconnection through an IWG

A carrier can provide virtual private line services on an MPLS network for multiple enterprises, by leveraging the traditional MPLS VPN technology. A carrier can build SD-WAN overlay tunnels over their own underlay network or a third-party underlay network to implement interconnection for enterprises' new or reconstructed SD-WAN sites.

As mentioned above, an enterprise's legacy MPLS sites and SD-WAN sites may need to communicate with each other. To implement this, a carrier can provide the IWG service, a site interconnection service, for all enterprise tenants. Specifically, a carrier deploys a multi-tenancy-capable IWG, which has links to and can communicate with both the carrier's legacy MPLS VPN and SD-WAN overlay network domains. Enterprise tenants can subscribe to the IWG service on the SDN controller as required in a method of subscribing to managed services.

The IWG functions as the autonomous system boundary router (ASBR) in the IP overlay VPN network domain of SD-WAN and is required to communicate with the ASBR (usually a PE) in the traditional MPLS VPN network. Two interworking modes are available: Option A and Option B, which are similar to those in the traditional inter-AS MPLS VPN solution.

5.7.2.1 Option A

In the traditional MPLS VPN architecture, the inter-AS VPN Option A mode is implemented as follows:

- ASBRs manage VPN routes through dedicated interfaces for VPNs that traverse different ASs.
- Option A does not need special configurations and MPLS does not need to run between ASBRs. ASBRs of two ASs are directly connected and function as PEs in the ASs. Each ASBR views the peer ASBR as its CE and creates a VRF instance for each VPN. Each ASBR also advertises IP routes to the peer ASBR through EBGP or Interior Gateway Protocol (IGP).

The implementation of interconnection through an IWG in Option A mode is similar and is described as follows:

1. For each VPN of each tenant, create a VRF instance on the IWG of the SD-WAN network domain and legacy MPLS network domain. Also create a pair of logical sub-interfaces (Ethernet sub-interfaces, VLAN sub-interfaces, or VXLAN tunnel sub-interfaces) on the back-to-back physical interconnection link.
2. Bind these sub-interfaces to the corresponding VRF instances on the IWG and PE. The number of sub-interfaces to be created depends on the number of VPNs that need to communicate with each other.
3. Configure static routes or dynamic routing protocols (such as EBGP and OSPF) on sub-interfaces. This will allow routes to be exchanged between the SD-WAN network domain and legacy MPLS network domain.

Figure 5.52 shows the interconnection through an IWG in Option A mode.

Interconnection in Option A mode is simple, and VPN route advertisement is easy to control, featuring high security and reliability. However,

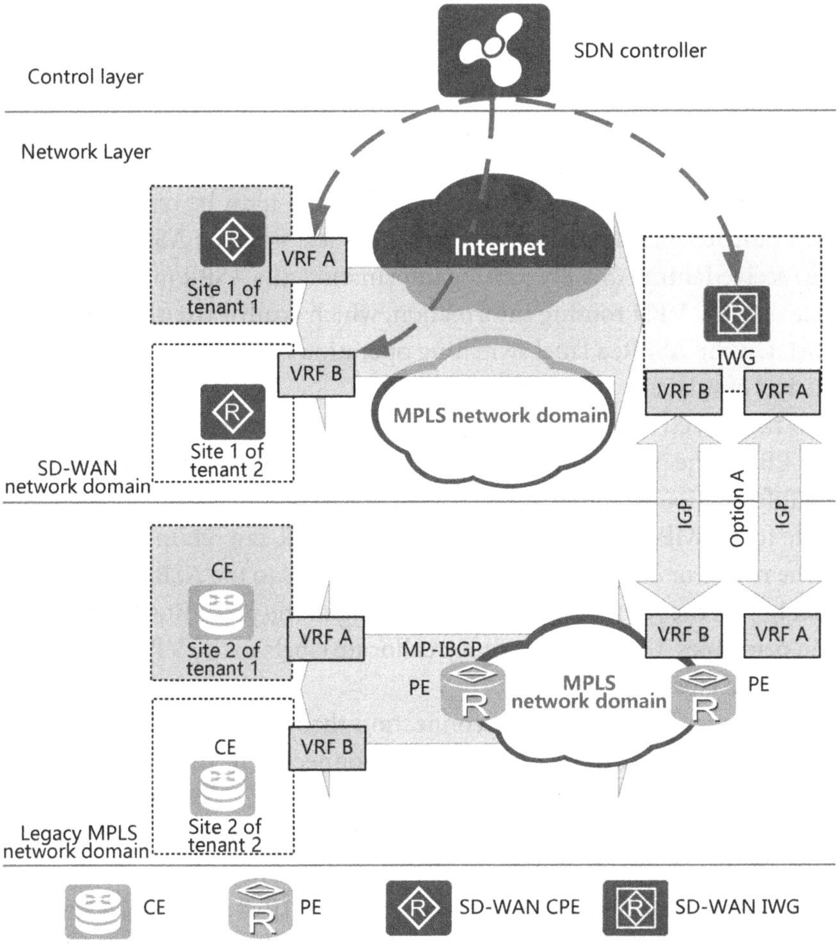

FIGURE 5.52 Interconnection through an IWG in Option A mode.

this mode has poor scalability and low service provisioning efficiency. This is due to the fact that deploying the IWG service, PEs in legacy MPLS VPNs need to be configured each time a new VPN is added. The IWG is orchestrated and configured on the SDN controller, and PEs — underlay backbone network devices — need to be configured using the NMS of the carrier's MPLS backbone network. This requires cross-department communication and coordination. Additionally, to avoid conflicts of key VPN configurations such as the route target (RT) policy, network design is required to be negotiated between both parties. The end result is low E2E service provisioning efficiency.

5.7.2.2 Option B

Figure 5.53 shows the implementation of inter-AS VPN Option B mode in the traditional MPLS VPN architecture.

ASBRs advertise labeled VPN-IPv4 routes to each other using Multi-Protocol EBGP (MP-EBGP). When advertising routes, MP-EBGP changes the next hop of the routes, upon which the label must be replaced locally according to label allocation principles. As such, when an ASBR advertises the received intra-AS VPN routing information, the ASBR must reallocate a label to the VPN routing information, which is advertised with the new label. On the ASBR, a label switching operation is performed between the old and new labels.

When receiving the VPN routing information advertised through MP-EBGP, the peer ASBR saves the information locally. It then advertises the information to the PEs in the local AS. When the ASBR advertises a route to the MP-IBGP peers in the AS, the ASBR can retain the next hop of the routes or change the next hop of the routes to the ASBR itself. If the next hop of the routes is changed, according to the preceding label allocation principles, the ASBR needs to reallocate labels to the VPN routes and switch the labels locally.

The implementation of interconnection through an IWG in Option B mode is similar. The IWG is physically connected to the PE of the peer MPLS VPN through a direct link and runs MP-EBGP between them. The IWG converts EVPN service routes in the SD-WAN network domain into VPN IPv4 routes and sends the VPN IPv4 routes to the peer PE through MP-EBGP. Additionally, the IWG receives VPN IPv4 routes of legacy sites from the peer end and converts the routes into EVPN routes. It then sends

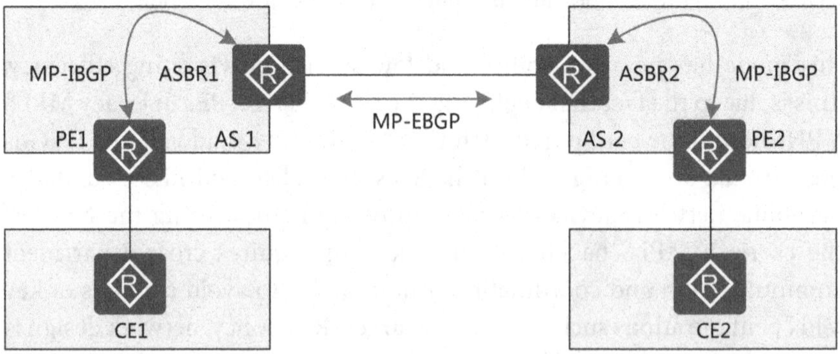

FIGURE 5.53 Inter-AS VPN Option B.

the EVPN routes to sites in the SD-WAN network domain, implementing service interworking.

Figure 5.54 shows the interconnection through an IWG in Option B mode.

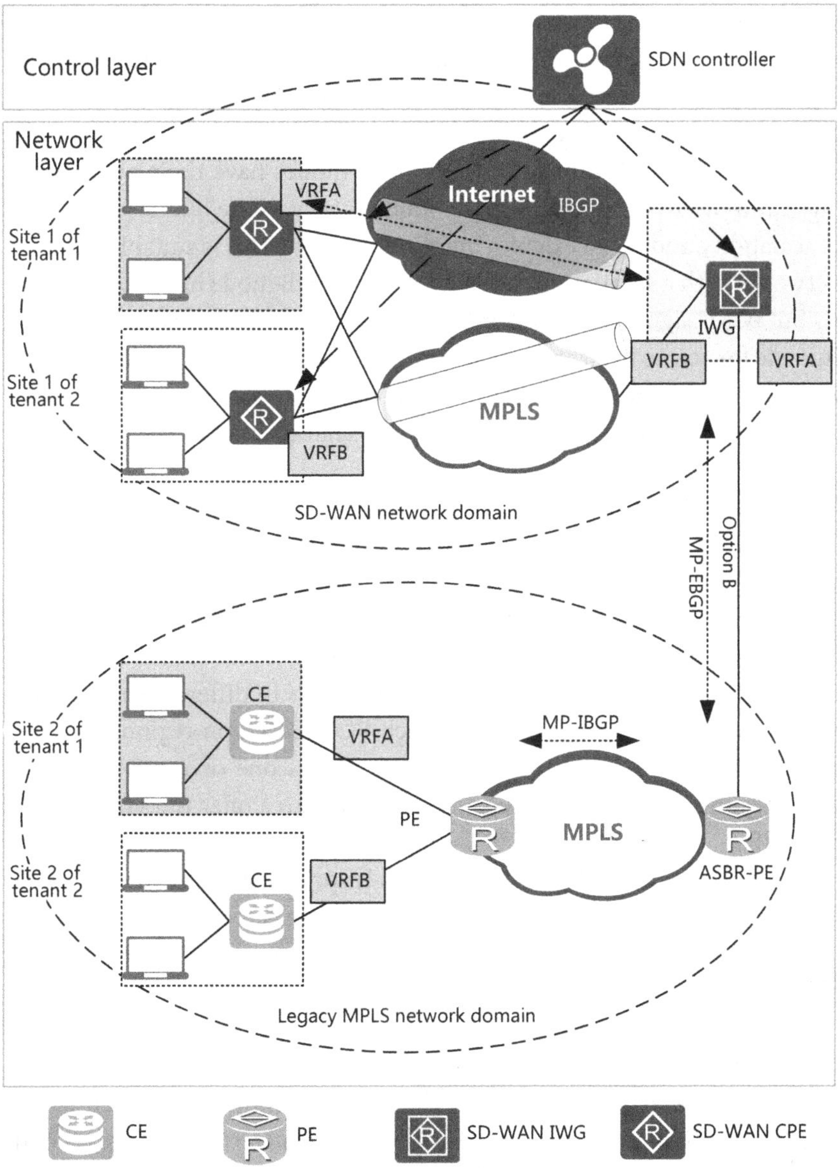

FIGURE 5.54 Interconnection through an IWG in Option B mode.

Interconnection in Option B mode features simple configurations and high service provisioning efficiency. All VPNs can reuse the MP-EBGP peer relationship configured between the IWG and PE, after one-time configuration on the ASBR-PE during deployment. PEs are not required to be configured even when new VPNs are added, which maximizes automated service provisioning efficiency. However, route advertisement in this mode is complex. VPN routes of different enterprises are flooded between ASs. To control and isolate routes between tenants, RT policies are required, complicating route management and control and carriers' O&M.

In conclusion, Option A and Option B modes have their own advantages and disadvantages. Option A mode is easy to use but has a low level of scalability and service deployment automation. This is suitable for small networks with a small number of VPNs. Option B mode has good scalability but poses significant challenges to O&M due to complex technologies, suitable for scenarios where a large number of VPNs need to communicate with each other and VPNs change frequently. Carriers should select solutions established on specific scenarios and characteristics.

5.7.2.3 Reliability Design

The IWG is a key device for interconnection between the SD-WAN network domain and legacy MPLS network domain, requiring redundancy and high service reliability mechanisms. Figure 5.55 shows the IWG reliability design.

A carrier has multiple POP equipment rooms in different regions, and one or more IWGs are deployed at each POP. Based on regions and reliability requirements, each access area contains one or more POPs and therefore multiple IWGs. IWGs in each access area offer the same level of reliability.

Tenants select two IWG access areas in primary/secondary mode for each site on the SDN controller. Based on the selected IWG access areas, the SDN controller selects two optimal-performance IWGs working in primary/secondary mode.

The IWGs learn the routes of legacy VPNs from the peer PEs. When routes are advertised to the SD-WAN network domain, the SDN controller orchestrates the routes. This is done so that routes advertised by the primary IWG have a higher priority than those advertised by the secondary IWG. In this way, SD-WAN preferentially selects the primary IWG

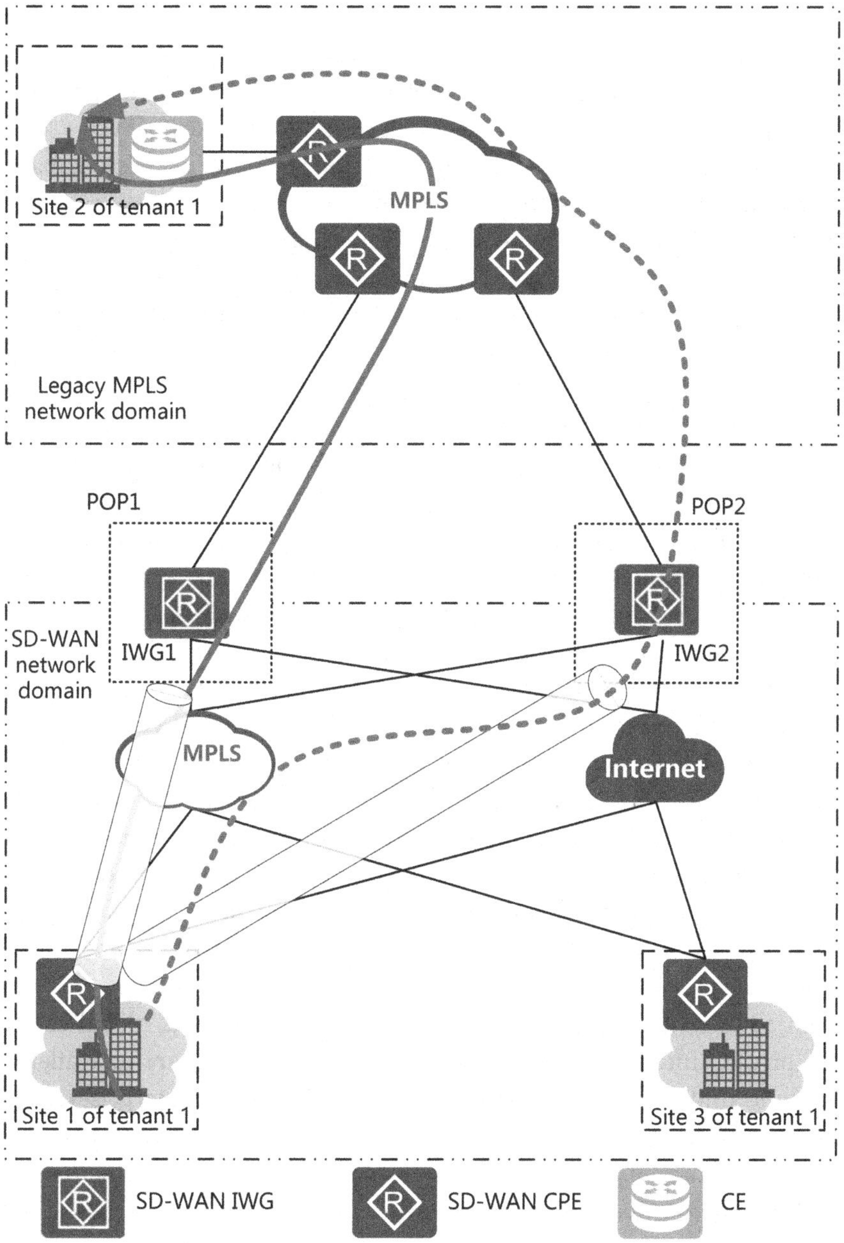

FIGURE 5.55 IWG reliability design.

to access legacy networks. Similarly, IWGs leverage BGP routing policies to control the priority of routes advertised to the legacy MPLS network domain. This is implemented so that routes learned from the primary IWG take precedence over those learned from the secondary IWG.

When the primary IWG fails, the RR in the SD-WAN network domain detects that BGP between the SD-WAN network domain and the primary IWG fails. The RR then sends route withdrawal messages to all sites in the SD-WAN network domain, upon which routes advertised by the secondary IWG are used by the SD-WAN network domain to access legacy networks.

5.8 POP NETWORKING

A growing number of enterprises are expanding their business across regions or carrying out their globalization strategies. This is driven by economic digitalization and globalization. Enterprises inevitably need to interconnect branches, synchronize data, and access cloud and SaaS applications across wider areas. Against this backdrop, carriers are pressingly challenged by how to provide fast, secure, and high-quality cross-region WAN interconnection for enterprises. For example, to adapt to the globalization trend and follow the "One Belt, One Road" initiative, increasing numbers of Chinese enterprises expand services to the global market to seek their fortunes in markets outside China. Quick construction of cross-region or cross-country WANs is of vital importance to enterprises.

Enterprises often interconnect their branch sites to the headquarters across countries or regions using private lines provided by global carriers or the Internet, neither of which, however, can fully meet enterprises' requirements.

In addition to high costs and long construction period, global carriers' private lines have insufficient network coverage and are unavailable in some underdeveloped areas. To address this issue, enterprises have to deploy microwave devices, which are costly but offer limited network bandwidth. In contrast, Internet access across regions or countries involves connectivity between Internet networks from multiple carriers and cannot offer guaranteed transmission quality. This occurs even though the Internet has wide coverage, is easy to access, and has increasingly better service quality in recent years.

Cross-region enterprises also have similar problems. For example, large logistics enterprises in China may have thousands of branches, which are distributed in regions with unbalanced economic development and WAN coverage and quality. These branches may use WANs provided by different carriers for interconnection, deteriorating connectivity quality. The end result is poor WAN application experience.

Cross-region and cross-country enterprises are challenged by balancing network costs and access experience/efficiency.

The answer to these challenges is the SD-WAN POP networking sub-solution.

5.8.1 Concepts and Principles

To deliver high-quality cross-country or cross-region interconnection for enterprise sites, the POP networking solution needs to solve the following problems:

- Build a high-quality cross-region or cross-country WAN backbone network. This delivers high-quality transmission and experience for enterprise WAN applications.
- Enable fast "last-mile" access to the backbone network, regardless of enterprise site locations.

High-quality cross-region or cross-country WAN backbone networks are exactly the strength of cross-region carriers/MSPs.

With relevance to fast "last-mile" access to the backbone network, SD-WAN is the optimal solution for its capabilities of automated overlay network orchestration and SDN-based fast service provisioning. Leveraging SD-WAN, carriers/MSPs can better provide high-quality cross-region or cross-country networks for enterprise sites.

Moreover, we have two issues to tackle: POP gateway deployment and quick access to POP networks.

First, carriers/MSPs that have a cross-region or cross-country private line backbone network establish POPs at the edge of their backbone networks and deploy SD-WAN gateways in the POPs. These gateways are also referred to as POP gateways. Typically, CPEs act as POP gateways, which can be software nodes or fixed-configuration hardware devices. The POP gateways are physically connected to edge devices on backbone networks over the underlay network. This forms an SD-WAN POP network across regions or countries, as shown in Figure 5.56.

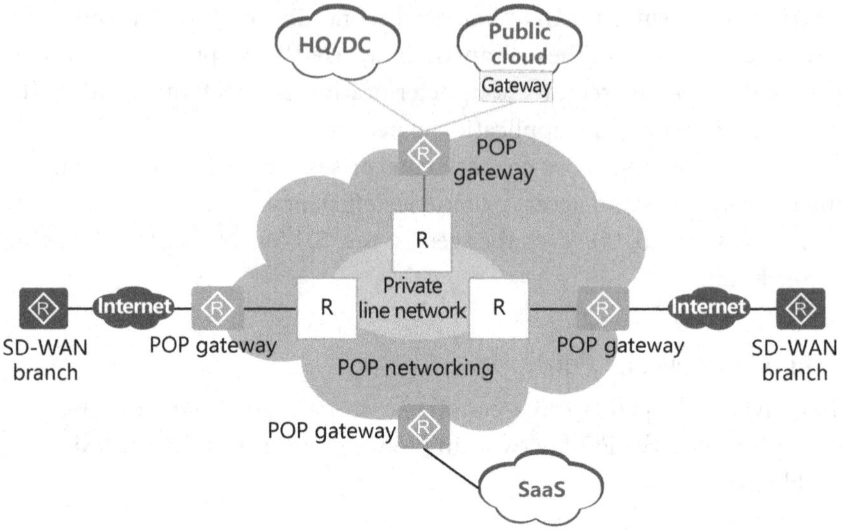

FIGURE 5.56 POP networking.

The other issue to tackle would be quick access from a broad range of key applications at widely distributed WAN sites to POP networks. A POP network usually has a limited coverage. When it comes to branch interconnection, we need to achieve fast "last-mile" access to the POP network. In SD-WAN, this is implemented using Internet links provided by local carriers. SD-WAN CPEs are deployed at enterprise branch sites. Under the centralized management and control of the SDN controller, these CPEs establish IP overlay tunnels with the nearest POP gateways, and traffic between enterprise branches distributed across countries is transmitted through the POP network. This achieves high-quality interconnection between the branches.

Carriers, MSPs, and public cloud providers have high-quality cross-region backbone networks and are eager to expand services across regions. However, they cannot fully cover the "last-mile" access for enterprises. The combination of the POP networking and SD-WAN is the best choice for this situation and as a result, all parties benefit.

A major difference between carriers and device vendors is that carriers provide network services, predominantly network resources. Carriers need to establish POP on their existing networks to build carrier-class SD-WAN for customer-oriented marketing and promotion. This enables customers to build SD-WAN networks. However, if a carrier does not establish POP on their existing networks, the carrier is merely an integrator and can only carry out projects one by one, without any standardization. Therefore, such a carrier will fail in large-scale market promotion.

Carriers want their POP networks to support multi-vendor compatibility and multi-tenant capabilities. This is done in order to accommodate requirements of common customers as well as those of high-end customers. Another key capability is intelligent traffic steering between POPs.

To sum up, carriers have inherent advantages in providing basic line resources and network-wide service delivery capabilities required by SD-WAN services. Therefore, the POP networking sub-solution is well suited for carriers/MSPs to launch promotion.

5.8.2 Solution Design

The SD-WAN POP networking sub-solution consists of the SDN controller, POP gateways, and SD-WAN sites to be interconnected, as shown in Figure 5.57. The SDN controller centrally performs orchestration and management to create POP gateways and interconnect POP gateways with each other and with sites.

5.8.2.1 POP Gateway

A POP gateway can be considered as one type of SD-WAN gateways and has common features of SD-WAN gateways.

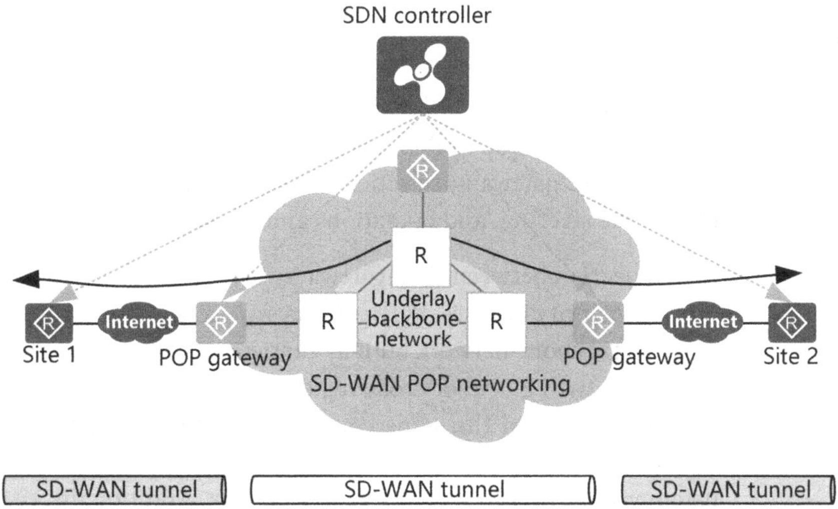

FIGURE 5.57 SD-WAN POP networking.

A POP gateway is a multi-tenant device, which is shared by tenants and does not belong to any specific tenant. On the SDN controller, the POP gateway is created and maintained by an MSP administrator and is invisible to tenants. Moreover, POP gateways are transit nodes on the network. It is the entrance for branch sites to access the POP network and connects to the POP gateways in other areas through the cross-domain underlay backbone network.

The SDN controller centrally controls and manages POP gateways and orchestrates and provisions all services on POP gateways. In addition to the configuration required for ZTP of POP gateways, configurations are required to be completed on the SDN controller. This is completed for interconnection between POP gateways and interconnection between POP gateways and SD-WAN sites.

A POP gateway may be a hardware- or software-based device. Software-based devices are more flexible and are ideal for POP in the computing virtualization environment or on public clouds.

5.8.2.2 Site

Branch sites, headquarters/DC sites, and public cloud sites are typical SD-WAN sites that need to be interconnected through POP networking.

POP networking is provided for tenants as an MSP service. Although POP gateways are invisible to tenants, the tenants are required to select sites or departments at sites for which the POP networking service is provided.

1. Key technologies

   The key issues to tackle in POP networking include: interconnection between POP gateways, selection of the optimal POP gateway for CPEs, traffic steering, and operations and charging.

   a. Interconnection between POP gateways

      In the majority of cases, POP gateways are connected through the WAN backbone network built by carriers or MSPs. For example, an MSP with global or cross-region MPLS VPNs can deploy the POP gateway in the same equipment room as that of PEs and directly connect it to the nearest PE or Multi-VPN-Instance Customer Edge (MCE) through a physical link. Additionally, the MSP can deploy back-to-back interconnection in Option A mode on the control plane, create VLAN sub-interfaces or VXLAN tunnels on devices at both ends, and exchange VPN routes through BGP. This implements service interworking.

In addition, carriers or MSPs that do not have their own WAN backbone networks can lease third-party Internet private lines for POP gateway interconnection. The Internet is flexible but offers unguaranteed quality. To transmit traffic over the optimal path on the POP network, configure special traffic steering.

b. Selection of the optimal POP gateway for CPEs

After CPEs at branch sites go online, they need to select an optimal POP gateway and establish an SD-WAN tunnel based on instructions from the SDN controller. In this manner, the CPEs interconnect with the selected POP gateway so as to connect to the POP network.

The SDN controller offers a wide variety of algorithms for selecting the optimal POP gateway. An MSP administrator can also statically specify the POP gateway to be interconnected with branch sites based on physical locations and distances of POP gateways. Each access area may contain one or more POP gateways. The SDN controller selects an idle POP gateway for a branch site as the primary POP gateway based on the POP gateway load, including the number of forwarding resources and tenants.

In addition, after the CPE at a site goes online, the SDN controller notifies the CPE of a POP gateway group. The CPE detects all POP gateways in this group and notifies the SDN controller of the detection result. Then, the SDN controller selects an optimal POP gateway based on the forwarding performance and load for the CPE as the primary POP gateway. If network conditions change, the primary POP gateway can be dynamically reselected.

The SDN controller needs to specify a secondary POP gateway for the branch sites, to enhance the reliability of the POP networking for branch sites. In normal cases, branch sites forward traffic through the primary POP gateway. When the primary POP gateway or the corresponding link is faulty, traffic is switched to the secondary POP gateway. Different sites can select different POP gateways as their primary and secondary gateways. In this way, all POP gateways work and load balance traffic.

c. Traffic steering

POP gateways are usually connected through a global WAN backbone network provided by carriers or MSPs. In SD-WAN, how can we ensure the network quality between POP gateways? Typically, the POP backbone network is lightly loaded. For example, on the MPLS backbone network of a carrier, high network transmission quality can be ensured simply by directly forwarding packets based on routes. In the case that the POP backbone network is not lightly loaded, select forwarding paths with guaranteed SLA for important applications, which can be implemented using either of the following methods:

Method 1: If the POP backbone network supports traffic steering based on application requirements, the optimal path can be selected based on conditions such as the packet loss rate, latency, and bandwidth. To implement this, the POP backbone network needs to function as the underlay layer to provide the MPLS Traffic Engineering or SRv6 function.

Method 2: If carriers or MSPs build POP networks based on a third-party Internet service and the optimal path cannot be selected through the Internet, SD-WAN implements traffic steering.

d. Operations and charging

POP backbone network resources are precious. The SD-WAN POP networking sub-solution provides effective traffic and QoS methods to maximize the utilization of POP network resources. The SDN controller can rate-limit traffic on the overlay tunnel that connects each site to the POP gateway, or limit the bandwidth of links for connecting backbone network devices to the POP gateway. MSPs can charge their customers by month based on the bandwidth or based on the actual traffic that traverses the POP network.

2. Overall service process

The overall service configuration process of the SD-WAN POP networking is described as follows:

a. Create a POP network

   i. An MSP administrator creates a POP gateway on the SDN controller and sets ZTP parameters.

   ii. The MSP administrator sets parameters for interconnection between the POP gateway and the POP backbone network, including sub-interfaces and BGP.

   ii. The MSP administrator powers on the POP gateway and performs ZTP. The POP gateway proactively registers with the SDN controller. After successful registration, the SDN controller automatically delivers configurations to the POP gateway.

   iv. The POP gateway goes online successfully.

b. Bring CPEs at branch sites online and select the desired POP networking service

   i. The SDN controller allocates the nearest POP gateway to CPEs at branch sites.

   ii. Each CPE establishes an SD-WAN overlay tunnel with the POP gateway.

   iii. Branch sites are successfully connected to the POP network and can communicate with each other through the POP network.

5.8.3 Solution Highlights

The SD-WAN POP networking solution has the following highlights:

- Cost-effective cross-domain high-quality access to multiple services

  In addition to branch sites, key services such as public clouds and SaaS may also be deployed across regions. The DC where these key applications are located usually has abundant private line resources. The POP network enables high-speed interconnection between enterprise branch sites and cloud services. These include public clouds and SaaS applications, facilitating access to key enterprise services that originally seemed to be impossible.

  For example, a multinational enterprise deploys its DC and research tools on AWS servers in the US. When the enterprise's branches in China carry out daily services, they need to synchronize data with the servers in the US. Due to the long distance and data transmission across carriers, Internet access performance is poor. The POP networking can effectively address this challenge. In this networking, POP gateways are deployed near AWS servers, offering network services with higher transmission quality than the Internet and lower overall costs than private lines.

- Internet used for "last-mile" access

  The Internet is a highly commercial WAN with the widest coverage across the globe. Driven by increasingly demanding requirements amid fierce market competition, the access-side network quality of the Internet has been greatly improved in recent years. According to surveys by professional organizations, the Internet provided by each carrier offers stable quality and low latency. This has a small impact on the long-distance network transmission. Consequently, the Internet can be combined with the overlay tunnel technology of SD-WAN to achieve "last-mile" access.

  This helps MSPs expand the service scope of their backbone networks and makes up for multinational branches that are out of carrier/MSP networks' reach, improving enterprise customer satisfaction and facilitating carriers/MSPs in market expansion.

- Rapid provisioning

  Branches of multinational enterprises are widely distributed, leading to high WAN O&M costs. Addressing this issue is extremely straightforward with SD-WAN. The SDN controller implements centralized management and control. With the SDN controller, key services can be provisioned automatically and remotely, including creating POP networks and establishing overlay tunnels between branch sites and POP gateways. This enables branch interconnection and fast provisioning of key services for multinational enterprises.

5.9 FLEXIBLE CONNECTION TO PUBLIC CLOUDS

With the rapid development of cloud computing technologies, the public cloud has evolved to provide relatively comprehensive network and IT services, typically represented by infrastructure as a service (IaaS, encompassing computing, network, storage, and security), platform as a service

(PaaS, including the microservice engine, cloud container engine, and AI development platform), and software as a service (SaaS, such as ERP).

Compared with traditional enterprise IT construction modes, the public cloud is vastly superior in terms of service provisioning, on-demand resource provisioning (scale-in or scale-out), resource utilization, and O&M costs. As such, enterprises are increasingly deploying their mission-critical IT service systems and applications on the public cloud, or in many cases on multiple public clouds, in order to meet the needs of flexible resource expansion and service reliability enhancement. Consequently, SD-WAN must provide network interconnections between enterprise sites and public clouds, and among public clouds, in addition to the regular network interconnections between branch sites and headquarters/DC sites.

In short, the cloud and the network are inseparable and their integration is essential. SD-WAN serves as the critical link to achieve such enterprise cloud-network synergy.

5.9.1 Cloud Concepts and Basic Principles

Before analyzing the requirements for an SD-WAN-to-public cloud connection and designing an optimal solution, we must first understand what the public cloud is and how it works. The following uses the AWS architecture as an example to describe the key concepts and architectural features of common public clouds, as shown in Figure 5.58.

The key concepts and architectural features of the public cloud are described as follows:

- Region

  A region is a separate geographical area based on which DCs are deployed. A public cloud serves customers in different regions around the globe, where DCs are usually deployed. For example, HUAWEI CLOUD has been deployed in regions such as North China, South China, Asia Pacific, and Europe. Enterprise tenants then select nearby DCs based on their service requirements. As such, public cloud DCs within the same area can be collectively called a region.

- Availability Zone

  An Availability Zone (AZ) is a geographical area with independent power supply and cooling systems within a region. A region can have multiple AZs that are physically isolated, and a DC is generally treated

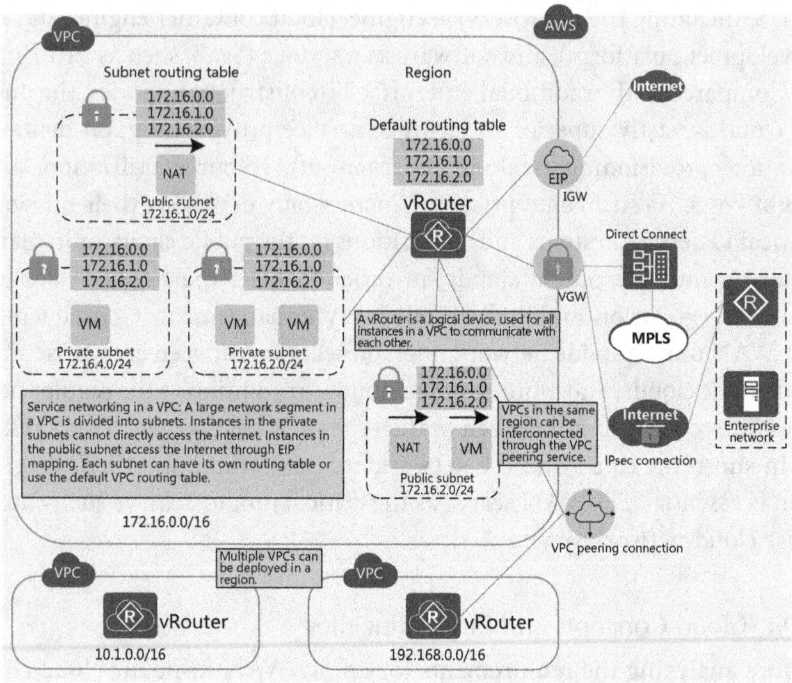

FIGURE 5.58 Key concepts and architectural features of the public cloud.

as an AZ. If enterprise applications require high reliability, resources are usually deployed in different AZs within the same region.

- VPC

  A VPC is a VN environment planned, configured, and managed by enterprises themselves, and serves as the "home" of enterprise tenants on the public cloud. Each VPC is an independent Layer 3 IP network, where enterprise tenants can plan network parameters, such as subnets, routes, security groups, and bandwidth, based on their requirements. VPCs are isolated from each other and can have their IP addresses overlapped.

- Elastic IP Address

  An Elastic IP Address (EIP) provides independent public IP address resources, including public IP addresses and public network egress bandwidth. EIPs can be associated with diversified instances such as Elastic Cloud Server (ECS) and NAT gateway instances.

- Internet Gateway

  As the access gateway of a VPC, an Internet Gateway (IGW) enables communication between VPCs and the Internet and provides mappings between EIPs and VPCs. When users access an EIP through the Internet, traffic is first sent to the IGW that will then find the corresponding VPC for the users, thereby implementing communication between VPCs and the Internet. Each VPC of an enterprise tenant has only one logical IGW.

- Direct Connect

  Direct Connect enables enterprises to establish connections from the local to public cloud VPCs through the private network. Direct Connect is a secure and reliable VPN service used to construct MPLS private lines.

  Most public clouds support both Internet access and Direct Connect access. In Internet access mode, both sites and VPCs have public IP addresses and can communicate with each other through any tunnel. In Direct Connect access mode, users must have private lines connected to the Direct Connect equipment room, which is usually provided by an Internet eXchange Provider (IXP). Tenants can deploy their own routers or rent them from public cloud service providers, and these routers are interconnected in the Direct Connect equipment room.

- Virtual Private Gateway

  Similar to an IGW, a Virtual Private Gateway (VGW) provided by a public cloud service provider also enables communication between tenant VPCs and the Internet. Despite its similarity to an IGW, a VGW is considered more secure and of higher quality due to the following:

  - Supports the standard IPsec VPN function and serves as the IPsec gateway for external devices, allowing external devices to connect to the VGW through IPsec tunnels to communicate with VPCs.

  - Connects to a private line network to implement high-quality interconnection with the external network through Direct Connect.

- VPC peering connection

  A VPC peering connection enables networks in different VPCs to communicate with each other. Generally, most public clouds can provide the connection service between VPCs in the same region, but do not support the VPC peering connection service across regions. In such cases, SD-WAN is required.

5.9.2 Cloud-Network Synergy Scenario Analysis

Now that we have gained a conceptual understanding of the public cloud architecture, let's take a look at the main service scenarios and requirements for SD-WAN-to-public cloud connection.

While the migration of enterprise services to the cloud has diversified cloud-network interconnection service scenarios, they can be generally classified into two different types: The first involves enterprise branch sites connecting to the public cloud for communication, in addition to the traditional communication between enterprise branch sites and their headquarters/DC sites. The second involves hybrid cloud services, whereby the private cloud and public cloud where mission-critical enterprise applications are deployed can communicate with each other, and the same is true for multiple heterogeneous public clouds.

Figure 5.59 shows the cloud-network synergy scenario.

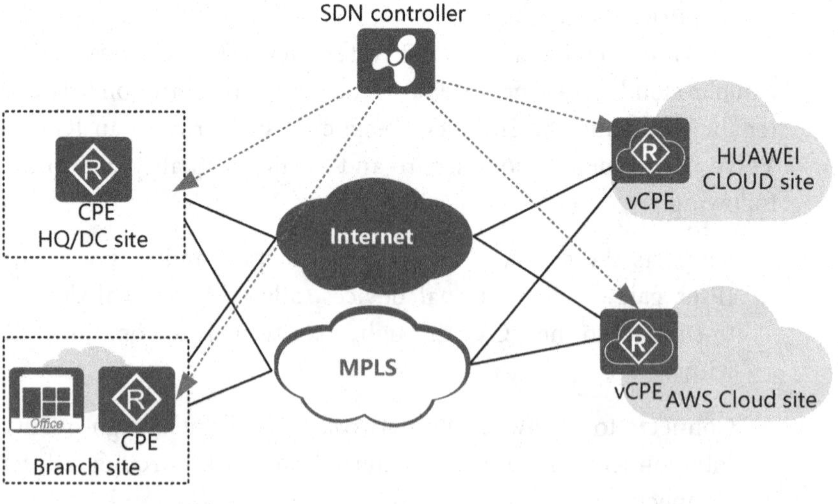

FIGURE 5.59 Cloud-network synergy.

5.9.2.1 Connecting Branch Sites to the Public Cloud

Migrating enterprise services to the cloud involves gradually deploying key application systems on the public cloud, such as enterprise office automation (OA), ERP, online supply chain, and data collection and storage systems. If enterprise branches require access to such systems deployed on the public cloud, they must first connect to the public cloud.

Based on service characteristics, connecting branch sites to the public cloud can be classified into the following three typical scenarios:

1. Simple connection to the public cloud

   Many small and midsize enterprises deploy only a limited range of services on the public cloud. For example, application systems are usually deployed in the same VPC of the public cloud, and enterprise branches centrally access application systems in the VPC. This does not pose high requirements on the service provisioning speed and network quality, so the primary goal is to implement communication between enterprise branches and the public cloud, as shown in Figure 5.60.

2. Enhanced connection to the public cloud

   In contrast to the above, many large enterprises deploy diverse service systems on the public cloud. Such deployment is complex and places high requirements on the service provisioning speed and cloud service experience. For example, an enterprise deploys multiple application systems in different VPCs of the public cloud, and enterprise branches need to access different VPCs as required. In addition, the application systems must also communicate with each other. As such, communication between VPCs is required.

   Figure 5.61 illustrates the network diagram in this scenario.

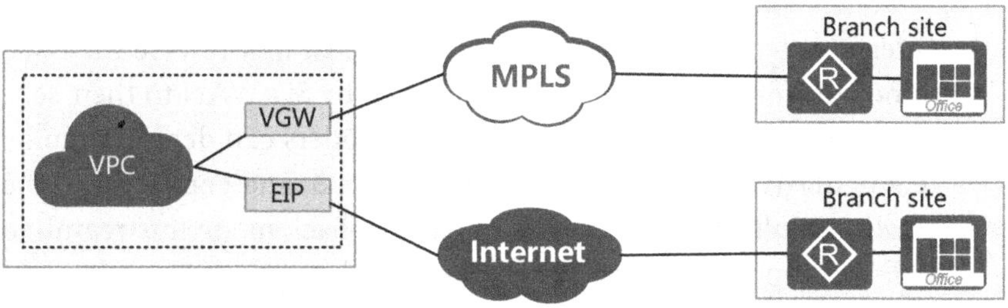

FIGURE 5.60 Simple connection to the public cloud.

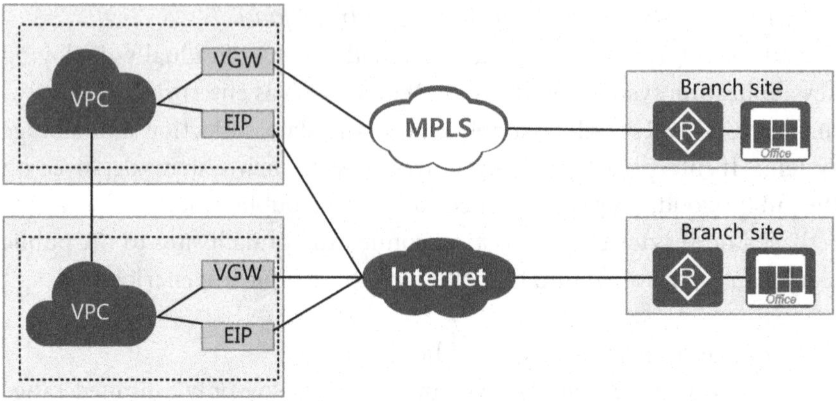

FIGURE 5.61 Enhanced connection to the public cloud.

In addition, to deliver an optimal application experience for the abovementioned application systems, high network quality must be ensured. Otherwise, the network congestion from a poor network environment prolongs latency and causes additional packet loss, which in turn adversely affects the application experience.

In order to overcome these problems, it is necessary to steer application traffic based on link quality to ensure an optimal application experience. For data storage systems or video systems deployed in a VPC, it is a must to quickly and stably transmit large files and videos.

3. Connection to the public cloud through self-built gateways

   Many SD-WAN service providers also build and operate their own public clouds and provide public cloud services for enterprises. Such service providers often take the logical step to connect SD-WAN to their own public cloud to form an E2E integrated solution.

   SD-WAN service providers can comprehensively integrate SD-WAN and cloud services, providing one-stop cloud services for enterprise tenants. To achieve this, they must first resolve the issue of how to connect the networks managed by SD-WAN to their self-built public clouds. SD-WAN service providers can deploy a multi-tenant SD-WAN gateway on the public cloud, and connect it to the existing multi-tenant gateways in back-to-back mode to streamline E2E cloud-network synergy services. In this way, enterprise users can enjoy a one-stop cloud experience while SD-WAN services are provisioned for enterprise tenants.

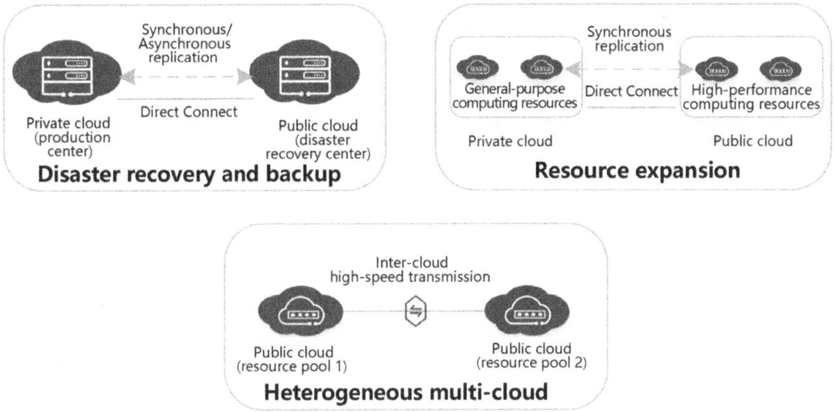

FIGURE 5.62 Hybrid cloud.

5.9.2.2 Hybrid Cloud

Hybrid clouds are primarily used for application data backup, disaster recovery, and elastic scaling on the public cloud, as shown in Figure 5.62. However, regardless of the scenario, we must address the problems of the interconnection between private and public clouds, the interconnection between multiple public clouds, and the centralized management of multi-cloud sites.

The general concept of SD-WAN deployment in hybrid cloud scenarios is to deploy edge devices or gateways at hybrid-cloud sites that must be interconnected, and implement on-demand and flexible interconnection of hybrid clouds under the unified management and control of the SDN controller.

5.9.3 Public Cloud Connection Solution

After analyzing the main scenarios of SD-WAN cloud-network synergy, we will discuss the three solutions for connecting branch sites to a public cloud, as shown in Figure 5.63.

In SD-WAN, the following three solutions are available for connecting branch sites to the public cloud:

- **IPsec VPN gateway solution**: A simple solution, whereby CPEs at the enterprise branch sites directly establish IPsec tunnels to connect to the VGWs of the public cloud. The IPsec configuration can be completed by the SDN controller.

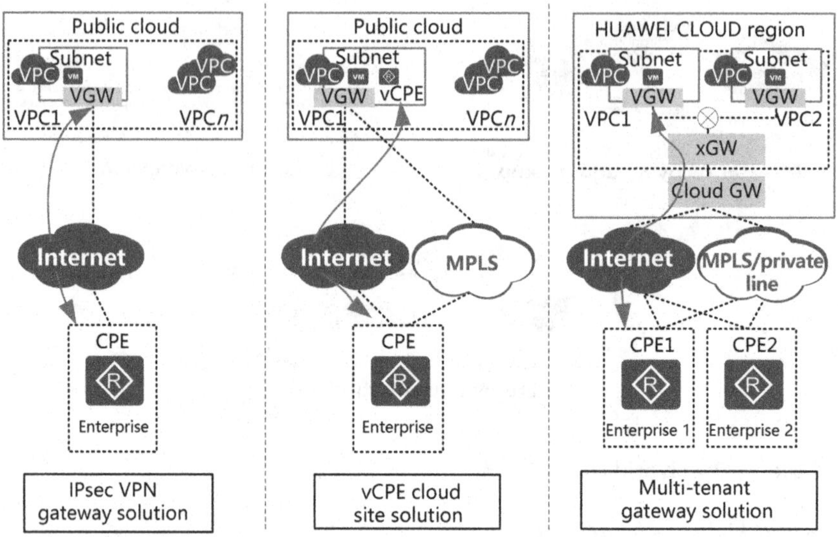

FIGURE 5.63 Connecting branch sites to a public cloud.

- **vCPE cloud site solution**: An enterprise deploys vCPEs in VPCs of the public cloud. Functioning as edges of VPC cloud sites, vCPEs are managed by the SDN controller and interconnected with enterprise sites through SD-WAN overlay tunnels.
- **Multi-tenant gateway solution**: The SD-WAN gateways (cloud gateways) are deployed on the public cloud and connect to gateways on the public cloud in a unified manner under the management of the SDN controller.

5.9.3.1 IPsec VPN Gateway Solution

After a tenant creates a VPC on the public cloud, the VPC can communicate with external sites (such as local branches and DCs) deployed by the tenant in either of the following ways:

1. Internet-based IPsec VPN

   In the IPsec VPN solution, IPsec VPN tunnels are established on both CPEs and VGWs at the external sites of an enterprise, and the Internet Key Exchange (IKE) protocol is used to implement communication over the tunnels. In addition, a BGP peer relationship is established on each tunnel to exchange service routes between the two ends. Figure 5.64 shows the networking of this solution.

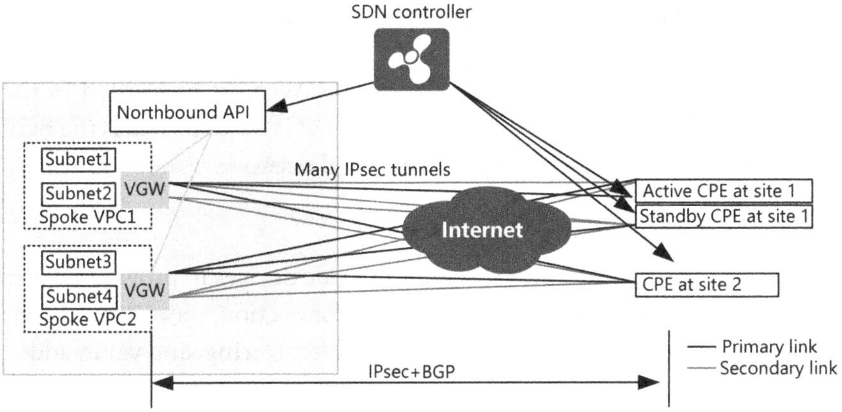

FIGURE 5.64 Internet-based IPsec VPN solution.

Two methods of service provisioning are available. The first is loose coupling, whereby CPEs are configured on the SDN controller and VGWs are configured on the public cloud platform. The second method uses the SDN controller to uniformly orchestrate the creation of IPsec VPNs, meaning the SDN controller not only directly configures IPsec VPNs for CPEs at the sites but also invokes related northbound APIs of the public cloud to indirectly configure IPsec VPNs.

2. Direct Connect based on MPLS private lines (or other private lines)

   In the VPC Direct Connect solution, after private lines are connected to the Direct Connect equipment room of the public cloud, the tenant must invoke APIs on the public cloud management console to create virtual interfaces (VIFs) and bind them to the VGWs of specific VPCs. Multiple dedicated VIFs can be created for one physical private line, and they are isolated by VLAN. A VGW creates BGP peer relationships between VPCs and sites for route exchange.

This solution features simplified operations and low costs and is ideal for small and midsize enterprises. Customers are typically only required to pay a monthly rental for VPN connections and traffic, and are charged only for the bandwidth in the outbound direction. This solution is suitable for enterprises seeking a simple connection to the public cloud. Specifically, the application systems of an enterprise are deployed in a VPC on the public cloud, and users at the enterprise's branch sites access the application systems in the VPC in a centralized manner. As such, this solution achieves fast provisioning and centralized management of network services.

Regardless of which solution is used, a VGW must be created in the VPC to be interconnected for the tenant. This VGW also provides the BGP function to exchange routes with the external network.

5.9.3.2 vCPE Cloud Site Solution

In some scenarios, customers require VPC-side devices to provide a wider array of functions, such as multi-VPC interconnection, more diverse tunnels (such as VXLAN), application-based traffic steering, and value-added network services (such as WAN acceleration and security services). They also demand more flexible routing protocol selection and route control. The vCPE cloud site solution is capable of fulfilling such requirements.

In this solution, vCPE components are deployed on the cloud in order to build cloud sites, which are orchestrated and managed by the SDN controller. The vCPE cloud site solution implements access control between enterprise branch sites and VPCs, quickly provisions application-based traffic steering and WAN acceleration, and implements unified control between enterprise branch sites and VPC networks.

A vCPE is essentially an SD-WAN CPE and offers all the same key features, such as multi-VPN interconnection, application-based traffic steering, and WAN acceleration. During deployment, we must first create one or a pair of vCPEs in a VPC and configure service subnets' VPC routes pointing to the vCPEs. Then, on a vCPE, we must configure static advertisement of intra-VPC routes to sites. All these operations can be automatically orchestrated and configured by invoking the northbound APIs of the public cloud through the SDN controller.

Based on the VPC connection mode, the vCPE cloud site solution can be classified as a transit VPC solution or a host VPC solution. We will use the AWS as an example to describe the transit VPC solution, and HUAWEI CLOUD to describe the host VPC solution. The vCPE cloud site solution is decoupled from the specific public cloud architecture and only needs to adapt to the APIs of the specific public cloud. As such, this solution is also applicable to public clouds of other service providers.

1. Transit VPC solution

   A transit VPC is an independent VPC that connects to service VPCs. This solution is recommended for SD-WAN sites to access the AWS Cloud. vCPEs are deployed in a transit VPC as enterprises' cloud sites, which are orchestrated and managed by the SDN controller. A transit VPC is also a hub site on the cloud where service VPCs are deployed, used to connect these service VPCs. Leveraging the transit VPC solution, enterprises can interconnect cloud sites with their SD-WAN sites, as illustrated in Figure 5.65.

The transit VPC solution implements cross-region interconnection of VPCs. For enterprises that have deployed VPCs in multiple regions on the cloud, a transit VPC can be deployed and used to connect to service VPCs in other regions on the cloud, as well as to remote branch sites.

In the transit VPC solution, vCPEs are introduced to flexibly control the interconnection between CPEs and VPCs based on vCPEs'

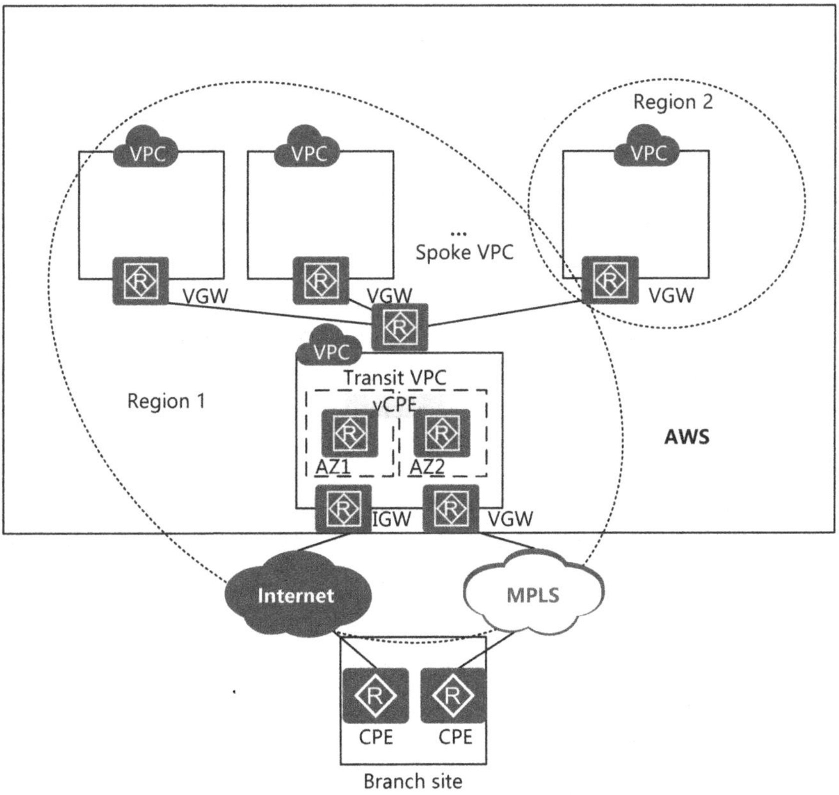

FIGURE 5.65 Transit VPC solution.

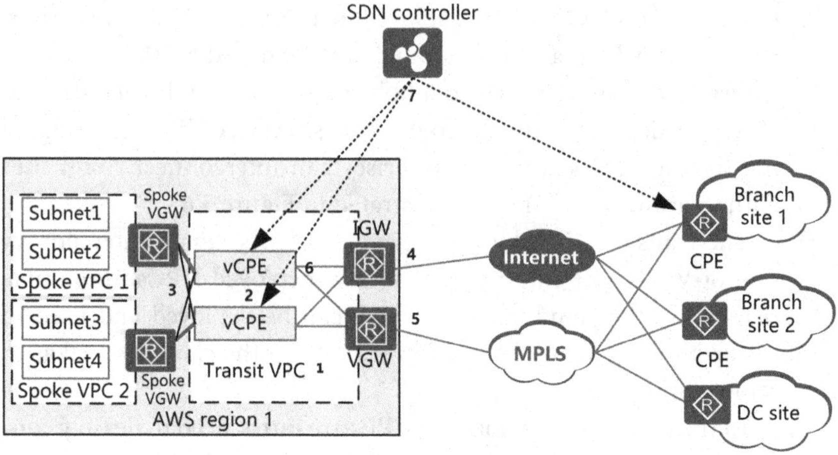

FIGURE 5.66 Working principles of the transit VPC solution.

routing and VPN functions. Figure 5.66 illustrates how the transit VPC solution works.

In the transit VPC solution, one or a pair of vCPEs are deployed in a dedicated VPC (specifically, a transit VPC) to connect remote sites to service VPCs in different regions on the cloud. The WAN side of vCPEs can connect to remote sites through the Internet or Direct Connect. On the LAN side, vCPEs establish IPsec VPN+BGP connections with VGWs in service VPCs to exchange routes and learn network-wide routes. As such, remote sites can communicate with VPCs, and features such as multi-VPN interconnection, intelligent traffic steering, and WAN acceleration are supported. The transit VPC solution operates as follows:

a. The transit VPC created by a tenant functions as an aggregation point of enterprise tenants and connects to the enterprise service VPCs, DC sites, and branch sites.

b. vCPEs are deployed in the transit VPC. The same number of WAN-side interfaces are added to each vCPE based on the number of WAN-side links, and a LAN-side interface is allocated to each vCPE.

c. In the transit VPC, EIPs are assigned for the WAN and LAN sides of the instances where vCPEs reside. The WAN side connects to the Internet through the IGW. On the LAN side, GRE interfaces are created to establish IPsec VPN+BGP connections with VGWs of the service VPCs, implementing mutual access between service VPCs and the transit VPC. In addition, service VPCs can use the multi-VPN interconnection feature of vCPEs to implement on-demand access control. Different service VPCs can belong to different VPNs.

d. The vCPE of the transit VPC on the AWS Cloud can access the Internet through the IGW, and the CPEs at non-cloud sites can also access the Internet. As a result, the SDN controller can uniformly orchestrate services for both on-cloud vCPEs and off-cloud CPEs.

e. When Direct Connect is deployed in the transit VPC, two links (IGW+VGW) can be established between vCPEs and CPEs, and the intelligent traffic steering feature can be used.

f. vCPEs support comprehensive QoS features for traffic control.

g. vCPEs can report interface traffic statistics to the SDN controller, which can then monitor and analyze traffic statistics on the cloud.

The transit VPC solution has the following advantages:

a. The use of a transit VPC reduces the impact on services of tenant service VPCs.

b. The use of vCPEs enables unified service management and orchestration by the SDN controller, ensuring consistent user operations. vCPEs provide diverse routing and policy features, contributing to high scalability.

c. As a centralized control point, a transit VPC simplifies the network topology, as a complex full-mesh topology is no longer required in this case.

d. A transit VPC provides visibility on applications and security and creates more space for the expansion of VASs.

Figure 5.67 shows an enterprise's SD-WAN networking with VPCs deployed on the cloud.

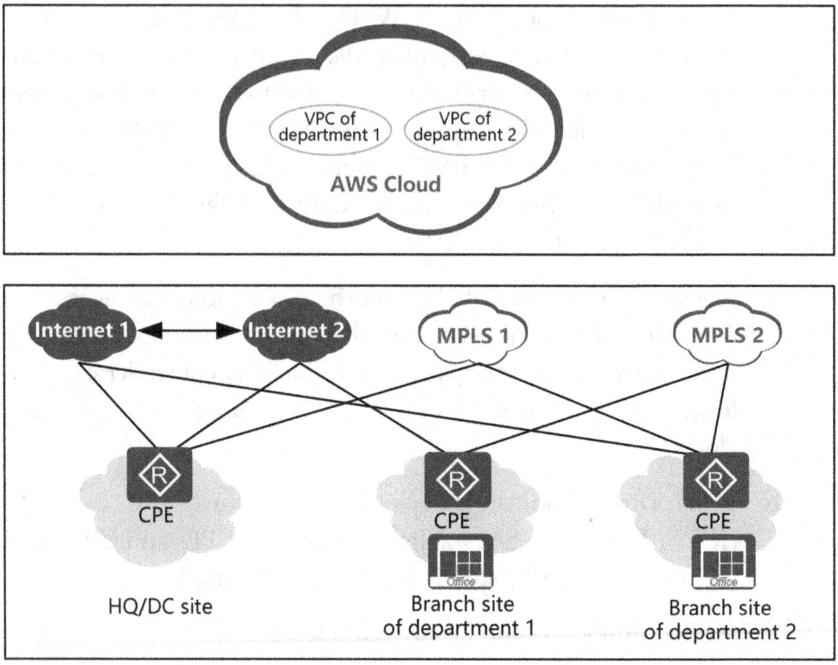

FIGURE 5.67 VPC deployment on the cloud.

The branch sites of different enterprise departments are connected through Internet 1, Internet 2, MPLS 1, and MPLS 2. Internet 1 and Internet 2 are interconnected. The enterprise applies for an independent VPC for each department on the AWS and has the following requirements regarding VPC communication: The VPC of each department must be able to communicate with branch sites over multiple networks, but VPCs of different departments must not communicate with each other.

Underlay network deployment

Transit VPCs are created, vCPEs are deployed, and an underlay network is constructed, as shown in Figure 5.68.

Three interfaces (two WAN interfaces and one LAN interface) are assigned for a vCPE to connect to three different subnets of the transit VPC. WAN interface 1 (an Internet interface) connects the vCPE to a public subnet. After an EIP is obtained for WAN

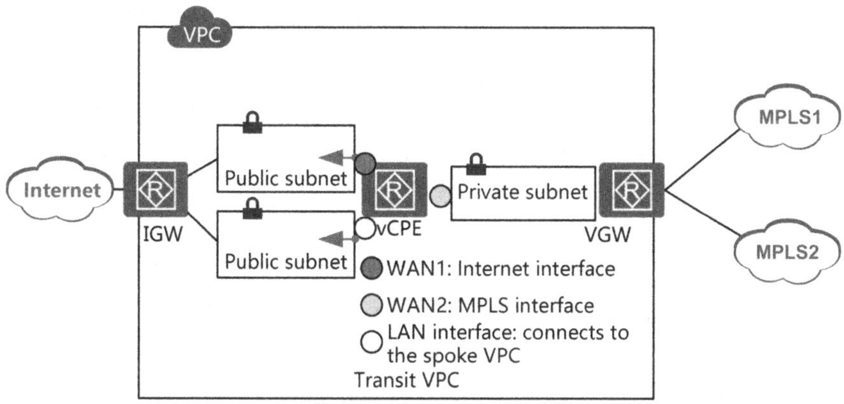

FIGURE 5.68 Underlay network deployment.

interface 1, the vCPE can access the Internet through WAN interface 1 and the IGW. WAN interface 1 can communicate with the Internet link of the enterprise CPE over the Internet. WAN interface 2 (an MPLS interface) connects the vCPE to a private subnet. The vCPE then connects to MPLS 1 and MPLS 2 through WAN interface 2 and the VGW. The VGW learns the routes of the WAN links that connect branch CPEs to MPLS 1 and MPLS 2, allowing communication between the vCPE and CPEs over the MPLS underlay network. The LAN interface connects the vCPE also to a public subnet. After an EIP is obtained for the LAN interface, the vCPE can access the Internet through the LAN interface and the IGW. This LAN interface allows the vCPE to communicate with the VGW of the service VPC through the Internet.

Overlay network deployment

vCPEs are configured as cloud sites and register with the SDN controller. The two WAN interfaces of each vCPE are bound to the RDs of the Internet and MPLS network, respectively. After vCPEs are added to the corresponding VPNs (departments) and the VPCs to be connected are specified, the SDN controller automatically orchestrates and provisions SD-WAN services, creates VRF entries on vCPEs, and establishes overlay networks to CPEs and VPCs based on the VRF entries.

The reliability design of the transit VPC solution involves the following:

a. **Active/standby vCPE deployment in multiple regions**: Two vCPEs can be deployed in active/standby mode in a transit VPC, and they can reside in different AZs. The VPC solution leverages the innate reliability feature of the AZ to provide fault isolation at the DC level and even the more refined device level.

b. **Independent EIP deployment for the active and standby vCPEs**: The active and standby vCPEs are each assigned an independent EIP, enabling network-level fault isolation.

c. **vCPE multi-subnet and multi-EIP deployment**: WAN interface 1 connecting a vCPE to the Internet, and WAN interface 2 and the LAN interface connecting the vCPE to the DC are bound with different subnets and EIPs to achieve subnet-level fault isolation.

Figure 5.69 shows the reliability design for service interconnection in the transit VPC solution.

By default, the VGW in a service VPC obtains two ECMP routes from two vCPEs through IPsec VPN tunnels. The SDN controller automatically selects the active and standby vCPEs for the service VPC. By adjusting route attributes (EBGP), the SDN controller selects a specific vCPE as the gateway in the outbound direction of the service VPC.

By default, a CPE site obtains two ECMP routes from two vCPEs through SD-WAN overlay tunnels. The SDN controller adjusts the route attributes (IBGP) and selects a specific vCPE as the gateway in the outbound direction of the CPE site.

The SDN controller centrally orchestrates the route attribute settings of vCPEs to ensure that the incoming traffic and the corresponding outgoing traffic are forwarded by the same vCPE. If a vCPE fails, the SDN controller can automatically switch traffic to the functional vCPE based on the BGP routing processing capability.

When multiple VPCs are deployed, the SDN controller can load balance traffic between two vCPEs.

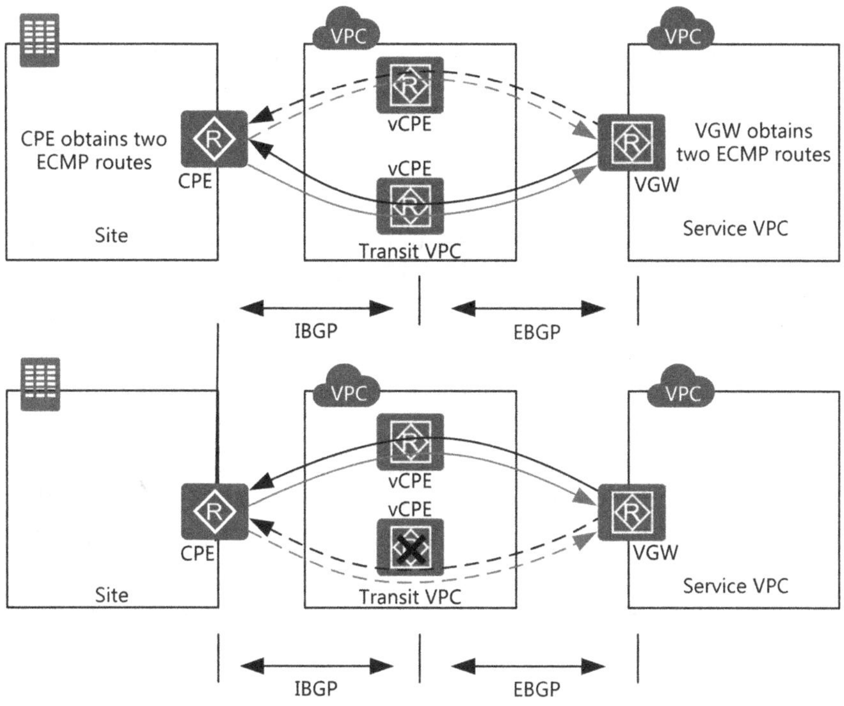

FIGURE 5.69 Service interconnection reliability design.

For security purposes, the SDN controller must be used to isolate subnets connected to vCPEs in a transit VPC based on security groups.

In the transit VPC solution, the SDN controller invokes the APIs provided by AWS to implement automated deployment. However, individually invoking AWS APIs to create and modify the large number of resources in a transit VPC leads to complex logic and difficult operations. As such, the SDN controller uses the AWS template to hand over the creation of transit VPC resources to the AWS. Template instantiation is an atomic operation, and the SDN controller is not required to perform a rollback if a single resource fails to be created, which greatly reduces implementation complexity and ensures reliability. The SDN controller injects the deployment configuration into vCPEs, which automatically load the deployment configuration during startup. The started vCPEs are then managed by the SDN controller.

2. Host VPC solution

A host VPC is a service VPC, where vCPEs are deployed as the cloud sites in an enterprise's SD-WAN and are uniformly managed and orchestrated by the SDN controller.

The host VPC solution also supports cross-region interconnection between VPCs. If an enterprise has deployed VPCs in multiple regions on the cloud, vCPEs can be deployed in each host VPC and the SDN controller will uniformly orchestrate services, implementing communication between host VPCs in different regions on the cloud and remote off-cloud branch sites.

After vCPEs are introduced to the host VPC solution, the routing and VPN functions of vCPEs can be used to flexibly control the interconnection between CPEs and VPCs, as shown in Figure 5.70.

In the host VPC solution, one or a pair of vCPEs are deployed in a service VPC to connect remote sites to the service VPC. The WAN side of vCPEs can connect to remote sites through the Internet or Direct Connect. On the LAN side, vCPEs connect to the subnets of host VPCs. VPC routes pointing to the LAN-side interface are configured in host VPCs so that traffic from host VPCs to remote sites passes through the vCPEs. As such, remote sites can communicate

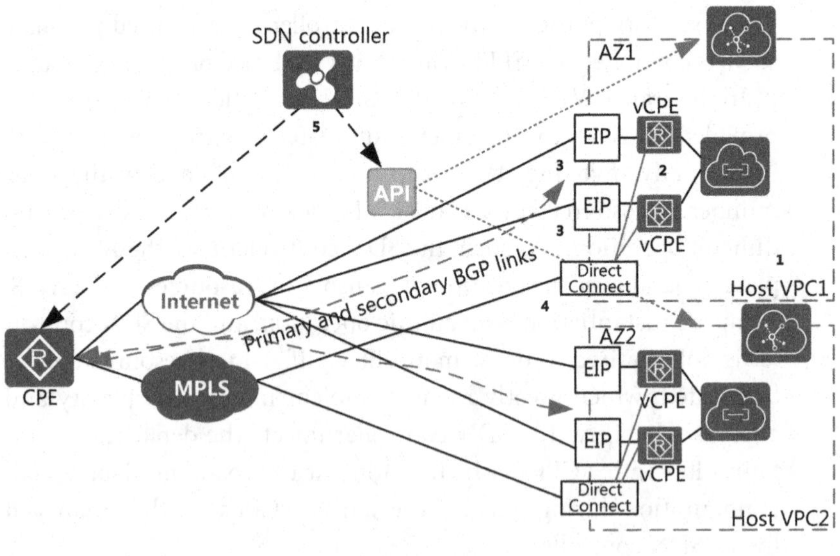

FIGURE 5.70 Host VPC solution principles.

with VPCs, and features such as multi-VPN interconnection, intelligent traffic steering, and WAN acceleration are supported.

The following describes the host VPC solution in detail:

a. In each host VPC, vCPEs are deployed as the egress gateways of the VPC and serve as the cloud sites in an enterprise's SD-WAN. To enhance reliability, a vCPE can be deployed in each of the two AZs, and the SDN controller uniformly orchestrates vCPE services, monitors vCPE status, and instructs VPC route switching.

b. Three interfaces (two WAN interfaces and one LAN interface) are assigned for a vCPE to connect to three different subnets of the host VPC where the vCPE resides. One WAN interface connects the vCPE to the Internet through the EIP, and the other WAN interface connects the vCPE to Direct Connect. The LAN interface connects to a subnet of the host VPC and serves as the aggregation point for connecting to other subnets. A summarized or default route is configured in a host VPC, and the next hop of other subnets' default or summarized route is set to the LAN interface of a vCPE in the host VPC. As a result, all egress traffic of the host VPC passes through the vCPE.

c. Based on the Internet, SD-WAN connections are established between different host VPCs through vCPEs, implementing communication between CPEs and host VPCs.

d. Based on Direct Connect, dual links (a Direct Connect link and an Internet link) can be established between host VPCs and CPEs.

e. The SDN controller monitors and analyzes traffic statistics on the cloud based on the interface traffic statistics reported by vCPEs.

To implement communication between a CPE and host VPCs, two overlay tunnels can be created in the service VPN to communicate with the CPE through the Internet and Direct Connect, and encrypt the tunnels using IPsec. BGP peer relationships are established over overlay tunnels for the CPE and vCPEs to advertise LAN-side routes to each other, enabling the CPE to learn the routes of VPC service subnets and vCPEs to learn the CPE routes.

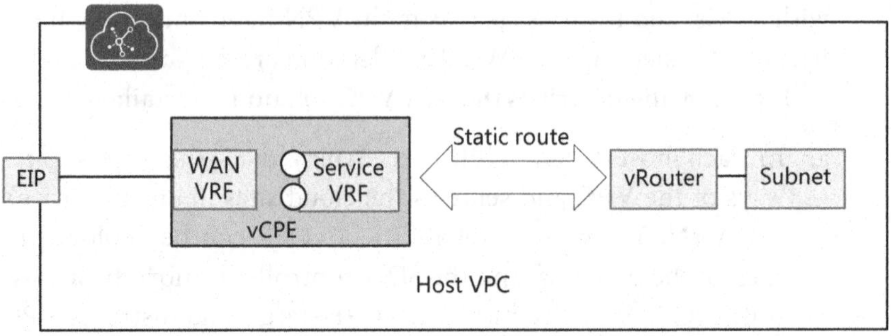

FIGURE 5.71 Service communication between a vCPE and the subnets in the VPC.

To implement service communication between a vCPE and the subnets in the VPC where the vCPE resides, when service traffic destined for the VPC enters the vCPE's WAN interface, the traffic is forwarded to the gateway of the vCPE's LAN interface based on the vCPE's routing information (specifically, the vRouter in the VPC). The vRouter forwards the traffic to the service address in the VPC based on the destination address.

Figure 5.71 shows how such service communication is implemented.

Figure 5.72 shows the reliability design for service interconnection in the host VPC solution.

It is recommended to deploy two vCPEs belonging to different AZs in a host VPC. The inherent reliability of AZs will provide fault isolation at the DC level and device level.

a. **vCPE multi-subnet deployment**: One WAN interface connecting a vCPE to the Internet, and the other WAN interface and the LAN interface connecting the vCPE to the DC are bound with different subnets and EIPs in order to implement subnet-level fault isolation.

b. When two vCPEs are deployed in a host VPC, CPEs at branch sites must be connected to each of the two vCPEs.

c. The SDN controller specifies the active and standby vCPEs for the host VPC, adjusts the route priority of the active vCPE to be higher than that of the standby vCPE, and sets the next hop of the host VPC route to the active vCPE so that the incoming traffic and the corresponding outgoing traffic both pass through the active vCPE.

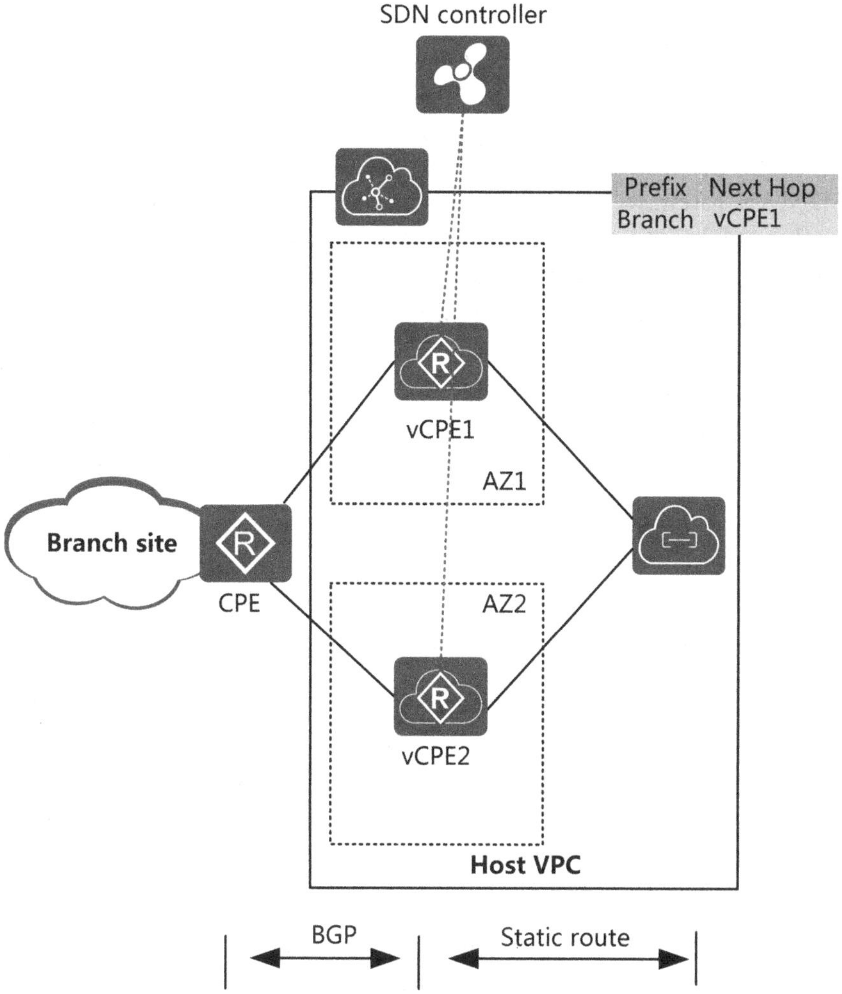

FIGURE 5.72 Service interconnection reliability design.

The SDN controller can monitor the vCPE status. If the active vCPE fails, the SDN controller sets the next hop of the host VPC's return route to the standby vCPE, implementing an active/standby switchover.

5.9.3.3 Multi-tenant Cloud Gateway Solution

While the IPsec VPN gateway solution and the vCPE cloud site solution described previously are suitable for public cloud connection in enterprise-built SD-WANs, the multi-tenant cloud gateway solution outlined below has instead been designed for cloud connection in SD-WANs operated by carriers. Each of these three solutions possesses unique technical strengths and weaknesses, and each is applicable to different service scenarios and public cloud architectures. We can flexibly choose one based on actual needs.

In the multi-tenant cloud gateway solution, one or more multi-tenant SD-WAN gateways (cloud gateways) are deployed on the public cloud. Under orchestration and control of the SDN controller, SD-WAN tunnels are created between branch CPEs and cloud gateways for communication. Additionally, interconnections between cloud gateways and tenant VGWs are configured as required.

For example, in back-to-back Option A mode, a VLANIF interface or VXLAN interface is configured for each tenant's VPN, and EBGP is run to implement interconnection between the enterprise branch network and public cloud, as shown in Figure 5.73.

The following describes how the multi-tenant cloud gateway solution is deployed:

- Cloud gateways are deployed on the public cloud.
- Two WAN interfaces can be assigned based on user access requirements. One connects users to the public cloud through the Internet and the other through Direct Connect. To facilitate description, we refer to them as WAN interface 1 (Internet) and WAN interface 2 (Direct Connect).
- It is recommended that WAN interface 1 be connected to the gateway at the Internet egress, and that WAN interface 2 be connected to the gateway at the Direct Connect egress.

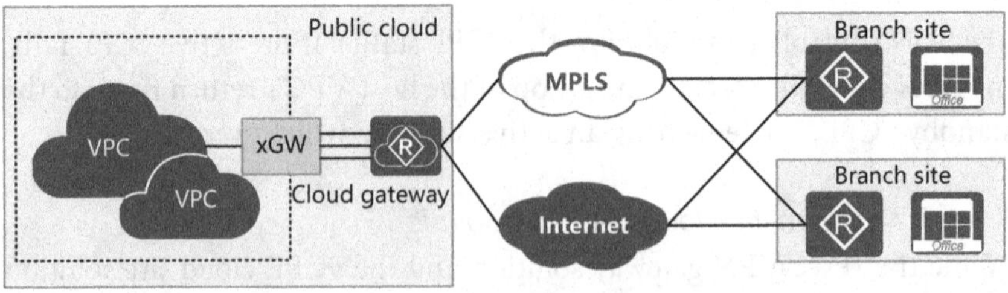

FIGURE 5.73 Multi-tenant cloud gateway solution.

- Two LAN interfaces (the same number as WAN interfaces) are assigned. One is connected to the Internet access switch, and the other to the Direct Connect access switch.
- Cloud gateways are managed and orchestrated through the SDN controller.
- In the MSP view of the SDN controller, a carrier can configure the interconnection mode for each VPN of a tenant, including interconnection with the Internet access switch and interconnection with the access switch of the MPLS private line. Through the VLANIF interface or the VXLAN interface, a tenant's VPC routes can be advertised to the cloud gateway along either the BGP or static route.
- A carrier can provision cloud gateway interconnection services for tenants in the MSP view of the SDN controller, and enterprise tenants can connect remote sites to cloud gateways through SD-WAN tunnels. The cloud gateway advertises a tenant's VPC routes to the remote sites of the tenant and advertises the remote site routes of the tenant to the access switch, implementing communication between remote sites and VPCs of the tenant. As route advertisement and learning are based on tenant VPNs, the networks of multiple tenants connected to the same cloud gateway are isolated, enabling this cloud gateway to serve multiple tenants.

5.9.4 Hybrid Cloud Solution

In traditional hybrid cloud scenarios, the cloud and the network are separated, and public cloud network resources cannot be flexibly scheduled based on service requirements.

In the SD-WAN hybrid cloud solution, various clouds are abstracted as special cloud sites that are managed and controlled by the SDN controller. IP overlay technology is leveraged to shield the differences between multi-cloud interconnection technologies, implementing fast, simple, and high-quality interconnection between different types of hybrid cloud networks.

In addition, open northbound APIs are provided for cloud services, integrating SD-WAN into the hybrid cloud service platform and achieving E2E unified management of cloud applications and in-depth cloud-network synergy.

5.9.4.1 Interconnection between Private and Public Clouds

Interconnection between private clouds and public clouds is the most common hybrid cloud scenario for enterprises, as shown in Figure 5.74. In this scenario, multiple SD-WAN solutions can be selected based on the SD-WAN business model and enterprise service characteristics.

If an enterprise builds its own SD-WAN, the vCPE cloud site solution can be used, whereby the SDN controller first plans and configures public cloud sites in offline mode and then remotely and automatically starts the vCPEs on the public cloud. Under the unified management and control of the SDN controller, network interconnection between private cloud sites and public cloud vCPE sites is implemented. If an enterprise leases multiple VPCs on the public cloud, the transit VPC or host VPC solution can be deployed.

If public cloud providers supply both public cloud services and SD-WAN services for enterprises, the multi-tenant cloud gateway solution is recommended. After one or more multi-tenant SD-WAN gateways are deployed on the public cloud, the public cloud usually provides a multi-tenant gateway to interconnect with the SD-WAN gateway(s) through back-to-back sub-interfaces or VXLAN tunnels (Option A). In addition, dynamic routing protocols such as BGP are deployed for route learning and service communication between the multi-tenant gateway and SD-WAN gateway(s). Under the unified management and control of the SDN controller, network interconnection between private cloud sites and public cloud SD-WAN gateway(s) is implemented, which ultimately achieves network interconnection between enterprise private and public clouds, and between enterprise sites and public clouds/private clouds.

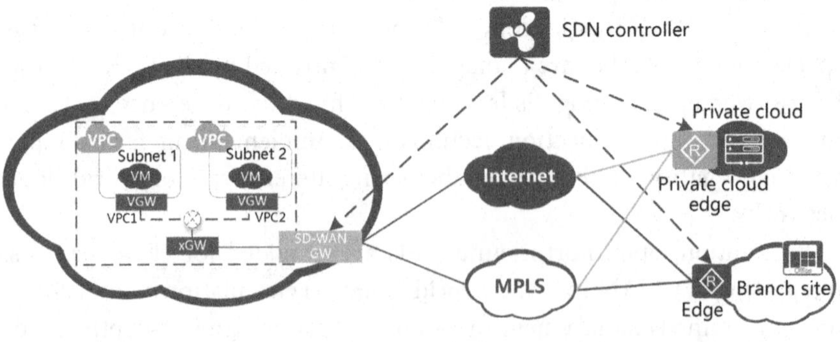

FIGURE 5.74 Interconnection between private and public clouds.

5.9.4.2 Heterogeneous Public Cloud Interconnection

Due to the limited resources and coverage capabilities of a single public cloud provider, enterprises typically deploy services on multiple public clouds, as shown in Figure 5.75.

As multiple public cloud providers are involved, the vCPE cloud site interconnection solution is preferred. The architecture used in this solution is decoupled from the basic architecture of the public cloud, and only lightweight interconnection between SD-WAN and the public cloud service platform is required, facilitating deployment.

Specifically, the SDN controller first plans and configures different public cloud sites in offline mode, and then remotely and automatically starts the vCPEs on different public clouds. Under the unified management and control of the SDN controller, network interconnection between different public cloud vCPE sites is automatically implemented. If an enterprise leases multiple VPCs on the same public cloud, the transit VPC solution can also be deployed to implement interconnection between VPCs.

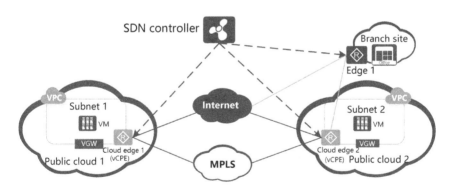

FIGURE 5.75 Heterogeneous public cloud interconnection.

CHAPTER 6

Guaranteed Application Experience

WITH THE GLOBALIZATION OF ENTERPRISE SERVICES, a growing number of enterprises are having to interconnect their branches across countries and regions using a broad range of WAN link types, including Ethernet cables, transoceanic cables, wireless networks, and satellite links.

For such complex networks with long-distance WAN links and a large number of intermediate nodes, guaranteeing link quality is extremely challenging. Issues such as packet loss, delay, and jitter are common, restricting enterprise development.

Enterprises with hybrid WAN links therefore require guaranteed application availability and consistent experience. It is for such purposes that the SD-WAN application experience assurance sub-solution is designed.

6.1 ASSURANCE SOLUTION

Today's enterprises have extensive applications, which generally fall into categories such as production, collaboration, cloud, and entertainment. Such diverse applications no doubt have varying requirements on bandwidth and link quality. For example, collaboration applications such as real-time video conferences require links to have extremely low packet loss rate, delay, and jitter. Once a packet loss occurs on a link, freeze frame and artifacts may occur. Another example is production applications, among which file transfer applications such as email and File Transfer Protocol (FTP) are able to withstand some packet loss but require high bandwidth.

To meet applications' diversified requirements, traditional WANs need to resolve the following issues:

- Applications of different priorities are carried on the same link but are not handled differently.

  This means that traffic of multiple applications cannot be distinguished, and traffic of low-priority applications cannot be dynamically adjusted to guarantee high-priority traffic. Here, let's take the voice call and FTP-based file transfer services as an example. These two services have different link quality requirements. If the destination addresses of the two applications are the same, traditional routing technologies cannot identify these applications, let alone distribute them to different links for transmission. If the traffic volume exceeds the link bandwidth, network congestion inevitably occurs, impacting even high-priority services such as voice traffic.

- When link quality deteriorates, dynamic routing cannot be implemented.

  Traditional IP networks determine packet paths using routing protocols, such as BGP and OSPF, which focus only on packet reachability, neglecting the impact of link quality on application availability. As a result, network O&M is separate from application experience management, and networks need to be statically configured by O&M personnel. When link quality deteriorates, O&M personnel have to manually adjust forwarding paths, which is time-consuming and error-prone.

- No effective measure is available when link quality deteriorates.

  The Internet cannot provide reliable transmission, with packet loss leading to application traffic being transmitted inefficiently. If an enterprise has no better links, there is no effective measure to improve transmission. The end result is inevitably poor application experience. For example, if the quality of all available links deteriorates and a large number of voice packets are lost, voice calls are intermittent or even cut off completely. Resolving such issues requires some sort of network optimization technology.

To tackle the preceding issues on traditional WANs, enterprises need a solution that can identify applications as well as guaranteeing their experience. Figure 6.1 shows the SD-WAN application experience assurance sub-solution, which consists of application identification, application-based traffic steering, QoS, and WAN optimization. These functions can be used separately or in various combinations.

1. Application identification

   Application identification is a technology that identifies application types in traffic transmitted on the network based on traffic characteristics. Enterprise applications have varying requirements on link quality, while their corresponding optimization measures also vary. Applications must be identified before the application experience assurance sub-solution can be implemented.

2. Application-based traffic steering

   Traditional network technologies are unable to dynamically select paths for applications based on their link quality requirements. Application-based traffic steering continuously monitors the status of multiple WAN links based on enterprise applications' priorities and requirements on link quality, and then selects the optimal link for transmission. This ensures high-value application experience upon WAN link congestion while ensuring high-quality links are fully utilized.

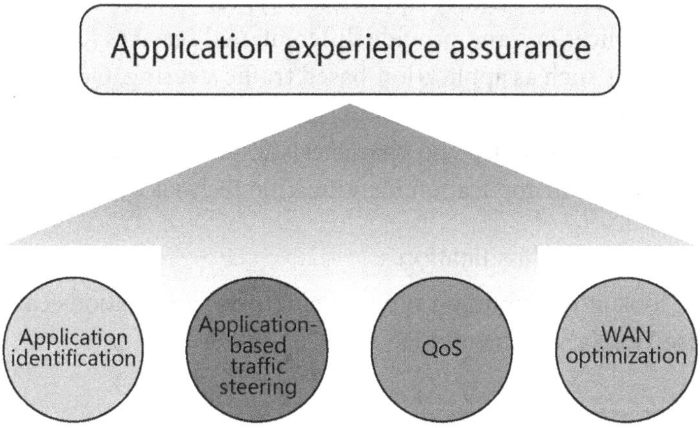

FIGURE 6.1 Application experience assurance sub-solution.

3. QoS

   Building on traditional QoS technologies such as traffic policing, traffic shaping, and queue scheduling, SD-WAN further expands the QoS functions to provide service awareness (SA), which provides differentiated services for enterprise applications. When multiple departments of an enterprise need to be isolated, the solution provides service quality assurance based on departments, forming hierarchical (service-department-site) quality assurance.

4. WAN optimization

   WAN optimization is a series of technologies used to improve the quality and efficiency of data transmission on WAN links, including packet loss mitigation technologies. WAN optimization technologies focus on how to obtain good enterprise application experience on low-quality links. An example technology is packet loss mitigation, which ensures that no freeze frame or artifacts occur when the link quality deteriorates or a large number of packets are lost. In addition, transmission and data optimization technologies are provided to improve data transmission efficiency.

Policies can be configured on demand to automatically apply traffic steering, QoS, and WAN optimization after applications are identified, thereby automatically ensuring application experience.

6.2 APPLICATION IDENTIFICATION

Once we are able to identify application types, we can better manage enterprise applications and provide differentiated services for them. That is, technologies such as application-based traffic steering, QoS, and WAN optimization are possible only after applications are identified.

Now, we will introduce the characteristics of enterprise applications before moving on to application identification technologies.

6.2.1 Application Classification

Let's first look at the common types of enterprise applications carried on enterprise WANs, as shown in Figure 6.2.

6.2.1.1 Production Applications

Production applications include Enterprise Resource Planning (ERP), customer relationship management (CRM), and email applications.

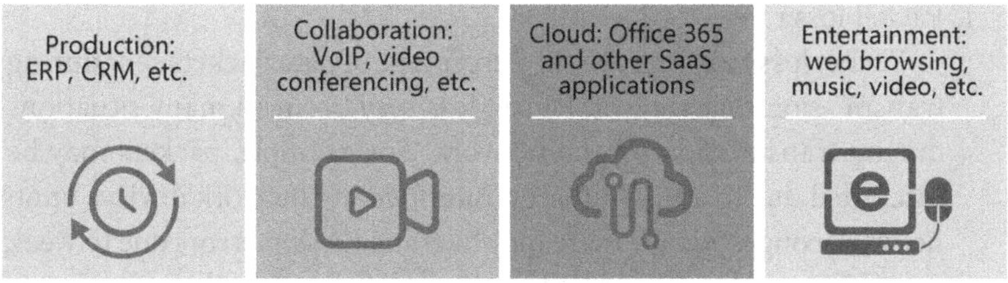

FIGURE 6.2 Common types of enterprise applications.

Such applications are typically deployed on servers in the headquarters or DC and are centrally accessed from branch sites across the WAN.

6.2.1.2 Collaboration Applications

Collaboration applications refer to voice over IP (VoIP) and video conferencing applications between individuals and organizations, and are generally used for office purposes. To meet performance requirements, such applications require data to be directly exchanged between branch sites after sessions are established.

6.2.1.3 Cloud Applications

As cloud computing becomes more widely used, a growing number of enterprise applications are evolving toward software as a service (SaaS). For example, instead of using locally installed Microsoft Office software, a growing number of enterprises are using browsers to access cloud-based Office 365 application servers, forming a traffic model that spans from enterprise branch sites to the SaaS cloud.

6.2.1.4 Entertainment Applications

Entertainment applications enable employees or guests to do things like listen to music and watch videos over the Internet.

As mentioned already, the application experience assurance subsolution needs to use different optimization technologies for different applications. But before we can apply an appropriate experience assurance technology to an application, we need to first analyze the characteristics of the application.

Metrics such as bandwidth, packet loss rate, delay, and jitter are crucial to enterprise applications during transmission over WANs. The following describes their effect in more detail.

1. Packet loss rate

   Packet loss rate refers to the percentage of data packets sent during transmission that are lost. Packet loss may occur in many situations during transmission on the network. For example, packets may be discarded due to a full buffer on intermediate network devices upon network congestion, or there may be transmission errors due to weak signals caused by signal blocking or interference. Such packet loss is unacceptable for video applications, as this will cause freeze frame and artifacts.

2. Delay

   Latency refers to the delay in transmitting packets on a link between two specified devices. It is related to the link transmission rate and the buffer usage of intermediate devices. If the latency is high, audio and video applications experience freezing or stuttering. To ensure a good audio or video experience, the latency of a two-way call must be below 100 ms.

3. Jitter

   Jitter is calculated by subtracting the interval between two consecutive packets being received from the interval between which they were sent. It represents the change of latency. During network transmission, the latency from the transmit end to the receive end varies between data packets. This occurs for many reasons; for example, packets are transmitted to the receive end over different paths, the CPU load of intermediate network devices changes, the buffer usage changes, or packets enter different queues.

   A common way for audio and video applications to cope with jitter is to buffer data. A buffering mechanism enables smooth audio and video experience even if packets reach the receive end at different rates. However, the buffering mechanism is ineffective if the jitter is excessively large: If the buffer size is increased, the waiting time increases, resulting in a long audio and video playback delay. On the other hand, if the size of data to be buffered is not increased, packet loss occurs, resulting in freeze frame or artifacts.

   Different measures need to be taken to ensure experience for different applications with varying requirements on link quality. Table 6.1 lists the link quality requirements of some typical applications.

TABLE 6.1 Link Quality Requirements of Typical Applications

Type	Typical Application	Priority	Bandwidth (kbit/s)	Packet Loss Rate (%)	Latency (ms)	Jitter (ms)
Production	ERP	Highest	30	1–2	50–100	-
	Email	High	40	5–10	200–750	-
	File sharing	Medium	100	2–5	200–750	-
Collaboration	VoIP	Highest	80	1–2	100–200	25–40
	Video conferencing	High	4000	0–1	50–150	15–30
	Screen sharing	Medium	200	1–3	100–150	-
Cloud	SaaS	High	50	1–2	100–200	-
	Other applications	Medium	30	2–5	100–400	-
Entertainment	Social networking	Low	400	2–5	1000–2000	-
	News	Low	200	5–10	1000–2000	-

Now that we have covered different types of applications and their requirements, let's look at how packets are identified and classified. Many application identification methods are available, including traditional methods such as packet identification based on the 5-tuple, traffic characteristics, and packet payload; and special methods such as packet identification based on the domain name system (DNS) and correlation identification.

Differing in whether applications can be identified upon receipt of the first packet, application identification technologies are classified into: first-packet identification (FPI) and SA. Figure 6.3 shows these application identification technologies. Specifically, FPI can identify an application when the first packet arrives, whereas SA can identify an application only after in-depth analysis on signatures of multiple packets.

As mentioned, in SD-WAN, service processing can be performed only after applications are identified and classified. But how is this achieved? The following sections describe application identification technologies in detail.

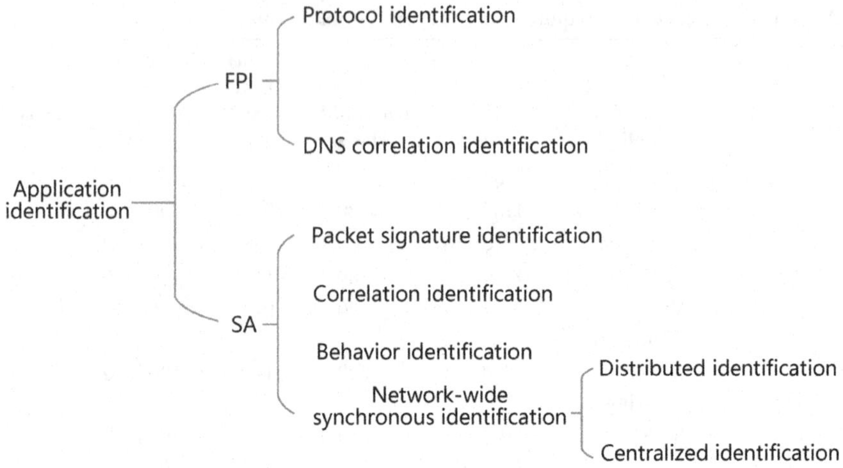

FIGURE 6.3 Application identification technologies.

6.2.2 FPI

As its name suggests, FPI is a technology that identifies an application based on the first packet of its data flow. With FPI, a network device can perform corresponding processing as soon as the first packet of a session arrives, thereby saving network resources.

FPI is applicable to the following scenarios:

- **Application-based traffic steering:** An application is correctly identified as soon as the first data packet is received, after which traffic is transmitted on the correct link. This prevents experience of an application from being affected due to link switching that occurs after the application is identified.
- **Cloud-based applications:** Traffic of SaaS applications is directly transmitted from branches to the Internet, rather than being diverted through the DC. This reduces network latency and costs.
- **Internet access by employees:** If a NAT device is deployed on the link through which application traffic passes and an application is not identified upon the receipt of the first packet, application packets will be discarded due to TCP handshake packets not being processed on the NAT device.
- **Security:** To block malicious applications, a security policy is matched when the first packet is received, thereby eliminating

security risks and excessive bandwidth consumption. In addition, FPI is independent of the application payload and can identify the application with encrypted traffic.

Typically, FPI is implemented using the following methods:

6.2.2.1 Protocol Identification

Protocol identification refers to the process of matching applications by extracting key information from data packets and searching the FPI table.

FPI identifies one or more fields in an IP, a TCP, or a UDP packet header. Information such as the 5-tuple and Differentiated Services Code Point (DSCP) value of packets can be recorded in the FPI table, which is queried to identify an application when the first packet of a data flow arrives.

Figure 6.4 shows the IP fields that can be matched in the FPI table, including type of service (ToS), source/destination IP address, protocol, and port number.

1. ToS

   The ToS field indicates the precedence of IP packets and requirements of services on the network.

   a. **Latency:** normal or low latency

   b. **Throughput:** normal or high throughput

   c. **Reliability:** normal or high reliability

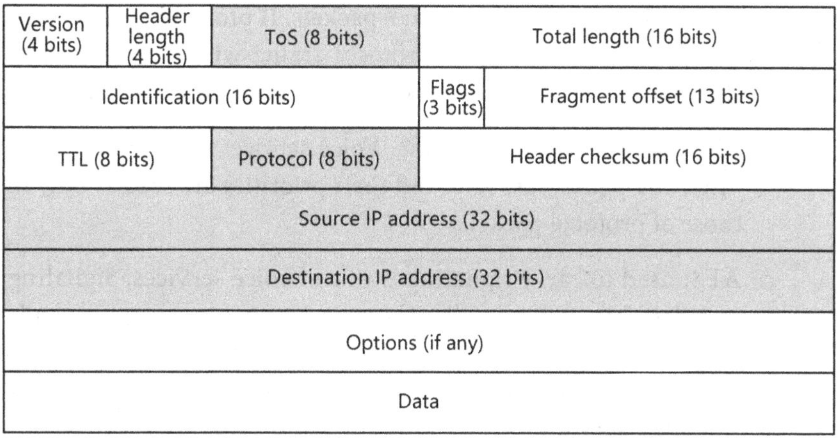

FIGURE 6.4 IP fields used by FPI.

In Request For Comments (RFC) 2474, the Differentiated Services (DiffServ) working group of IETF redefined bits 0–5 (six bits in total) in the ToS field of the IPv4 packet header as the DSCP field and renamed the ToS field the Differentiated Service (DS) field. The DS field selects a behavior based on the DSCP value. For this reason, many enterprises use DSCP values to classify enterprise services and provide differentiated services.

The DSCP value ranges from 0 to 63, with a larger value indicating a higher priority. To make it easier to understand DSCP values, they are classified into four types:

a. **Class selector (CS):** The binary format is aaa 000, of which the first three bits are variables. CS is further classified into CS1 to CS7.

b. **Expedited forwarding (EF):** The binary format is 101 110, and the value is 46.

c. **Assured forwarding (AF):** The binary format is aaa bb0. The AF is further classified into AF1, AF2, AF3, and AF4, each of which has three values.

d. **Best effort (BE):** The value is 0, which is the default value of the DS field.

Typically, DSCP values are allocated to enterprise services as follows:

a. **CS6 and CS7:** used by protocol packets. If protocol packets cannot be received properly, protocol traffic will be interrupted. For this reason, protocol packets have the highest priority.

b. **EF:** used to carry voice traffic. Voice services require low delay, jitter, and packet loss rate, and their priorities are second only to those of protocol packets.

c. **AF4:** used to carry signaling traffic of voice services. Signaling is utilized to control calls. Users can wait for several seconds before a call is connected, but a call being interrupted is not acceptable. Therefore, voice services take precedence over signaling services.

d. **AF3:** used to carry traffic of video conferencing services. For video conferencing services, both continuity and large throughput need to be guaranteed.

e. **AF2:** used to carry traffic of video on demand (VoD) services, which have lower requirements on real-time performance than video conferencing services and allow delay or buffering.

f. **AF1:** used to carry low-importance services, such as data backup and email.

g. **BE:** used to carry the least important services such as Internet access and entertainment services.

With the DSCP values, the priorities of communication services can be preliminarily classified, ensuring the quality and efficiency of important services. Therefore, SD-WAN can identify enterprise services based on DSCP values.

2. Source/destination IP address

Based on the source and destination IP addresses, we can identify where a packet is from and where it is destined for. IP addresses can be assigned to different devices, departments, companies, and regions or countries, and can therefore be used to identify service information.

For example, we can identify important services if packets contain IP addresses of servers where such services reside.

3. Protocol field

The protocol field in the IP packet header indicates the protocol used at the IP layer. By parsing the protocol field, we can identify the service type, such as Internet Control Message Protocol and DNS.

4. Port number

For TCP or UDP, we can further identify the port number in the TCP or UDP packet header. Figure 6.5 shows a typical example of the port number in the TCP packet header.

Many services or applications use fixed ports, allowing us to identify them based on these ports. Ports are divided into three segments:

a. **0–1023:** well-known ports, which are generally allocated to specific services. For example, port 21 is allocated to the FTP service.

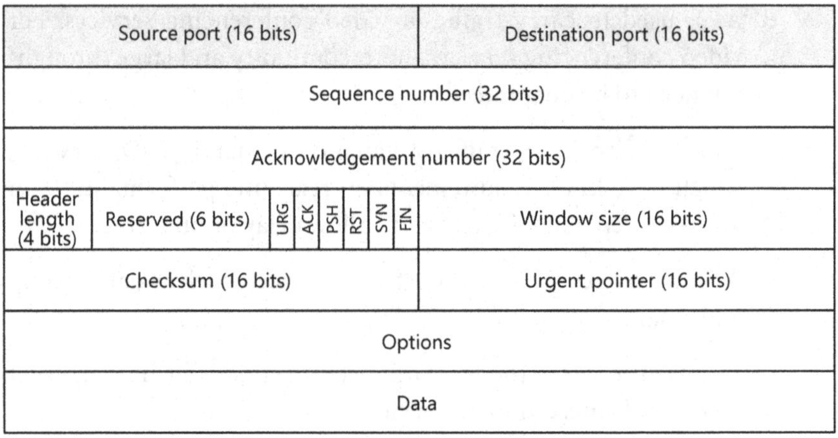

FIGURE 6.5 TCP fields used by FPI.

b. **1024–49151:** registered ports, most of which are not defined for specific services. Applications can use such ports based on the site requirements.

c. **49152–65535:** dynamic or private port numbers, which are not fixed for a specific enterprise application.

Common applications can be predefined in the FPI signature database, while applications that are not predefined in the FPI signature database can be customized.

FPI matches packets based on Layer 2–Layer 4 information. Given this, when customizing applications, ensure that the Layer 2–Layer 4 information varies between applications.

Searching the FPI table can effectively identify services with fixed port numbers or addresses. However, this mode can only identify application protocols due to limited identification precision. This means that if multiple applications use the same protocol, they cannot be distinguished. In this case, advanced identification methods are required.

For example, Hypertext Transfer Protocol (HTTP) generally uses port 80 for communication, which means that typical web applications use port 80. However, there are also special cases:

a. Websites of many web applications use port 80, resulting in the failure to identify the website of traffic based on the port number.

In this case, in-depth traffic identification is required to identify the specific website by matching keywords such as the host or URL field in HTTP requests.

b. To prevent them from being blocked by the firewall, some applications also use port 80. For example, if other ports are unavailable, some video call applications may use port 80. In this case, we need to check whether packets contain application keywords. If traffic of such applications is encrypted, we need to configure SA for in-depth identification.

c. Some applications may also use port 80 to provide RESTful APIs for clients to use or websites to access. Traffic of such applications can be identified by matching keywords.

d. Web applications do not necessarily use port 80. The administrator may configure port 8080 or, to enhance security, use Secure Socket Layer (SSL) port 443. If an application fails to be identified based on port 80, we need to check whether the application uses other ports.

6.2.2.2 DNS Correlation Identification

FPI also supports DNS correlation identification. If an application is defined based on the domain name, the CPE associates and caches the domain name (e.g., www.example.com) and the corresponding IP address (for example, 1.1.1.1) contained in the DNS response during DNS resolution. In this way, the application type can be identified based on the IP address of the first packet during the subsequent TCP handshakes, as shown in Figure 6.6.

Even though encrypted data is displayed as unrecognizable ciphertext, the DNS correlation identification process is the same. Therefore, DNS correlation identification also applies to encrypted data.

6.2.3 SA

As mentioned above, FPI can identify an application upon receipt of the first packet but is not applicable to all applications. For example, some applications hide their signatures in data flows of sessions, and therefore cannot be identified by FPI. In this case, SA is required.

SA is a technology that identifies applications by matching packet signatures. For a typical TCP-based application, the three-way handshake

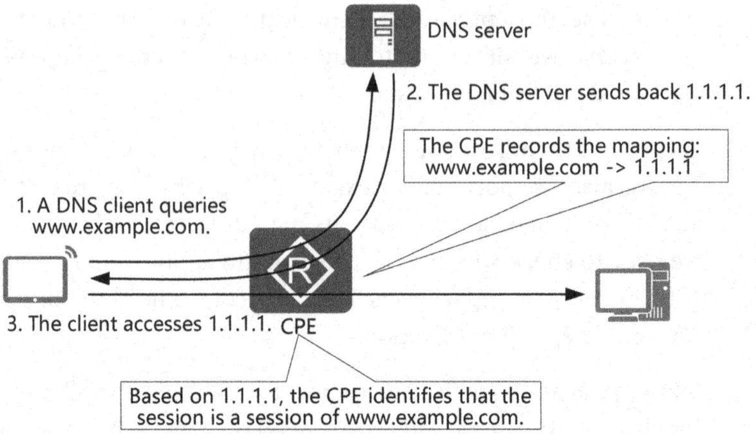

FIGURE 6.6 DNS correlation identification.

packets do not carry any payload. The application can be identified only after the handshake succeeds and the packets with the payload are transmitted.

SA is more precise than FPI. Instead of checking basic information, such as the DSCP value, protocol ID, IP address, and port number, SA checks and analyzes some keywords in the packet payload, packet sending rate, and length and sequence of multiple packets in a more comprehensive and precise manner in order to identify applications. This makes SA suitable for scenarios with multi-channel protocols or non-fixed port numbers.

In terms of identification methods, SA is further classified into packet signature identification, correlation identification, behavior identification, and network-wide synchronous identification.

6.2.3.1 Packet Signature Identification

For undisclosed protocols such as most P2P and VoIP protocols, vendors do not disclose their protocol details, which is due to many considerations. In this case, applications can be identified based on character sequences that have obvious characteristics in packet flows. The downside of this method is that it involves a heavy workload and protocols change rapidly, meaning that packet signatures must be constantly updated to ensure efficient and reliable identification.

If an application cannot be identified based on a single signature, multiple signatures can be combined for better effect. Packet signature identification includes single-packet identification and multi-packet identification, of which the latter identifies and records the identification results of multiple packets, resulting in higher resource overheads. Multi-packet identification is also unable to identify an application upon receipt of the first packet.

Signature identification needs to match the packet payload to identify keywords, but the payload is not transmitted in the TCP handshake phase. This is why signature identification cannot be performed in the TCP handshake phase.

6.2.3.2 Correlation Identification

To achieve better performance and architecture flexibility, many applications use separate protocol and data channels for communication, allowing the actual data transmission channel to be dynamically negotiated. As shown in Figure 6.7, when VoIP, P2P, and file transfer applications (such as FTP) are identified by port number, only protocol channels can be identified, with the actual data transmission channels for audio, video, and text services going unidentified.

In multimedia communication, the actual audio, video, and text channels are dynamically negotiated. The source and destination ports of these channels differ from the well-known ports used in the initial connection and are carried in negotiation packets. Before the application type can be identified, the negotiation packets need to be parsed to obtain the dynamically negotiated port numbers. A technology that identifies an application like this through collaborative detection on multiple channels is called correlation identification.

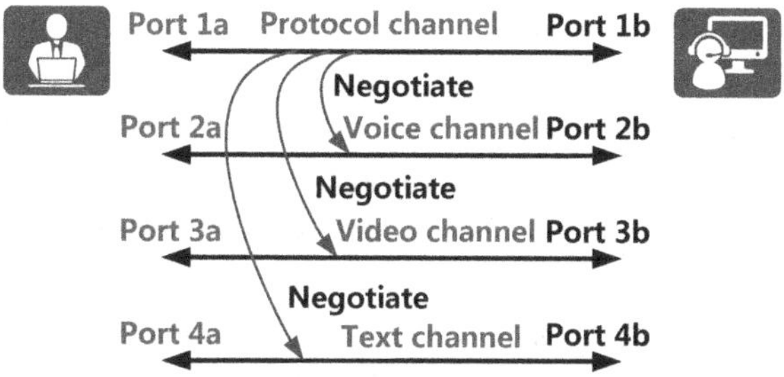

FIGURE 6.7 Dynamic port negotiation.

6.2.3.3 Behavior Identification

Complex applications with encrypted traffic cannot be identified based on ports or signatures. This is because ports are dynamically negotiated and signatures are encrypted, each time using a different encryption mode.

There are two methods for identifying such application traffic: decryption and identification; traffic behavior identification. The first method requires a proxy for traffic and a private key built into the enterprise's device. This method is complex to deploy and is not described in detail here.

Traffic behavior identification identifies applications based on the port range, packet length statistics, packet sending frequency, packet receive/transmit ratio, and distribution of destination addresses in packets. For example, for a VoIP application, the length of voice data packets is relatively stable, and the sending frequency is relatively constant; for a P2P application, the number of connections with a single IP address is large, and the port number varies between each connection; for a file sharing application, packets are large and stable in size. It is thanks to such behavior characteristics that applications can be identified.

Complex protocols do not have signature keywords for identification, and keywords cannot be extracted from encrypted packets. This is why behavior identification applies to such scenarios with complex protocols and encryption protocols.

6.2.3.4 Network-Wide Synchronous Identification

In SD-WAN, devices at multiple sites need to collaborate. As such, SD-WAN also supports network-wide synchronous identification, in addition to the preceding application identification technologies.

Network-wide synchronous identification can be implemented in distributed and centralized modes.

1. Distributed identification

   Similar to the FPI table, a dynamic mapping table records information such as the 5-tuple and DSCP value of packets, which is used to identify applications when the first packet is matched. Such information is dynamically added based on the result of SA. As illustrated in Figure 6.8, CPE1 imports the identification results (such as application A and application B) to the SDN controller, before distributing the results to other CPEs. Other CPEs can then reuse the identification results. This is known as distributed identification.

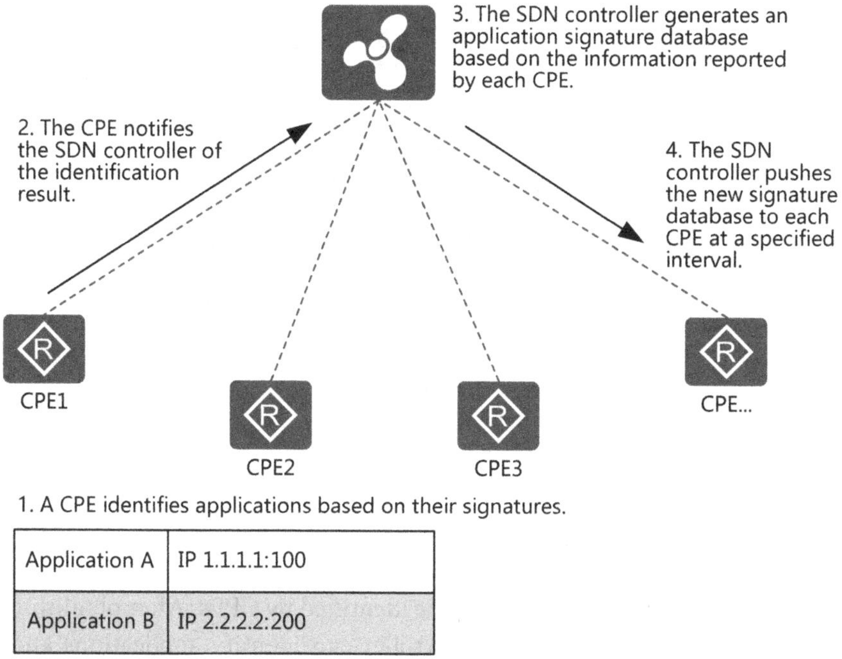
3. The SDN controller generates an application signature database based on the information reported by each CPE.
2. The CPE notifies the SDN controller of the identification result.
4. The SDN controller pushes the new signature database to each CPE at a specified interval.
CPE1
CPE2
CPE3
CPE...
1. A CPE identifies applications based on their signatures.
Application A
IP 1.1.1.1:100
Application B
IP 2.2.2.2:200

FIGURE 6.8 Distributed identification.

The address set for SaaS applications is large and changes frequently. This makes it difficult to maintain manually. By deploying the SDN controller on the public network, the vendor can maintain the application identification result and periodically push the result to each CPE. In this way, the optimal SaaS address can be pushed to CPEs in different regions, enabling CPEs to select the optimal link for data forwarding.

2. Centralized identification

   In SD-WAN, sometimes a device may fail to collect all traffic of an application. This is especially likely to occur at a site with multiple egresses, where incoming and outgoing traffic is transmitted along different paths and traffic is going in only one direction on the intermediate device.

   In this case, applications cannot be identified based purely on some of the application traffic that is transmitted on the intermediate device. For this reason, centralized identification over the entire network is required. In this mode, an SDN controller centrally identifies

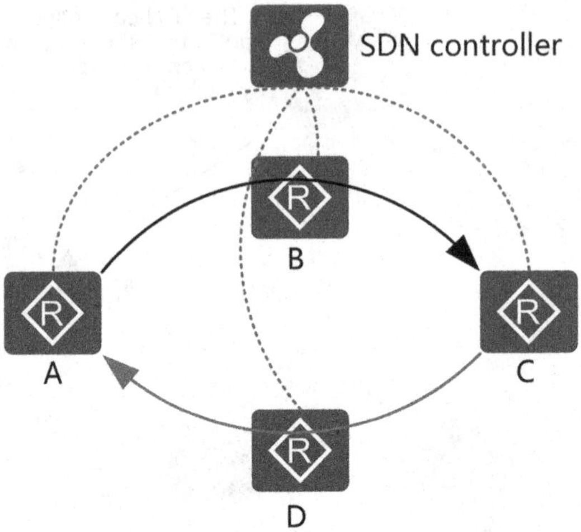

FIGURE 6.9 Centralized identification.

application traffic that cannot be identified by CPEs. After obtaining complete traffic, the SDN controller can identify applications and send the identification result to each CPE.

Figure 6.9 illustrates the centralized identification process.

In Figure 6.9, CPEs B and D can obtain traffic in only one direction and cannot identify applications. Instead, they send the traffic to be identified to the SDN controller for centralized identification, after which the SDN controller sends the identification result to the CPEs.

6.3 APPLICATION-BASED TRAFFIC STEERING

6.3.1 Traffic Steering Scenarios

Traffic steering on networks can be thought of like traffic management on a city's roads. After application traffic is identified, how should we steer traffic and what traffic steering policies can we use? From the application and global perspectives, SD-WAN supports three typical traffic steering scenarios: link quality-based traffic steering, load balancing-based traffic steering, and application priority-based traffic steering.

6.3.1.1 Link Quality-Based Traffic Steering

Applications have varying requirements on the link quality. Given this, we don't want to transmit all application traffic over the optimal link.

For example, if we transmit FTP traffic on an MPLS link, excessive bandwidth resources are consumed, while audio and video quality deteriorates. For this reason, enterprises usually transmit traffic of different applications over different links according to their available bandwidth resources and investments. For example, VoIP services are preferentially transmitted on an MPLS link, which offers good link quality but is high in cost; FTP traffic is preferentially transmitted over Internet links, which feature high available bandwidth and low cost but suffer from high latency and packet loss rate. Figure 6.10 shows an example of transmitting traffic of various applications over different links.

The WAN link used for transmitting application traffic is selected based on the link quality, which is measured using delay, packet loss rate, and jitter. In general, the link quality deteriorates as the amount of data transmitted increases. If the link quality deteriorates or even becomes unavailable, thereby failing to meet an application's requirements, the application traffic needs to be migrated to another link that can meet them. This can be achieved using link quality-based traffic steering, as shown in Figure 6.11.

To implement link quality-based traffic steering, you need to:

1. Configure requirements that various applications have on link quality so that only links meeting the requirements of a given application are used to transmit its traffic. SD-WAN provides predefined application templates, which specify the link quality requirements of common applications, making such configurations easy even for nonnetwork management professionals who understand little about applications' requirements on link quality.

   Table 6.2 lists the types of common link quality templates.

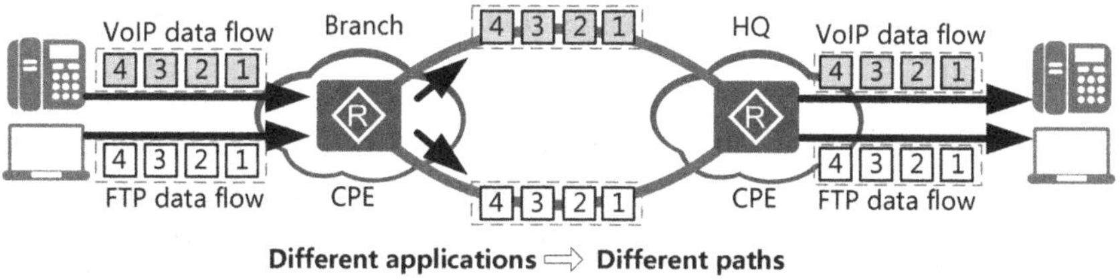

FIGURE 6.10 Transmitting traffic of various applications over different links.

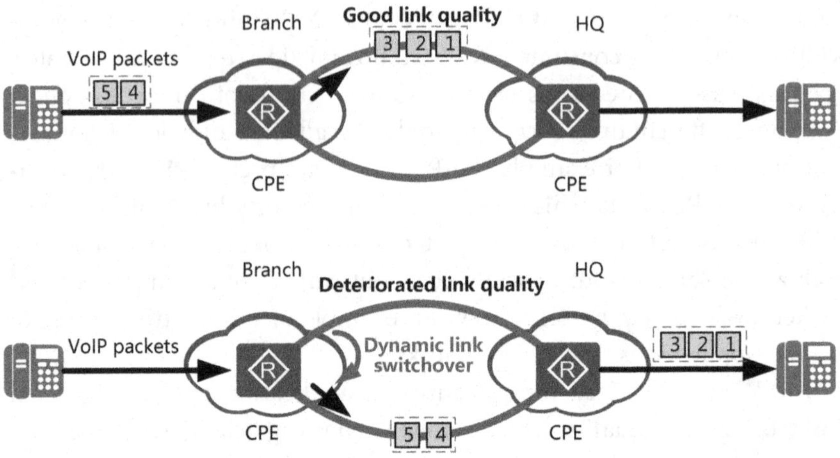

FIGURE 6.11 Link quality-based traffic steering.

TABLE 6.2 Types of Common Link Quality Templates

No.	Type	Description
1	Voice	Voice applications require low latency and packet loss rates but consume few bandwidth resources
2	Real-time video	Real-time videos require low latency and packet loss rates, and consume a large amount of bandwidth
3	Low-latency data	Some protocol data requires low latency
4	Common data	Common Internet data requires much bandwidth but has low requirements on packet loss rate and latency
5	Customized application	Link quality requirements of customized applications vary

2. Configure dynamic link quality detection to determine in real time whether the link quality meets application requirements. This requires SD-WAN to check the quality of all links in real time and involves advanced technologies that will be described in the next section.

3. When the link quality deteriorates or a link becomes unavailable (e.g., it is disconnected), traffic should be automatically switched to another link with a lower priority to ensure high network availability. When it comes to link switching, the link priority and link group need to be considered.

a. The link priority determines the preferred link for transmitting traffic. For example, to preferentially transmit VoIP traffic on an MPLS link and switch it to an Internet link only when the MPLS link quality deteriorates to the point that it cannot meet requirements, set the priority to 1 (primary) for the MPLS link and to 2 (secondary) for the Internet link.

b. A link group is used to define escape links. Enterprises often use an LTE link as the escape link. When links in other link groups are unavailable, SD-WAN diverts application traffic to the LTE network for transmission.

6.3.1.2 Link Load-Based Traffic Steering

In order to maximize bandwidth utilization, enterprises can configure link load-based traffic steering between multiple links to assign traffic to these links based on their bandwidths.

For example, in Figure 6.12, an enterprise has two MPLS links provided by different carriers: one 100 Mbit/s MPLS link and one 50 Mbit/s MPLS link. VoIP service traffic is preferentially transmitted on the two MPLS links. CPEs check and collect statistics on links' bandwidth utilization and applications' bandwidth consumption in real time. If both MPLS links meet VoIP service requirements, VoIP service traffic is load balanced between them according to their bandwidths, maximizing the bandwidth utilization.

6.3.1.3 Application Priority-Based Traffic Steering

If multiple types of service packets are transmitted on the same link, to ensure user experience, traffic of high-priority applications needs to be preferentially processed when congestion occurs. To this end, SD-WAN

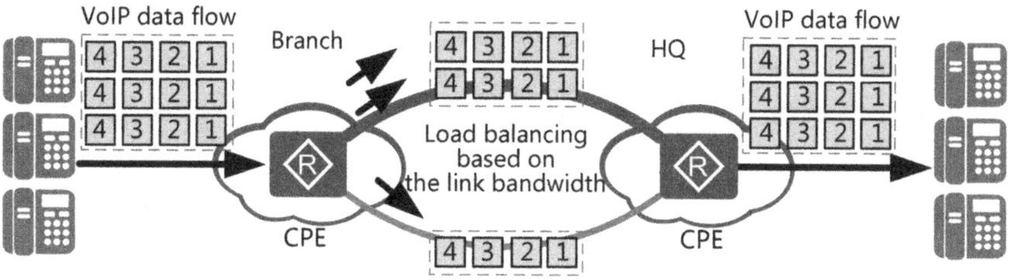

FIGURE 6.12 Link load-based traffic steering.

provides application priority-based traffic steering. For example, if both VoIP and FTP data flows are transmitted over MPLS links, when link resources are insufficient, VoIP services are preferentially guaranteed.

Both VoIP and FTP service traffic is preferentially transmitted on the MPLS link, with the Internet link being used as the secondary link and the VoIP service taking precedence over the FTP service. As VoIP and FTP service traffic increases, the MPLS link may become congested. When this occurs, FTP service traffic is gradually migrated to the Internet link, ensuring user experience of the VoIP service, as shown in Figure 6.13. In order to fully utilize the bandwidth of the MPLS link, when the MPLS link recovers, FTP service traffic can be gradually switched back to the MPLS link.

For application priority-based traffic steering, in addition to checking the link quality in real time, CPEs need to collect statistics on the bandwidth consumed by various applications. In this way, traffic of lower-priority applications can be migrated to other links upon link congestion.

In summary, SD-WAN needs to detect the application and link quality in real time and dynamically execute traffic steering policies when application or link quality requirements are not met. Without doubt, an important part of this is quality detection.

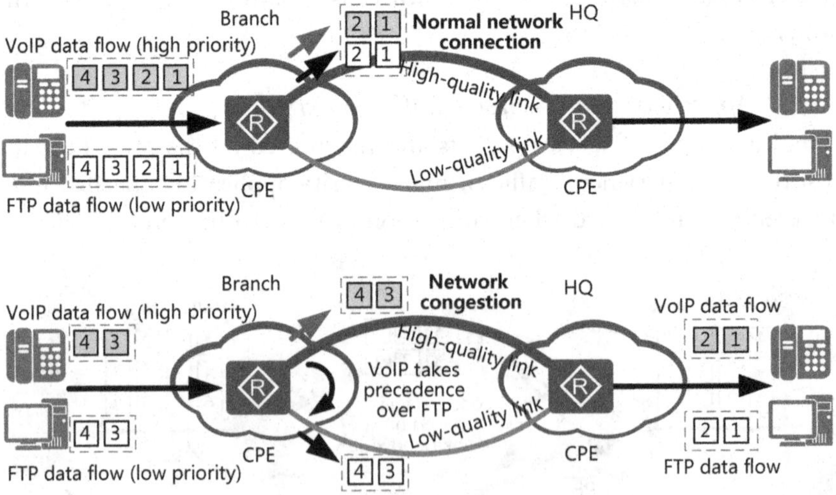

FIGURE 6.13 Application priority-based traffic steering.

6.3.2 Quality Detection

As mentioned in the previous section, during traffic steering, dynamic adjustments need to be made in real time based on the link quality. As such, real-time link quality detection is vital to high-quality application traffic steering.

Real-time application and link quality detection technologies can be either proactive or passive, both of which are explained in detail below.

- Proactive detection

  In proactive detection, the transmit and receive devices proactively send detection packets and calculate the packet loss rate, latency, and jitter based on the number of packets sent and received between the two ends, as well as the timestamp information. Fixed-interval proactive detection periodically obtains network quality information, leading to additional network bandwidth consumption and therefore higher latency. Packets used for proactive detection are not actual application packets, and therefore the detection result may not reflect the actual application quality.

  A typical proactive detection technology is Network Quality Analysis (NQA), which measures network performance and collects statistics on the response time, jitter, and packet loss rate. NQA monitors QoS indicators of the network in real time, effectively diagnosing and locating network faults.

  Additionally, NQA measures the performance of different protocols running on the network. This facilitates real-time collection of network performance counters, such as the total HTTP connection latency, TCP connection latency, DNS resolution latency, file transmission rate, FTP connection latency, and DNS resolution error rate.

  One way that NQA does this is by using an HTTP test case to test whether a client can establish a connection with a specified HTTP server. In this way, it determines whether the client provides the HTTP service and what the connection setup duration is. Figure 6.14 shows the NQA process.

  NQA can obtain the DNS resolution time, TCP connection setup duration, and HTTP transaction time. Both real-time and historical results are recorded in test cases, which can be checked using

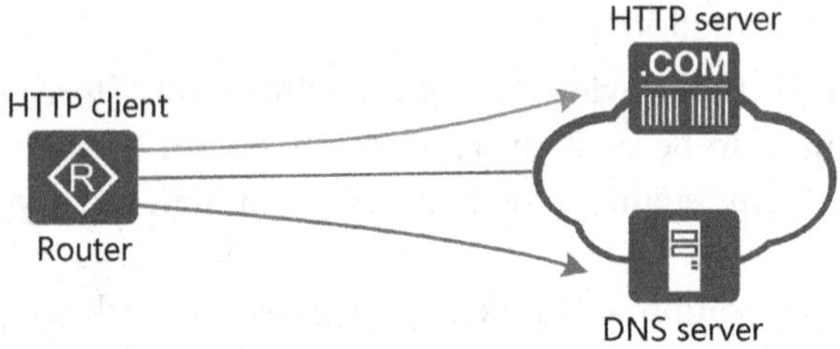

FIGURE 6.14 NQA.

commands and invoked by other services to check the link or service running status.

- Passive detection

  Passive detection is a new type of technology that differs greatly from NQA in that it enables detection information to be carried in packets that are transmitted, without sending out-of-band detection packets. It directly detects application packets, delivering high accuracy and better presenting the actual transmission quality of applications, without consuming additional bandwidth resources.

  IP Flow Performance Measurement (IP FPM) is an IP-based passive network performance detection solution. The solution collects statistics on E2E performance of Layer 3 networks by means of marking packets (also known as coloring). IP FPM measures service packets to assess IP network performance. It also monitors services on the IP network in real time on a per-tunnel basis and presents the application running status on each path. Figure 6.15 shows the IP FPM function.

A test instance is configured for each link to obtain the real-time link quality. When the link quality deteriorates, application traffic is switched in real time based on the applied traffic steering policy.

Passive detection technology obtains the real-time link quality based on service traffic transmitted on links. If no traffic is transmitted within a period of time, the real-time link quality cannot be obtained. Achieving this requires SD-WAN to combine IP FPM with NQA.

Link quality is assessed based on the packet loss rate, latency, and jitter. The following describes the detection technologies employed by SD-WAN from the perspectives of the preceding indicators.

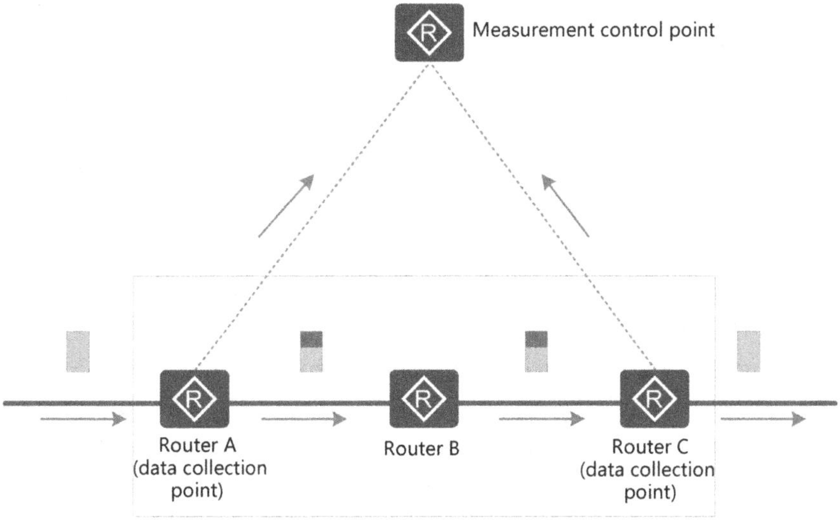

FIGURE 6.15 IP FPM.

6.3.2.1 Packet Loss Rate

Packet loss detection technologies include coloring all packets transmitted within a given period, sending keepalive packets between two ends of a tunnel, and proactively sending probe packets.

1. Coloring all packets transmitted within a given period

   Coloring refers to marking reserved packet bits in the IP packet header at one end of a tunnel and unmarking the packet at the other end of the tunnel. This is how statistics are transmitted in service packets. This technology can distinguish packets transmitted in different periods. For example, red packets are sent in period T1, whereas green packets are sent in period T2.

   Coloring is a passive detection technology. It directly detects service packets, without affecting services or occupying bandwidth resources. Packets are colored by period, improving detection accuracy.

   In Figure 6.16, statistics collected using the coloring technology show that SndA (number of sent packets) is 5 and RcvB (number of received packets) is 4 in period T1. After negotiation between the transmit and receive ends, we can determine that one packet was lost during transmission over the network in period T1.

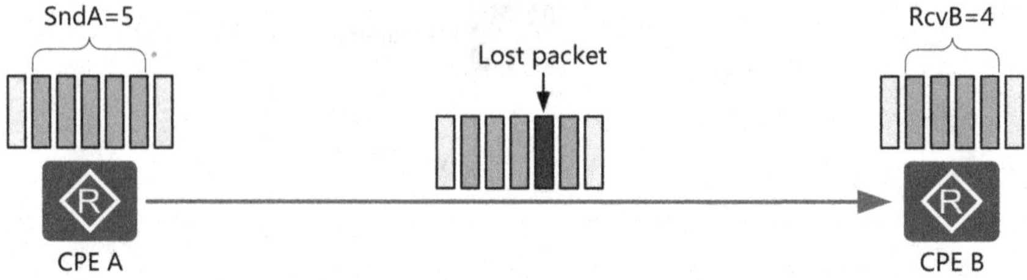

FIGURE 6.16 Packet loss detection through coloring.

2. Sending keepalive packets between two ends of a tunnel

   CPE A sends a keepalive packet containing SndA (number of sent packets) and TsA (packet sending timestamp) in period T1 to CPE B. After receiving the keepalive packet, CPE B sends RcvB (number of received packets) and TrB (packet receiving timestamp) in period T1 to CPE A. After statistics are synchronized between CPE A and CPE B, the two CPEs calculate the packet loss rate of CPE A. The formula for calculating the one-way packet loss rate is as follows:

$$\text{Loss A} = (\text{Snd A} - \text{Rcv B}) / \text{Snd A} \times 1000\%$$

   The packet loss rate of CPE B can be calculated similarly.

   The average value obtained after multiple rounds of calculation is used as a basis for traffic steering.

3. Proactively sending probe packets

   Not all applications constantly send packets. For example, if a user does not perform any operations in a web browser, no packets are transmitted. This prevents passive detection technologies from obtaining packet loss information of the network. Given this, if no service traffic is transmitted, the service and link quality cannot be detected by merely coloring packets. This is where proactive detection technologies come in.

   When detecting that no service traffic is transmitted or there is no traffic on a link, a CPE automatically sends probe packets at a low rate to check network connectivity and packet loss rate, without occupying excessive bandwidth. This is illustrated in Figure 6.17.

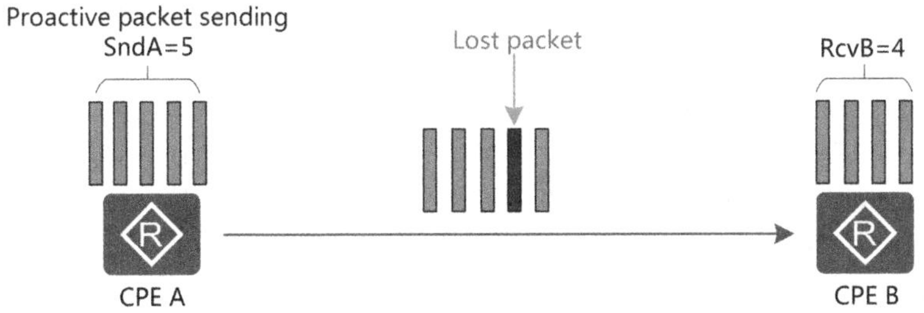

FIGURE 6.17 Proactively sending probe packets.

On real-world networks where service traffic is transmitted, we can combine proactive detection technologies with the keepalive-based detection technologies for service and link quality detection, without occupying extra bandwidth resources.

6.3.2.2 Two-Way Latency

Keepalive packets used for measuring the packet loss rate carry timestamps (sending time: Ts; receiving time: Tr), which are used to detect the latency, without the need to send other extra packets.

Figure 6.18 shows the formula for calculating the two-way transmission latency of a single packet (*M* is the time difference between CPE A and CPE B):

$$\text{Latency} = \text{TrB} - (\text{TsA} + M) + (\text{TrA} + M) - \text{TsB} = (\text{TrB} - \text{TsA}) + (\text{TrA} - \text{TsB})$$

Again, the average value obtained after multiple rounds of calculation is used as a basis for traffic steering. According to the preceding formula, the two-way latency is not related to the time difference between CPEs, and therefore does not depend on Network Time Protocol (NTP).

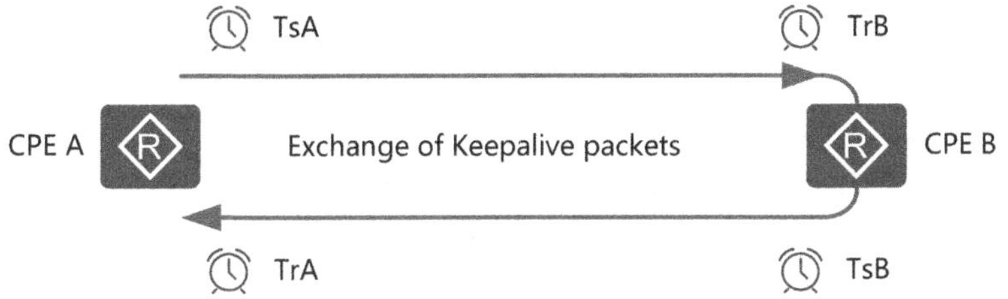

FIGURE 6.18 Two-way latency measurement.

6.3.2.3 Jitter

As mentioned earlier in the chapter, jitter is the variation of latency. No extra packet needs to be sent to obtain the jitter.

In SD-WAN, jitter is the standard deviation of all two-way latency values within a period and reflects the deviation from the average value within the period.

The jitter is calculated as follows:

$$\text{pdv} = \sqrt{\frac{1}{N}\sum_{i=M}^{N+M-1}\left(\text{Delay}[i] - \text{avrDelay}\{N, M\}\right)^2}$$

where

pdv is the jitter to be calculated.

N is the number of times that latency is measured within a period.

M is the Mth time that the latency is calculated.

Delay[i] is the latency measured for the ith time.

avrDelay{N, M} is the average value of N times of latency statistics.

6.3.3 Traffic Steering Policy

To properly steer traffic on SD-WAN, configure traffic steering policies first. When a CPE detects changes in link quality, it applies a new traffic steering policy to subsequent traffic. This section describes how to configure, update, and use traffic steering policies.

6.3.3.1 Configuring Traffic Steering Policies

Traffic steering policies are created based on traffic classifiers and need to be applied to sites. Traffic classifiers are used to classify traffic before steering. A traffic classifier can be defined based on one or more of the following items: IP 5-tuple, application/application group, and DSCP value.

A traffic steering policy defines the following items:

1. Requirements of applications on the link quality

   Specify the packet loss rate, latency, and jitter thresholds. When a threshold is reached, a link switchover is triggered.

2. Link group

   Specify the available links in normal situations and escape links. If a link in the primary link group is available, links in the secondary link group will not be used.

3. Links in a link group

   Specify the links in a link group and their priorities. Links can have the same or different priorities. If links have the same priority, traffic is load balanced between them.

4. Whether to perform load balancing

   If load balancing is configured, a load balancing algorithm based on the weights of same-priority link bandwidths is used to load balance traffic between the links. This algorithm enables links with higher bandwidth to take more traffic.

5. Application priority

   Different types of traffic can be steered based on their priorities. If congestion occurs, the traffic of low-priority applications will be migrated to other links to guarantee the experience of high-priority applications.

6.3.3.2 Updating Traffic Steering Policies

After traffic is identified, it is steered based on the link quality, which changes dynamically. As such, link information needs to be updated dynamically, including the link quality, whether the link is faulty, and link bandwidth. This information is used in traffic steering policies, and it indicates the link quality requirements, preferred link, and link status of applications.

When link status information is updated, the device traverses all available links in the traffic steering policy and checks whether the quality of each available link meets applications' requirements. If a link fails to meet the specific requirements, the device will set its status to unavailable.

After the link status is updated, the device migrates traffic on unqualified links to qualified links, and it will not assign subsequent application traffic to the unqualified links within a specific period of time.

Traffic steering based on link quality is achieved through this process.

6.3.3.3 Applying Traffic Steering Policies

When a device receives a packet, it identifies the application; searches for the set of links that can reach the destination site based on the destination IP address; determines the links that meet quality and bandwidth requirements; and selects the link for transmitting traffic based on the link priority, whether load balancing should be enabled, and application priority.

1. Link priority

   In a traffic steering policy, links available to applications and link priorities are specified. During traffic steering, links are selected in descending order of priority.

   If load balancing is configured and multiple links with the same priority are available, traffic will be load balanced between these links based on the weights of their bandwidths.

2. Application priority

   The device collects traffic statistics on each link to determine its congestion status. Bandwidth resources are allocated to applications based on their priorities. Depending on the congestion status, application traffic within the allocated bandwidth resources can be transmitted, enabling the device to transmit the traffic of a wide range of applications based on their priorities.

Traffic steering solutions on CPEs are generally summarized as follows:

1. Overall, CPEs steer traffic based on the configured link priorities, which generally do not dynamically change. During traffic steering, links with the highest priority are preferentially selected for applications. If the link quality changes, only qualified links will be selected based on the traffic steering policy.

2. When CPEs forward packets, they check the congestion status of each link to determine whether the link is available to an application. Bandwidth resources are allocated to applications based on their priorities, implementing priority-based dynamic traffic steering.

3. If load balancing is configured and multiple qualified links are available, traffic will be load balanced between these links based on the weights of link bandwidths.

Flexible traffic steering on a per-application basis is implemented on CPEs through this process.

6.3.4 Configuration Practices

Taking into account the implementation of application identification, link quality detection, and traffic steering, let's comprehensively cover traffic steering practices based on these technologies.

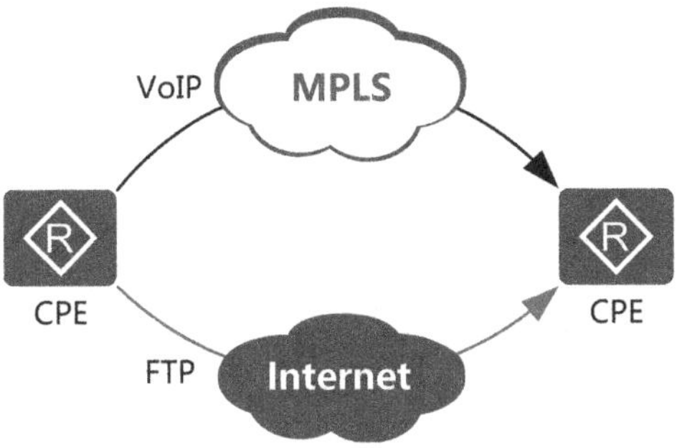

FIGURE 6.19 Traffic steering configuration for an enterprise using both the MPLS and Internet links.

The following example is of an enterprise that has both the MPLS and Internet links. As shown in Figure 6.19, the voice service uses the MPLS link as the primary link, with the Internet link serving as the secondary link; the FTP service uses the Internet link as the primary link, with the MPLS link serving as the secondary link.

Table 6.3 lists the key parameters of a traffic steering policy.

After you plan the parameters in the preceding table, create two traffic steering policies. Configure different switchover conditions for different applications on the controller; configure the MPLS link as the primary

TABLE 6.3 Key Parameters of a Traffic Steering Policy

Parameter	Value
Traffic Steering Policy for a Voice Application	
Policy name	voip_traffic_steering
Traffic classifier	voip_traffic_classification (voice service identification)
Policy priority	10 (high-priority application)
Switchover condition	Voice (link quality requirements of the voice service)
Transport network priority	1: MPLS (primary link) 2: Internet (secondary link)
FTP Application Policy	
Policy name	ftp_traffic_steering
Traffic classifier	ftp_traffic_classification (FTP service identification)
Policy priority	1 (low-priority application)
Switchover condition	Data (link quality requirements of the FTP service)
Transport network priority	1: Internet (primary link) 2: MPLS (secondary link)

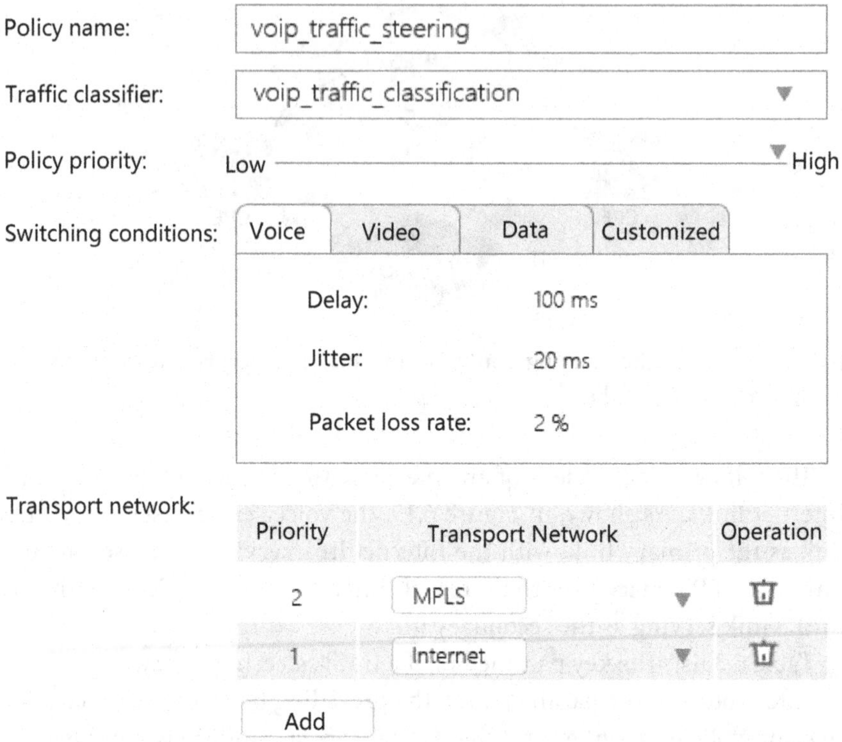

FIGURE 6.20 Configuring a voice service policy.

link (priority: 1), and the Internet link as the secondary link (priority: 2) for the voice service; and configure the Internet link as the primary link (priority: 1), and the MPLS link as the secondary link (priority: 2) for the FTP service. Set the priorities of the voice service policy and FTP service policy to 10 (high priority) and 1 (low priority), respectively. This ensures the FTP service does not preempt the bandwidth resources of the voice service when they are transmitted on the same link. The FTP service and voice service have similar configurations. Figure 6.20 uses the configuration of the voice service as an example.

6.4 QoS SOLUTION

On traditional networks, QoS has proven to be critical to the convergence of voice, video, and data services. In SD-WAN, QoS is essential to offering differentiated services for enterprises.

6.4.1 Overall Solution

From the traffic model perspective, the QoS solution in SD-WAN covers site-to-site and site-to-gateway rate limiting.

6.4.1.1 Site-to-Site Rate Limiting

Site-to-site rate limiting is the most common QoS application scenario in SD-WAN. It can work for both inbound and outbound directions. As enterprise sites usually have limited bandwidth, deploy QoS at the egress of a site to preferentially transmit high-priority service traffic over a congested link.

6.4.1.2 Site-to-Gateway Rate Limiting

Site-to-gateway rate limiting includes site-to-IWG and site-to-POP gateway rate limiting, and can also work for both inbound and outbound directions. Carriers offer legacy network access services for enterprises through the IWG, as well as multi-tenant and multi-site services on multi-domain large networks built across backbone networks through the POP gateways. To achieve this, carriers pose requirements on rate limiting and guaranteed QoS for various types of users.

Figure 6.21 shows the overall QoS scenario.

The following describes QoS requirements and solutions in these scenarios.

6.4.2 Site-to-Site Rate Limiting

Enterprises have limited egress bandwidth. Although we can transmit application traffic on a proper link through traffic steering, congestion and packet loss may still occur if the traffic volume reaches the maximum bandwidth of the egress link. To address this issue, we can apply QoS policies to ensure stable running of services from the application and departmental perspectives. Figure 6.22 shows the site-to-site rate-limiting scenario.

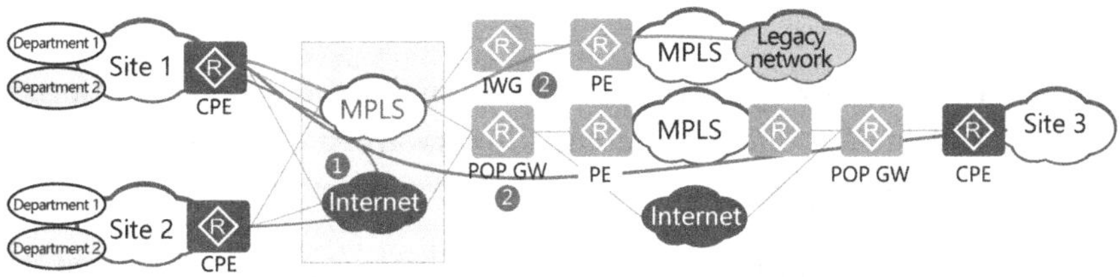

FIGURE 6.21 Overall QoS scenario.

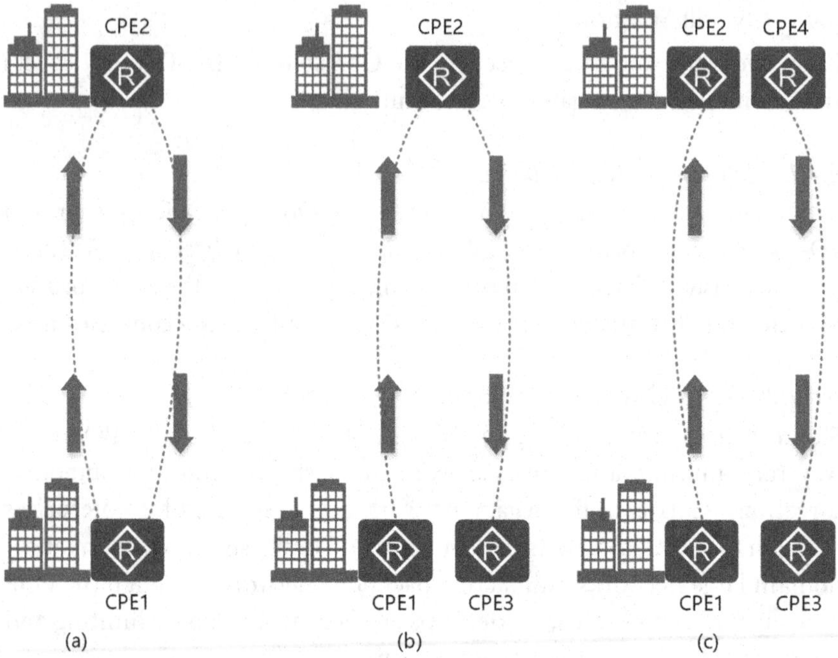

FIGURE 6.22 Site-to-site rate limiting.

As enterprises purchase bandwidth from carriers on a per-link basis, the volume of traffic transmitted on a link should not exceed the link bandwidth.

Site-to-site communication poses the following requirements on QoS:

- Provides differentiated priorities and bandwidth guarantees for applications.

  The experience of important applications should be preferentially guaranteed on a link with limited egress bandwidth. For example, applications must be configured as follows to meet QoS requirements:

  - Instant messaging (IM) applications have the highest priority and their traffic is preferentially forwarded even when the network is congested.
  - Video applications have the second highest priority, followed by web page access.

    - Services such as email do not preempt the bandwidth of other services.

- Provides differentiated bandwidth guarantees for an enterprise's departments.

  Enterprises usually have multiple departments, requiring traffic to be isolated and differentiated bandwidths. This means that the basic bandwidth of each department must be guaranteed. If a department does not occupy its full bandwidth, the idle bandwidth can be used by other departments.

- Limits the bandwidth of the egress link.

  As carriers charge enterprises by link bandwidth, traffic transmitted on a link should not exceed the bandwidth. To ensure this, limit the bandwidth of the egress link.

  For example, the following requirements should be met on a physical link with a total egress bandwidth of 100 Mbit/s:

    - Department 1 takes 40% of the physical link bandwidth, and department 2 takes 60%, with basic bandwidth being guaranteed for both departments.
    - Both departments can use each other's idle bandwidth if their own bandwidth is insufficient.
    - The ratio of bandwidth used for local Internet access and inter-site access is 4:6 for department 1 and 3:7 for department 2. If congestion occurs on the physical link, the minimum bandwidth will be guaranteed for the two types of access based on the configured ratios.

QoS implementation varies in inbound and outbound directions. Specifically, queue scheduling is used in the outbound direction, whereas Committed Access Rate (CAR) is used in the inbound direction. The following details QoS implementation in both directions.

6.4.2.1 QoS Implementation in the Outbound Direction

Enterprises may need to provide differentiated QoS levels for applications, as well as different guaranteed bandwidth for different departments. SD-WAN achieves this through Hierarchical Quality of Service (HQoS),

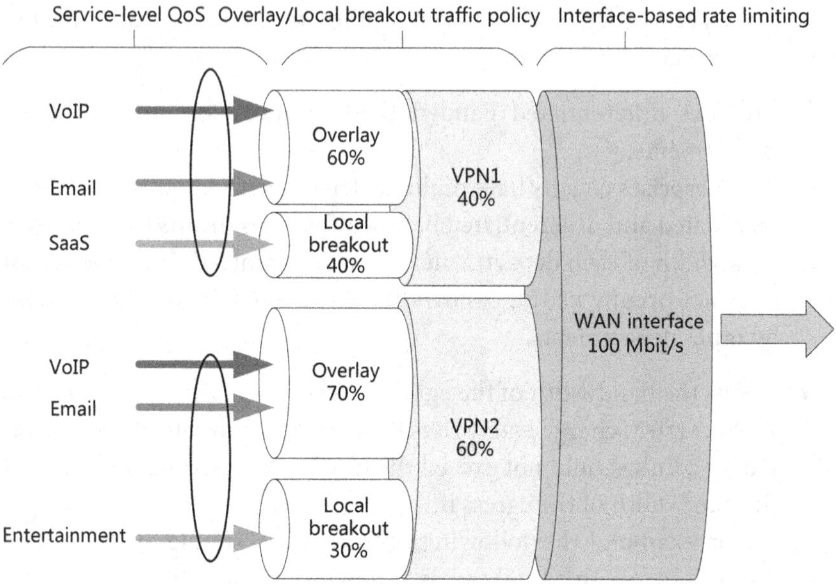

FIGURE 6.23 HQoS service model.

which isolates inter-departmental traffic through VPNs, as shown in Figure 6.23.

This service model involves the service-level QoS, VPN traffic policy (overlay/local breakout traffic policy), and interface-based traffic limiting. Table 6.4 lists the key parameters.

1. Service-level QoS

   On the SDN controller, configure a QoS policy for each VPN and apply the policy to control the quality of various services for each user in the VPN, as shown in Figure 6.24.

   To configure a QoS policy, perform the following steps:

   Step 1: Create a traffic classifier. When the traffic of VPN users (e.g., voice, data, office application, and Internet access traffic) needs to be classified, a traffic classifier must be created for each type of traffic object, which can be one or any combination of the following items: IP 5-tuple, application/application group, and DSCP value.

   Step 2: Define actions in a traffic behavior. Select a traffic classifier for each traffic policy, and specify the QoS actions to be

TABLE 6.4 HQoS Parameter Settings

Parameter	Value
Service-Level QoS	
Policy name (VoIP service for department 1 is used as an example)	voip_traffic_department1
Traffic classifier	voip_traffic_classification (voice service identification)
Policy priority	10 (high-priority application)
Queue priority	Highest
Guaranteed bandwidth for a queue	40% (40% of the department's bandwidth is assigned to the VoIP service)
Bandwidth limit	40 Mbit/s (Up to 40 Mbit/s bandwidth is available)
DSCP re-marking	46 (DSCP value of LAN-side packets)
VPN Traffic Policy (Overlay/Local Breakout Traffic Policy)	
Percentage of bandwidth for VPN 1	40% (The guaranteed bandwidth of department 1 is 40% of the total interface bandwidth)
Percentage of bandwidth for VPN 2	60% (The guaranteed bandwidth of department 2 is 60% of the total interface bandwidth)
Percentage of local breakout traffic in VPN 1	40% (Local breakout traffic accounts for up to 40% of department 1's bandwidth)
Percentage of local breakout traffic in VPN 2	30% (Local breakout traffic accounts for up to 30% of department 2's bandwidth)
Interface-Based Rate Limiting	

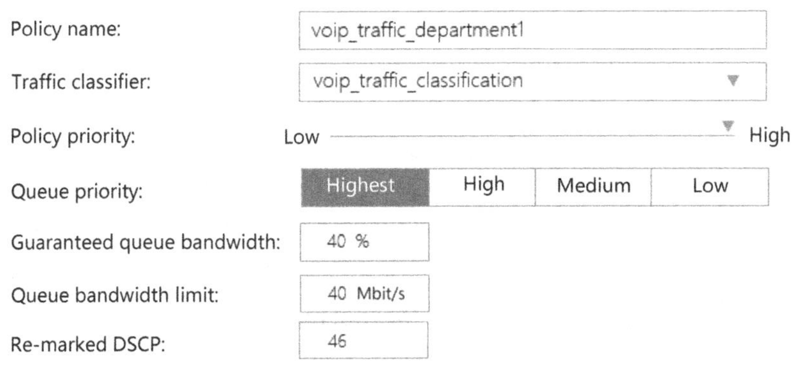

FIGURE 6.24 Configuring a QoS policy.

performed for traffic that the traffic policy is applied to, including queue scheduling, bandwidth limiting (traffic policing and traffic shaping), and DSCP re-marking.

Queue scheduling is a traffic control mechanism that provides high QoS levels for latency-sensitive services by placing packets into the following queues in descending order of priority: Low-Latency Queuing (LLQ), EF, AF, and BE. A minimum bandwidth is guaranteed for each queue based on the configured bandwidth. If traffic does not match a QoS policy, it is placed into the BE queue.

These queues have the following characteristics:

a. **EF queue:** This queue has a high priority and can satisfy the requirements of low-latency services. The traffic of one or more types of services can be placed into the EF queue, with each type of traffic taking configurable proportions of the queue's bandwidth. During packet scheduling, packets in the EF queue are always scheduled first, and the packets in other queues are sent only when the EF queue is empty or the bandwidth exceeds the maximum amount reserved for the EF queue. If congestion does not occur on an interface, the EF queue can occupy the idle bandwidth of AF and BE queues.

b. **LLQ queue:** LLQ is a special EF queue with a lower latency and provides guaranteed quality for latency-demanding applications such as VoIP. However, if congestion occurs, the volume of traffic in this queue cannot exceed the maximum bandwidth. Otherwise, excess traffic will be discarded.

c. **AF queue:** Each AF queue corresponds to one packet type. You can set the bandwidth occupied by each packet type to match the needs of key data services. The system sends packets out of the AF queue based on the bandwidth configured for each packet type, ensuring fair scheduling. The AF queue can occupy the remaining bandwidth on an interface based on the configured weight, and even if congestion occurs on the interface, the minimum bandwidth can be guaranteed for each packet type.

d. **BE queue:** When a packet does not match a QoS policy, the packet is placed into the BE queue by default. Leveraging Weighted Fair Queue (WFQ) scheduling, the BE queue performs per-flow queue scheduling for traffic based on its priority.

Bandwidth limiting includes traffic policing and traffic shaping. Traffic policing discards excess traffic to limit traffic within an appropriate range, reserve network resources, and guarantee application quality; whereas traffic shaping proactively adjusts the rate of traffic sent from an interface. When an inbound interface on a downstream device has a lower rate than an outbound interface on an upstream device, or burst traffic occurs, traffic congestion may occur on the inbound interface of the downstream device. Therefore, traffic shaping can be configured on the outbound interface of the upstream device to ensure outgoing traffic is sent at even rates and avoid congestion. Traffic shaping uses the token bucket mechanism to control traffic. Specifically, when packets are sent at a high speed, traffic is cached in the buffer and evenly sent out based on this mechanism.

DSCP re-marking can be configured as follows:

a. On the LAN side, DSCP re-marking is configured on an inbound interface to re-mark the DSCP value of an IP packet entering a CPE. If the packet is subsequently forwarded through the overlay tunnel, the outer DSCP value of the packet will be copied from the inner DSCP value of the IP packet. That is, both the inner and outer DSCP values of the IP packet are the re-marked value.

b. On the WAN side, DSCP re-marking is configured on an outbound interface to re-mark the DSCP value of an IP packet that is forwarded out through the outbound interface on the WAN. If an overlay tunnel header is added to a packet, only the DSCP value in the outer IP packet header is re-marked. That is, the DSCP values in inner and outer IP packet headers may be different, and the outer DSCP value is the re-marked value.

If DSCP re-marking is configured on both the LAN side and WAN side, the DSCP value will be re-marked for an IP packet when the packet enters the CPE, and when it is forwarded out through an outbound interface on the WAN. In this case, for an IP packet encapsulated through the overlay tunnel, the inner DSCP value of the IP packet is the re-marked DSCP value configured on the LAN side, and the outer DSCP value is the re-marked DSCP value configured on the WAN side. The DSCP value of an IP packet that is routed out through local breakout is the re-marked DSCP value configured on the WAN side.

2. VPN traffic policy (overlay/local breakout traffic policy)

   All the traffic of a VPN is considered to be one traffic object, and it enters the AF queue for scheduling. As previously mentioned, the AF queue can share the remaining bandwidth based on the weight, and its minimum bandwidth can be guaranteed for each packet type even if congestion occurs on the interface. In this way, each VPN has its own guaranteed minimum bandwidth if link congestion occurs, and shares the idle bandwidth of other VPNs on the link if congestion does not occur.

   If Internet traffic is routed out through local breakout in a VPN, you can also specify the percentage of such traffic in relation to the total bandwidth of the VPN and allocate the remaining bandwidth for inter-site traffic in the VPN, as shown in Figure 6.25.

3. Interface-based rate limiting

   If Internet traffic routed out through local breakout or overlay traffic is directly sent from a WAN interface to the underlay network, you can rate-limit traffic on the WAN interface based on its total bandwidth. The interface bandwidth is the service bandwidth purchased from a carrier and needs to be specified for the WAN link as its base bandwidth for HQoS.

6.4.2.2 *QoS Implementation in the Inbound Direction*

Inbound interface-based rate limiting is similar to outbound interface-based rate limiting, and their difference lies in application scenarios. Outbound interface-based rate limiting is typically used in enterprise user

No.	VPN	Bandwidth Ratio	Internet Traffic	Operation
1	VPN1	40 %	40 %	
2	VPN2	60 %	30 %	

Add Remaining bandwidth: 0%

FIGURE 6.25 Specifying the percentage of such traffic in relation to the total bandwidth.

scenarios where the bandwidth of high-priority applications needs to be guaranteed, despite limited bandwidth. However, incoming traffic does not need to be rate-limited because carriers do this before it arrives at an enterprise site.

As such, inbound interface-based rate limiting is mainly used in carrier resale scenarios where CPEs are provided and managed by the carrier. The carrier can rate-limit traffic on CPEs instead of PEs.

Traffic from multiple sites may be simultaneously sent to the same site, and traffic with a higher priority must be preferentially sent to the site to guarantee its bandwidth. Figure 6.26 shows inbound interface-based rate limiting.

Assume that an enterprise with a 100 Mbit/s link has the following requirements:

- The bandwidth needs to be guaranteed for each VPN (40 Mbit/s for VPN 1 and 60 Mbit/s for VPN 2).
- The idle bandwidth of one VPN can be used by the other VPN.
- Each service in a VPN has a guaranteed minimum bandwidth and can use the remaining bandwidth of other services.

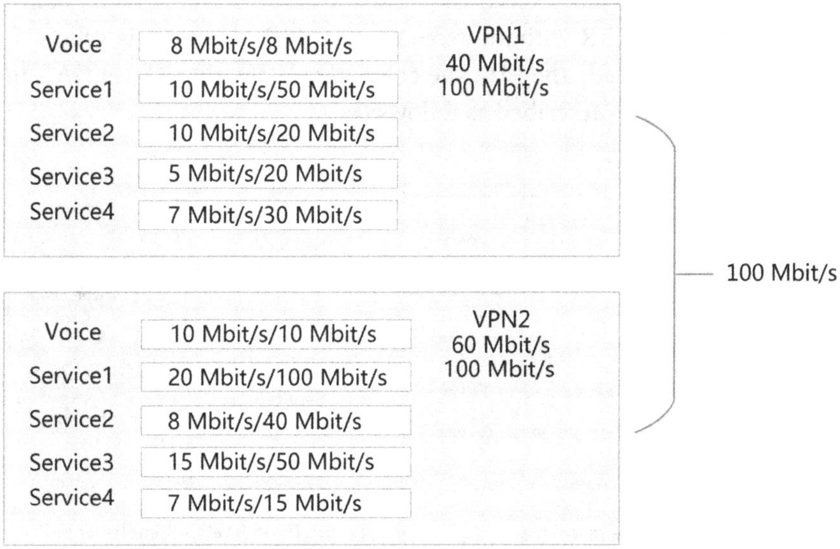

FIGURE 6.26 Inbound interface-based rate limiting.

HQoS can be used to meet the preceding requirements, and its configuration is similar to that of outbound interface-based rate limiting. Inbound and outbound interface-based rate limiting are implemented through different methods. Specifically, outbound interface-based rate limiting is implemented based on queues; whereas inbound interface-based rate limiting uses hierarchical CAR, which discards unnecessary packets more quickly. Figure 6.27 shows the implementation of inbound interface-based rate limiting, which is described as follows:

1. When receiving a service packet, the inbound interface marks that CAR needs to be performed to rate-limit the packet. However, CAR cannot be immediately performed for the packet because the packet has been encapsulated over the SD-WAN overlay tunnel.
2. The packet is decapsulated to obtain the original packet and determine which VPN it belongs to.
3. Application identification technologies are used to identify the application of the packet.

The color-aware mode is used to implement three-level CAR for rate limiting (Table 6.5).

If the traffic does not exceed the CIR, the color is green. If the bandwidth exceeds the CIR without exceeding the PIR, the color is yellow. If the bandwidth exceeds the PIR, the color is red. The three-level CAR in color-aware mode is described as follows:

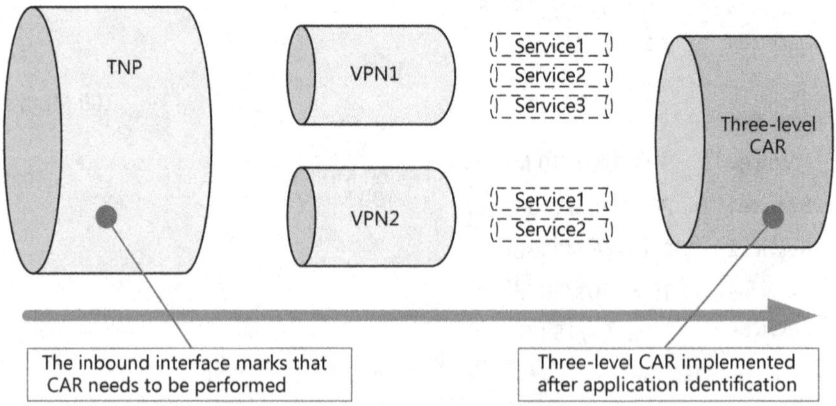

FIGURE 6.27 Inbound interface-based rate limiting.

TABLE 6.5 Rule for Implementing Three-Level CAR in Color-Aware Mode

Service	VPN	Initial Result	Total Bandwidth	Result
(Color Blind)	**(Color Blind)**	**(Color Aware)**	**(Color Blind)**	**(Color Aware)**
Green	Green	Green	Green	Pass
	Yellow	Green (CIR token overdraft)	Yellow	Pass
	Red	Green (CIR/PIR token overdraft)	Red	Pass (CIR/PIR token overdraft)
Yellow	Green	Green		
	Yellow	Yellow	Green	Pass
			Yellow	Pass
			Red	Drop
	Red	Red	Green	Drop
Red	Green	Red	Yellow	
	Yellow	Red	Red	
	Red	Red		

- If the level-1 CAR result is green, the packet is allowed to pass.
- If the level-1 CAR result is red, the packet is dropped.
- If the level-1 CAR result is yellow, whether the packet is permitted depends on the level-2 and level-3 CAR results:
    - If the level-2 CAR result is green, the packet is allowed to pass.
    - If the level-2 CAR result is red, the packet is dropped.
    - If the level-2 CAR result is yellow and the level-3 CAR result is green or yellow, the packet is permitted. If the level-2 CAR result is yellow but the level-3 result is red, the packet is dropped.

6.4.3 Site-to-Gateway Rate Limiting

Site-to-gateway rate limiting includes site-to-IWG rate limiting and site-to-POP gateway rate limiting.

6.4.3.1 Site-to-IWG Rate Limiting

A site can be deployed with a single or dual CPEs to access legacy sites. In the IWG scenario, after traffic is sent from a CPE to the IWG, the traffic is decapsulated and then sent to the PE.

In this scenario, SD-WAN supports the following rate limiting modes:

1. Tenant- or site-based rate limiting
   The IWG service can be purchased on a per-tenant or per-site basis. In SD-WAN, the QoS solution needs to limit the IWG bandwidth available to all sites under a tenant or the bandwidth available to each site.

2. Department-based rate limiting
   The IWG decapsulates packets transmitted over the SD-WAN overlay tunnel. On the IWG, departmental information and internal applications can both be viewed, enabling department-based rate limiting.

3. Application-based rate limiting
   In the IWG scenario, application-based rate limiting can also be provided.

If a carrier provides the IWG service to an enterprise, tenant- or site-based rate limiting is recommended. If an enterprise manages the IWG service, application- or department-based rate limiting is recommended.

As shown in Figure 6.28, when a site accesses the IWG, one IWG node implements rate limiting, regardless of whether a single CPE or dual CPEs are deployed at the site.

When a site accesses the IWG, traffic can be rate-limited at positions 1–7. The recommended rate-limiting positions are as follows:

Preferentially rate-limit upstream and downstream traffic at positions 3 and 4, respectively. Rate limiting at positions 3 and 4 is applicable to scenarios where a site uses two CPEs to connect to the IWG through multiple links. In this case, traffic on multiple links at the same site can be centrally rate-limited, guaranteeing the total overlay bandwidth, regardless of bandwidth allocation to multiple links on the underlay network. If traffic is rate-limited at position 1 or 4 and two CPEs are deployed at the site, multiple links cannot share bandwidth.

If rate limiting at positions 3 and 4 is infeasible, rate-limit upstream and downstream traffic at positions 5 and 6, respectively. Queue

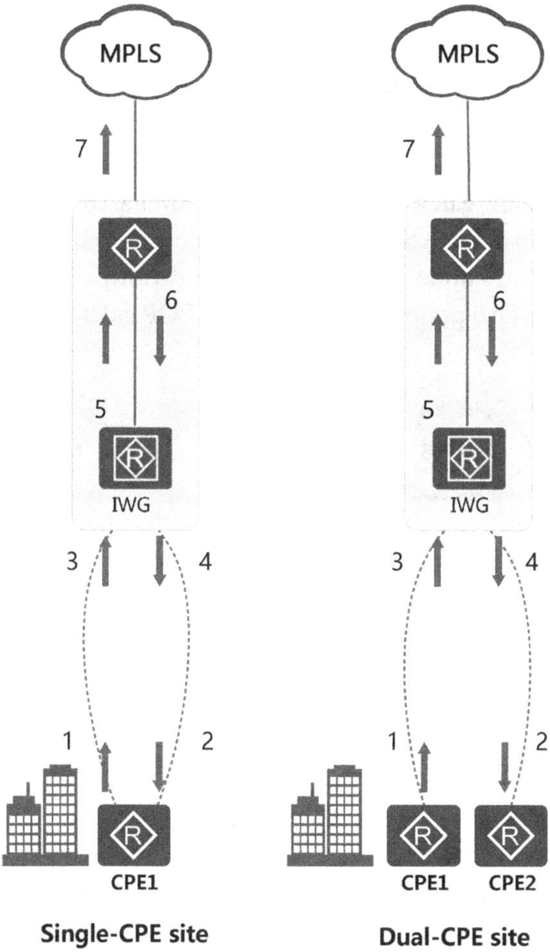

FIGURE 6.28 Site-to-IWG rate limiting.

scheduling and traffic shaping can be used for rate limiting at position 5, whereas only CAR is used at position 3. However, rate-limiting traffic at position 5 consumes more of the IWG's computing and forwarding resources.

Rate limiting at position 6 consumes less network resources than at position 4 because excess traffic is discarded earlier. It is important to note that PEs are carriers' devices and are therefore inaccessible to enterprise users.

6.4.3.2 Site-to-POP Gateway Rate Limiting

Figure 6.29 shows a site-to-POP gateway rate-limiting scenario where a site, configured with either a single or dual CPEs, accesses other sites through the POP gateway.

In this scenario, only tenant- and site-based site rate limiting are supported. The tenant gateway service can be purchased on a per-tenant or per-site basis. In SD-WAN, the QoS solution needs to limit the bandwidth available to all sites under a tenant or the bandwidth available to each site. Rate limiting is also performed on only one POP gateway.

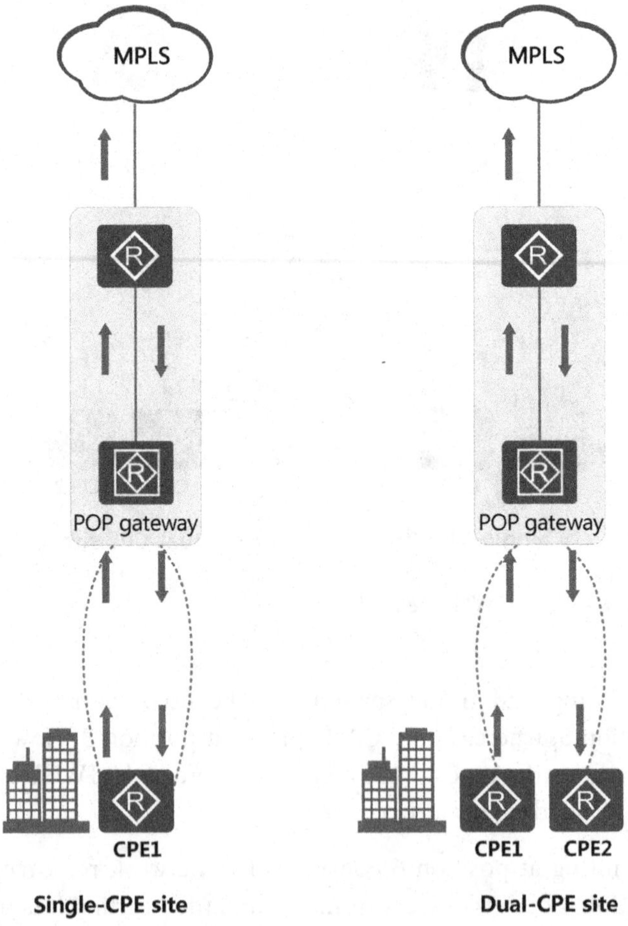

FIGURE 6.29 Site-to-POP gateway rate limiting.

6.5 WAN OPTIMIZATION

When network quality is poor, to guarantee the experience of key services, traffic steering selects the optimal link for transmission, and QoS discards traffic. Both traffic steering and QoS cannot proactively improve network quality.

Unlike traffic steering and QoS, WAN optimization technology enhances applications' adaptability to link quality and guarantees application experience even if link quality deteriorates.

On traditional enterprise networks, branches are interconnected. If branches are interconnected through WANs, the communication distance will be extended, and the packet loss rate, delay, and jitter will increase after multiple hops over a broad range of links. Consequently, the following problems will occur:

- Freeze frame in audio and video services

  The number of network nodes through which audio and video conference packets traverse increases with the distance between sites, resulting in a higher delay and packet loss rate. Users are very sensitive to this delay and packet loss rate as a high delay results in asynchronous audio and video experience, and a high packet loss rate results in intermittent voice experience as well as freeze frame or artifacts in video applications.

- Slow application running

  As enterprises move to the cloud, cloud branches, cloud computing, and cloud applications are becoming increasingly popular. As such, an increasing amount of data needs to be transmitted through WAN links, resulting in possible delay and packet loss, which in turn cause TCP-based applications to run slowly. In specific, when high-resolution photos or videos are backed up, a large amount of data needs to be transmitted, deteriorating transmission experience.

- Fast bandwidth consumption

  A tremendous amount of bandwidth is required for enterprise branches to transmit duplicate data between them, and the amount increases with the number of branches. For example, sending group emails, distributing applications and patches to devices, and playing on-demand videos consume a large amount of bandwidth.

6.5.1 Conventional WAN Optimization Solutions

Traditional WAN optimization, also referred to as WAN acceleration, focuses on improving the application access speed on WANs. SD-WAN also focuses on the service quality of applications, especially smooth video experience and network reliability. Currently, professional WAN optimization vendors are dedicated to data deduplication and compression as well as application acceleration technologies.

A WAN has lower bandwidth as well as a higher packet loss rate, delay, and jitter than a LAN. Therefore, numerous protocols and applications cannot adapt to WAN characteristics, resulting in performance deterioration and poor user experience.

WAN optimization typically employs the following technologies to improve user experience:

6.5.1.1 Packet Loss Mitigation

Packet loss on the network severely impacts IM applications. If an application retransmits packets after detecting packet loss on the network, freeze frame will occur; if the application does not retransmit packets, data will be lost.

Packet replication and FEC are common packet loss mitigation technologies. Both technologies transmit redundant packets to recover from packet loss, without the need to retransmit packets. The differences between them are as follows:

- Packet replication technology increases packet redundancy by replicating one or more copies of original packets and transmits redundant packets over multiple paths to further enhance packet loss mitigation. The receive end sorts and deduplicates the received packets.
- FEC uses the Reed–Solomon (RS) algorithm to calculate the redundant packets for original packets. It features higher recovery capabilities than packet replication because one FEC redundancy packet can be used to correct errors for and recover multiple original packets.

6.5.1.2 Transmission Optimization

TCP and UDP are common transport layer protocols. The process of sending and receiving packets based on UDP is simple and does not require optimization. Therefore, vendors perform transmission optimization mainly for TCP.

TCP was developed many years ago, when the packet loss-based bandwidth detection algorithm was used. This algorithm continuously increases the packet sending rate and cannot detect network bandwidth until when packet loss occurs. However, it is no longer suitable for modern networks due to the high delay and packet loss rate caused by long-distance transmission and the introduction of wireless networks. The characteristics of modern networks mean that TCP transmission performance will remain low even if network bandwidth is high.

To address this problem, the industry players have developed numerous TCP optimization methods, which can be classified into the following types:

1. Packet loss-based algorithm optimization

   This method uses a more radical algorithm to recover lost packets, and to a certain extent, it can mitigate the impact of packet loss on transmission efficiency. However, it cannot resolve the problem regarding the packet loss-based algorithm as network congestion may still worsen. As a result, bandwidth cannot be fully utilized if delay increases and packet loss occurs.

2. Latency-based algorithm optimization

   When network congestion occurs, packets that are being transmitted occupy the buffer of the network device. In this case, latency increases because packets need to wait in the buffer. This algorithm takes into account the fact that congestion occurs if latency increases and requests the transmit end not to increase the packet transmission rate.

   This method prevents packet loss caused by occasional transmission errors from affecting transmission performance. In addition, this method does not extend the delay, as it prevents the transmission rate from increasing when a bottleneck occurs. However, this method is not applicable to networks where latency varies significantly.

3. Comprehensive computing-based optimization

   Both of the preceding methods are used in limited scenarios and cannot fully utilize bandwidth.

Advanced TCP optimization algorithms are designed to compensate for these shortcomings by comprehensively implementing traffic control based on both packet loss and latency information. For example, the Bottleneck Bandwidth and Round-trip propagation time (BBR) algorithm calculates an optimal packet sending rate by separately measuring the bandwidth and latency of a bottleneck device.

The Fill up the Pipe (FillP) algorithm performs further optimization at both ends through technologies such as transmission rate negotiation, proactive packet loss requested by the receive end, and intelligent flow control. Through these optimizations, this algorithm accelerates the transmission rate of TCP applications and reduces latency, thereby resolving the issue of applications running slowly.

6.5.1.3 Data Optimization

Data optimization technologies deduplicate and compress data to minimize redundant data. This accelerates data transmission and reduces the volume of traffic transmitted on networks, conserving bandwidth resources.

1. Data deduplication

   Data deduplication can be configured at the transmit end, receive end, or both. It minimizes the amount of redundant data by buffering files or the data blocks in files.

   Buffering files is recommended for web pages and files whose content does not change; whereas buffering data blocks in files is recommended for scenarios where only some of the file content changes.

   Buffering data blocks is the most common data optimization method, and it is based on data blocking technology as well as data block search algorithms.

For security purposes, data must be encrypted before being stored in storage devices. Before optimizing encrypted data, the optimization device decrypts it. After optimization, the device encrypts the data and sends it to the receive end device, which then decrypts and restores the data, encrypts it, and sends it to the actual receive end device. The data, certificates, and private keys involved in this process must be encrypted for storage. The private keys and certificates can be dynamically obtained from a Certificate Authority (CA).

2. Data compression

    Redundant data strings may be transmitted, for example, aaaa, which occupies 4 bytes. If we use 4a to mark aaaa, only 2 bytes will be occupied. This is a simple example of compression. The industry has conducted extensive research on compression technologies. Depending on whether original information can be fully restored, compression algorithms can be classified into lossless or lossy compression algorithms.

    Lossy compression algorithms, which are applicable to applications, compromise image quality and audio resolution; therefore, network acceleration devices usually use lossless compression algorithms, including Run Length Encoding (RLE), Huffman, and Lempel-Ziv (LZ77).

6.5.1.4 Application Protocol Optimization

The preceding optimization technologies are only underlying technologies that implement optimization based on connections. If we combine them with application behavior-based optimization, better optimization effects can be achieved. This leads to more complex application-specific optimization algorithms. Typical application protocol optimization technologies include:

1. Protocol optimization technologies for web pages and web-based applications

    Many applications are web-based, including enterprise-built applications, inter-site applications, and Internet access applications.

    Web-based applications transmit texts and images based on HTTP or HTTPS, among which a large amount of content is duplicate. For example, a user accesses the same website multiple times or multiple users access the same website concurrently. In these scenarios, data optimization technologies can be used to reduce data volume and accelerate access speed.

2. Protocol optimization technologies for file transfer and database access

    The Server Message Block (SMB) and Common Internet File System (CIFS) protocols and some database access protocols are designed only for LANs. If they are used on the Internet, a large number of interactions will be involved due to high latency, and it will therefore take a long period of time to open a web page. To overcome this issue, a WAN acceleration device can combine these requests and centrally respond to them as a proxy, eliminating the need for frequent interactions and accelerating file access, which in turn enhances the file transfer performance.

3. Video optimization technologies

   Video applications can be classified into three types based on their behavior: IM, live streaming, and VoD. Different video application types have varying characteristics and optimization methods. For example, packet loss mitigation is ideal for IM video applications, whereas bandwidth usage needs to be optimized for live streaming and VoD applications.

4. SaaS application access optimization technologies

   SaaS applications are web-based applications deployed on the cloud, to which optimization technologies for web-based applications are also applicable. Traffic steering can also be optimized for common SaaS applications (such as Office 365, cloud disk, and ERP).

In summary, optimization technologies can be classified into optimization at either end and optimization at both ends. BBR is a typical example of the former and is generally deployed at the site where servers are deployed to accelerate the download speed for all users. Both-end optimization requires deploying CPEs at both ends and enabling the WAN optimization function. The typical technologies for performing optimization at both ends include the FEC algorithm for packet loss mitigation, FillP algorithm for transmission optimization, and data compression and deduplication.

The following sections provide detailed descriptions of the application scenarios, implementations, and methods of optimization technologies.

6.5.2 Packet Loss Mitigation Technologies

An enterprise may have many latency-sensitive applications, including IM applications, such as voice calls and video conferences. These tend to use UDP as the transport layer protocol to ensure low latency and minimize the number of TCP handshakes and retransmissions. However, UDP cannot deliver reliable transmission like TCP. So, when packet loss occurs on the network, application quality deteriorates.

WAN optimization offers a series of packet loss mitigation technologies, including single-path FEC, single-path Adaptive-Forward Error Correction (A-FEC), multi-path packet replication (dual-fed and selective receiving), and multi-path FEC.

6.5.2.1 FEC Algorithm: Recovers Lost Packets

Let's use video communication to show the importance of packet loss mitigation. As mentioned above, when videos are transmitted on a link with poor quality, packet loss usually occurs. This results in freeze frames or artifacts, affecting user experience.

Why does this happen? The answer lies in the characteristics of video communication. During a video call, an intra-coded frame (I-frame) — a complete image — is sent at certain intervals. In the meantime, a predicted frame (P-frame) only shows the changes in the image from the previous frame to reduce the occupied bandwidth, as shown in Figure 6.30.

Exceptions in the video stream (such as bit error or packet loss) can cause errors when decoding a P-frame. As a result, the images after the error P-frame do not show up correctly, as shown in Figure 6.31. To solve this issue, the I-frame needs to be re-transmitted.

When an error occurs due to packet loss on the network, the error P-frame and all subsequent P-frames before the next I-frame show up

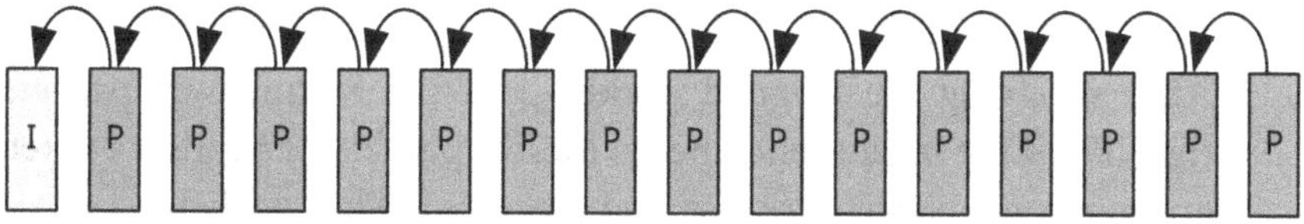

FIGURE 6.30 Video stream.

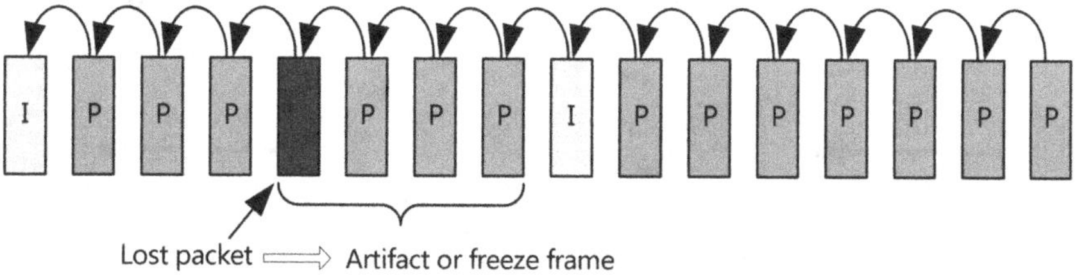

FIGURE 6.31 Packet loss in video streams.

abnormally. If the video terminal continues to forcibly output the image, artifacts occur. Meanwhile, if the video terminal stops image output, the frame freezes for a long time. As such, packet loss has a significant impact on video communication.

In order to transmit data in real time, UDP is generally used for transmission. However, UDP is not as reliable as TCP. Therefore, it depends on network reliability. An Internet link also has a higher packet loss rate than an MPLS link, which further deteriorates voice and video experience. Therefore, we need to mitigate packet loss for voice and video applications.

FEC technology can reduce packet loss on a network by intercepting specified data streams through a proxy and generating redundancy packets encoded according to specific algorithms. The redundancy packets carry information about protected original packets. The CPE at the transmit end sends the original VoIP packets and the redundancy packets to the receive end. The CPE at the receive end verifies these packets. If packet loss occurs during transmission, the original lost packets are restored using the redundancy packets, as shown in Figure 6.32.

The FEC implementation process is as follows:

1. The CPE at the transmit end receives original packets from Bob's phone, optimizes the packets, and groups accumulated packets.
2. The CPE at the transmit end performs FEC encoding on the original packets in each group, and generates an FEC redundancy packet.
3. After receiving the original packets and the redundancy packet, the CPE at the receive end checks for lost packets, rebuilds a packet group if some packets are lost, and recovers the lost packets.

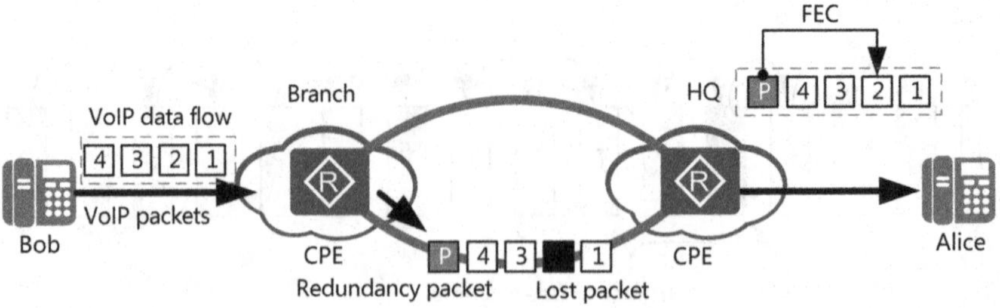

FIGURE 6.32 FEC implementation.

4. The CPE at the receive end performs FEC decoding on the received group of packets, both original and redundancy, to obtain the original packets.

5. The CPE at the receive end sends the original packets to Alice.

FEC generally uses XOR or RS encoding, where RS encoding better serves packet loss mitigation. Figure 6.33 shows the RS encoding process.

To illustrate encoding at the encoding end, we use five original packets in matrix *D* as an example. Matrix *G* is a generator matrix. Matrix *D* and matrix *G* are multiplied to obtain the packet redundancy matrix *R*.

Let's assume that packets D1, D4, and R2 are lost during transmission. After removing the corresponding rows, we get matrix *G*′, where the equation is still true, as shown in Figure 6.34.

The equation is still true after both sides of the equation are multiplied by the inverse matrix of *G*′, as shown in Figure 6.35.

The product of G'^{-1} multiplied by G' is I, and the product of I multiplied by D is D. Therefore, only the original matrix D is displayed on the left side, as shown in Figure 6.36.

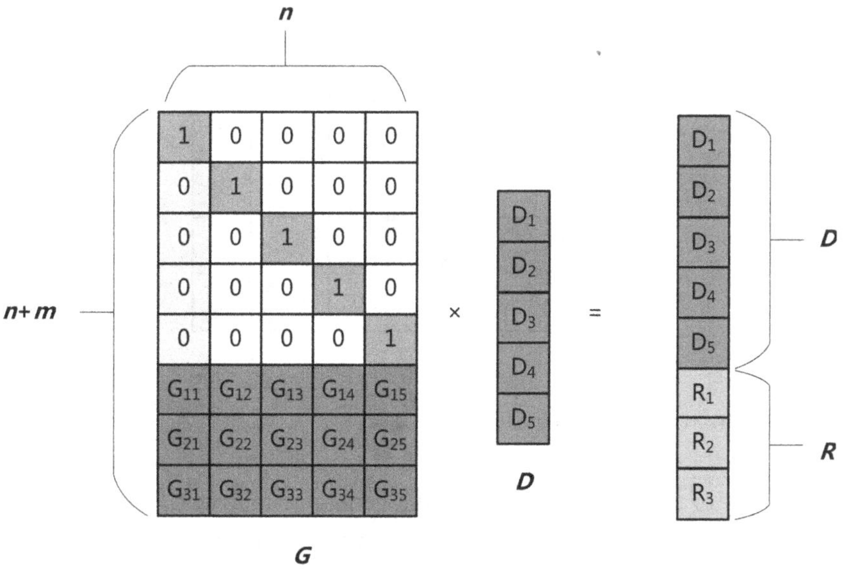

FIGURE 6.33 Encoding at the encoding end.

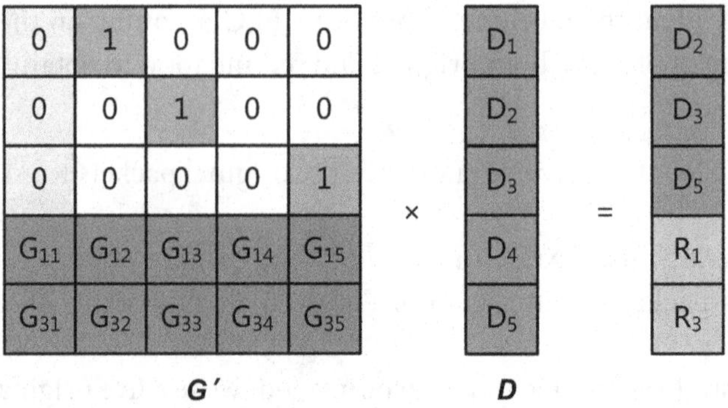

FIGURE 6.34 Matrix obtained after packet loss.

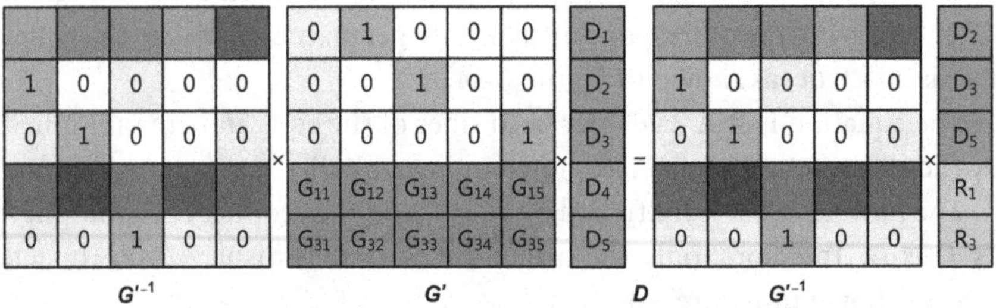

FIGURE 6.35 Obtaining the original matrix.

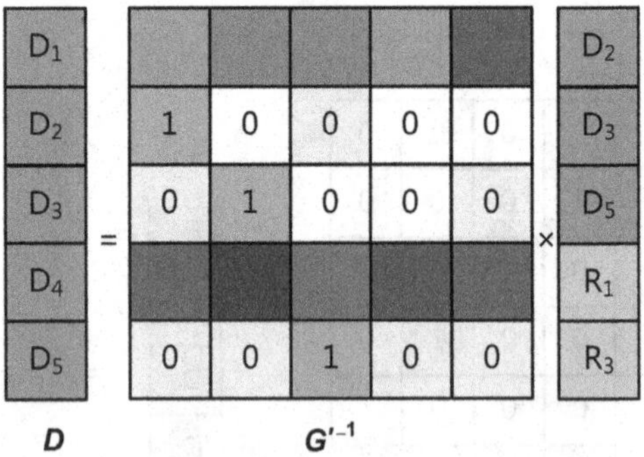

FIGURE 6.36 Original matrix.

The key of the algorithm is to find a generator matrix G, which makes any submatrix reversible. Does such a matrix exist? Of course! Both the Vandermonde matrix and the Cauchy matrix can do this. The RS algorithm originally used the Vandermonde matrix, but has since turned to the Cauchy matrix because it involves less computation in the inversion process.

The algorithm shows that if the number of randomly lost packets per unit time does not exceed the number of redundancy packets, the loss of original or redundancy packets does not affect the restoration of original packets.

In summary, FEC has the following features:

- Compared with the TCP-based retransmission mechanism, FEC does not need to retransmit packets, ensuring a real-time service experience.
- FEC is based on the IP protocol and is very likely to successfully restore lost packets for common applications.

FEC is typically applied to video conferencing. For example, an enterprise has two branch sites (A and B) and a DC, where a conference server is deployed. When branches A and B hold a video conference, video streams need to be forwarded through the server at the DC. They must also go over the Internet, which means jitter and packet loss are inevitable. Video conferences are sensitive to packet loss, so we need to enable FEC on egress CPEs at the DC and the two branch sites, to mitigate packet loss for video traffic that traverses the CPEs, as shown in Figure 6.37. It is recommended to configure FEC for these sites on the SDN controller.

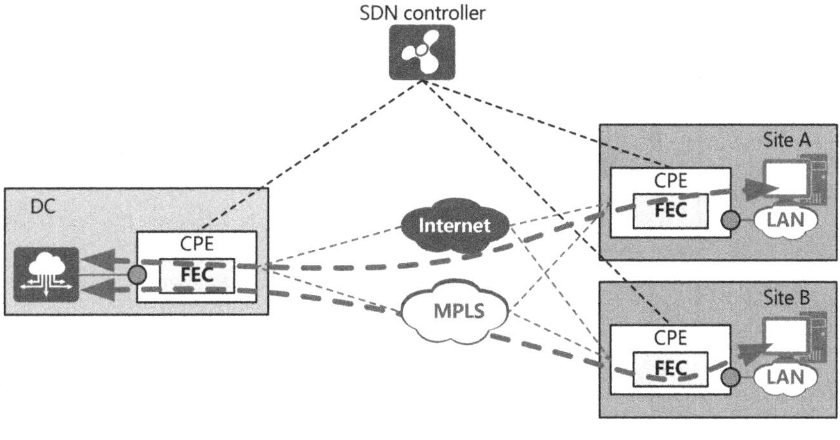

FIGURE 6.37 Typical application scenario of FEC.

6.5.2.2 A-FEC: Automatically Adjusts the Redundancy Rate

In FEC, we can manually adjust the proportion of redundancy packets to restore lost packets in the case of different packet loss rates. However, configuring a fixed redundancy rate will lead to the following issues:

1. Waste of bandwidth resources

   A configured redundancy means it is unable to dynamically adapt to changing network conditions. If the packet loss rate is low, transmitting excess redundancy packets will waste bandwidth resources. In addition, when network congestion occurs, excessive redundancy packets will further exacerbate the network load, thereby affecting other services.

2. Ineffectiveness to constant packet loss

   If the network is congested, packet loss may occur constantly. If packets are lost consecutively within a short period of time, the actual packet loss rate increases dramatically. In this case, the manually-configured fixed redundancy rate will be ineffective, because insufficient FEC redundancy packets can be generated in a short period of time, meaning that packets will still fail to be restored. Figure 6.38 shows FEC with a fixed redundancy rate.

   A-FEC can solve the above issues by automatically adjusting the FEC redundancy rate, so as to save bandwidth resources when the packet loss rate is low. Further, it adaptively increases the redundancy

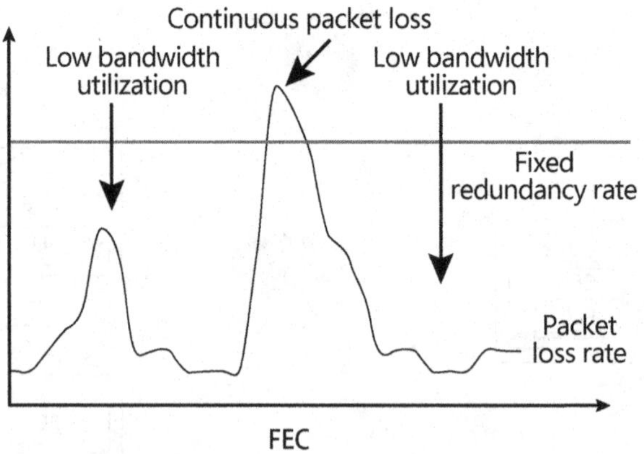

FIGURE 6.38 Fixed redundancy rate in FEC.

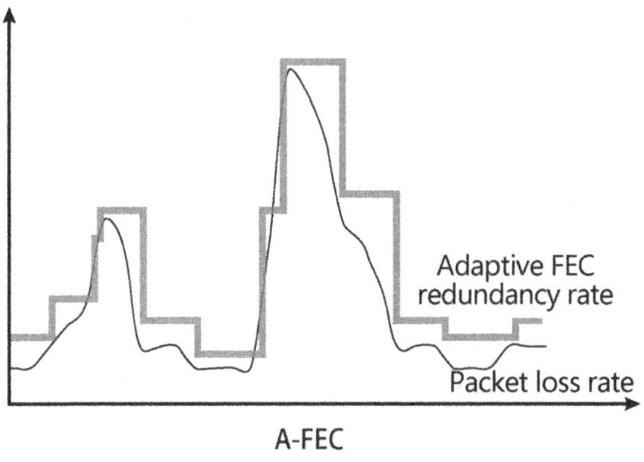

FIGURE 6.39 Dynamic redundancy rate adjustment in A-FEC.

rate when the packet loss rate increases dramatically within a short period of time. Figure 6.39 shows the dynamic redundancy rate adjustment in A-FEC.

A-FEC is implemented using the FEC-ACK packet that is sent from the receive end to the transmit end. This packet reflects the packet loss rate and constant packet loss on the network in real time. The FEC encoding end adjusts the FEC redundancy rate according to the FEC-ACK packet, reducing or even eliminating the impact of packet loss on data transmission.

The following factors should be taken into account when determining the optimal redundancy rate:

- **Maximum number of consecutively lost packets:** The number of redundancy packets for each code block must be greater than the maximum number of consecutively lost packets.
- **Packet loss rate reported by the decoding end:** The redundancy rate at the encoding end must be greater than the packet loss rate.
- **Historical packet loss rate:** The packet loss rate within a recent period is stored using the sliding window mechanism. The historical packet loss rate is calculated using the weighted average algorithm, which focuses on the recent packet loss rate. This ensures smooth packet loss rate change.

- **Security coefficient:** The encoding end has a certain delay in sensing the network packet loss rate. If the network packet loss rate increases, the packet loss rate reported by the decoding end and the historical packet loss rate are lower than the actual packet loss rate of the network. To make up for this, we need to evaluate the packet loss rate trend and reserve a margin.

When calculating the redundancy rate, we must use the maximum value for each of the above to minimize possible packet loss.

6.5.2.3 Multi-Path Packet Replication: Mitigates Packet Loss

FEC and A-FEC are implemented by transmitting redundancy packets on the same path. However, on a congested network, transmitting redundancy packets will further aggravate network congestion and even affect other services. Although A-FEC reduces the number of redundancy packets through dynamic adjustment, it still cannot avoid worsening link congestion.

In scenarios where multiple links are available, multi-path packet replication can reduce the impact of packet loss on packet loss-sensitive applications without aggravating link congestion.

Leveraging multi-path packet replication technology, the CPE at the transmit end makes multiple copies of the original and redundancy packets of the VoIP service. It then sends the copied packets through multiple links. When combined with intelligent traffic steering, multi-path packet replication selects two or more optimal links for transmission, as shown in Figure 6.40.

When receiving data packets transmitted on the links, the CPE at the receive end performs operations such as sorting and deduplication on the data packets.

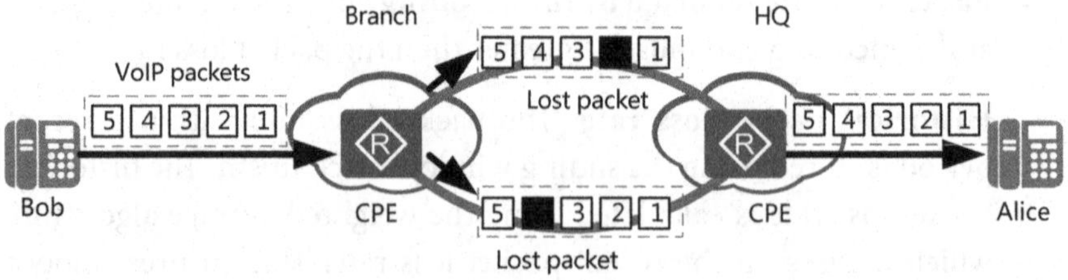

FIGURE 6.40 Multi-path packet replication.

As long as not all copies of the data packet are lost, it is possible to restore the original packet.

6.5.2.4 Multi-Path FEC: Mitigates Packet Loss

Multi-path packet replication technology can solve the issues associated with single-path FEC and A-FEC by fully utilizing bandwidth available on all links, yet without worsening network congestion. However, it also leads to the following issues:

- There is a high redundancy rate, because data is replicated 1:1 on other links, which occupies bandwidth resources.
- If there are only a few links and all copies of the same data packet are lost, the data packet cannot be restored.

Multi-path FEC technology can transmit original data packets of the VoIP service and FEC redundancy packets over different links, as shown in Figure 6.41. Combined with intelligent traffic steering, multi-path FEC transmits redundancy packets over a high-quality link, ensuring that the original packets can be restored when packet loss occurs.

To conserve bandwidth resources, the FEC redundancy packets, which are fewer than the original data packets, are transmitted on the redundant link. Additionally, multi-path FEC does not perform 1:1 mapping between the FEC original packets and the redundancy packets. This prevents failure to restore data packets resulting from losing all copies in multi-path packet replication. Instead, if original data packets are lost, FEC redundancy packets can be used to restore them as long as the redundancy rate is not exceeded. Ultimately, applications achieve higher packet loss mitigation capability.

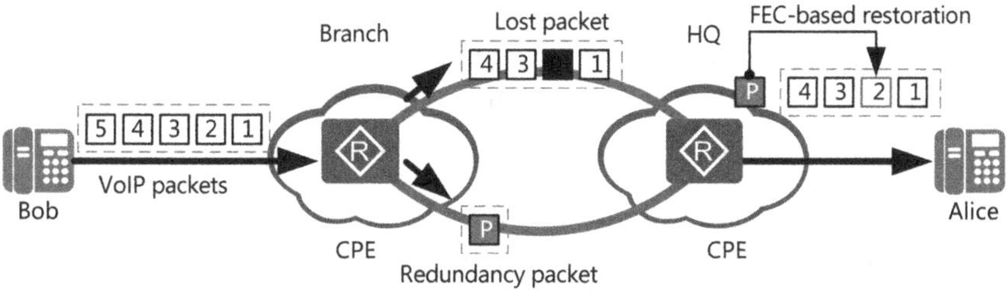

FIGURE 6.41 Multi-path FEC.

6.5.3 Transmission Optimization Technologies

6.5.3.1 Why Is Optimization for TCP-Based Transmission Necessary?

We have all impatiently waited as a website takes forever to load or simply fails to open. Everyone has felt the frustration of downloading or transferring large video files at a slow speed, even when using supposed 100 Mbit/s or even 1000 Mbit/s links.

This mainly happens because of TCP's network congestion control algorithm based on packet loss. This algorithm has been used since the 1970s and worked just fine for the low-speed links at the time. However, modern network technologies have been rapidly developing. The network adapter rate has evolved from Mbit/s to Gbit/s, and memory has grown from KB to GB. The packet loss-based congestion control algorithm cannot adapt to high-bandwidth networks with a certain packet loss rate, failing to maximize network efficiency. As such, we need to optimize TCP due to the following reasons:

1. On a link with a certain packet loss rate, TCP cannot fully utilize bandwidth resources.

   In the congestion control algorithm, TCP determines whether a link is congested based on packet loss. Packet loss due to transmission errors is common on networks and is not caused by link congestion. Packet loss that occurs during transmission using TCP algorithms will lead to the following issues:

   a. In the startup phase, the TCP algorithm uses the slow-start algorithm, where the sending rate grows exponentially. Network congestion occurs when the window size excessively increases.

   b. The TCP congestion avoidance algorithm uses the linear increase mode. As such, the window size increases slowly and cannot reach the maximum network bandwidth quickly.

   c. The packet loss-based TCP congestion control algorithm often overreacts to packet loss. When packet loss occurs, the congestion window decreases by 50%, as shown in Figure 6.42.

   Therefore, on a link with packet loss, the TCP transmission rate is always limited, failing to fully utilize bandwidth resources.

2. In the packet loss-based TCP algorithm for calculating the window size, the buffer of the device with a bandwidth bottleneck may be

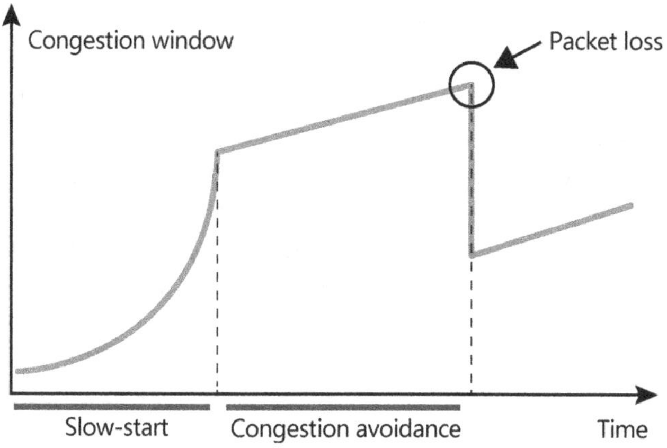

FIGURE 6.42 TCP rate severely affected by packet loss.

exhausted, which further increases the network delay. Additionally, the network's capacity to send packets depends on the capacity of the device with the minimum bandwidth.

On a network, a device usually has a buffer that resolves burst congestion. When the TCP algorithm attempts to increase the window size and determine the maximum bandwidth, the volume of data that is being transmitted on the link but is not received by the peer end keeps increasing.

To address these issues, the ICT industry has come up with several TCP optimization algorithms, which are classified into single-end and dual-end optimization algorithms, as described in Table 6.6.

TABLE 6.6 TCP Optimization Algorithm Categories

Category	Advantages	Applicable Scenario
Single-end optimization	Can be deployed for all TCP/IP applications on the network, without being deployed on the peer end Features extensive adaptability and enables flexible deployment	Scenarios where optimization cannot be implemented at the peer end Example: Internet access
Dual-end optimization	Supports negotiation of proprietary transmission protocols Delivers better optimization than the single-end optimization algorithms	Multibranch scenarios, where optimization is deployed at the egress of each branch

1. Single-end optimization for TCP

   Single-end optimization for TCP uses efficient congestion control algorithms to improve the TCP throughput while retaining the fairness and friendliness of TCP. The basic idea of congestion control is to adjust the TCP transmission rate at the transmit end, based on congestion feedback information obtained from the network.

   Based on the type of information they use, TCP optimization technologies can be classified into the following types:

   Type 1: packet loss-based TCP optimization algorithms

   These algorithms determine whether congestion occurs and adjust the transmission rate based on packet loss. Unlike traditional TCP algorithms, they increase the initial congestion control window size and radically restore the window if they determine that packet loss-based congestion is occurring. This reduces the impact of congestion on the transmission rate.

   Reno, New Reno, and Cubic are typical packet loss–based congestion control algorithms.

   However, these algorithms have the following limitations:

   a. Packet loss does not always mean congestion. For example, in a wireless scenario, packet loss on an air interface may be caused by interference instead of link congestion.

   b. Delay increases due to buffering. When measuring bandwidth, the packet loss–based congestion control algorithm occupies the buffer of the device with a bandwidth bottleneck, increasing delay.

   Type 2: delay-based TCP optimization algorithms

   These algorithms evaluate the level of congestion and the change in delay, adjusting the transmission rate accordingly. These are more suited to modern networks, as they decrease the transmission rate immediately when congestion occurs. This prevents congestion from increasing and reduces or even eliminates packet loss. In addition, delay-based TCP optimization algorithms do not consider packet loss as a necessary result of congestion. Instead, they can still maintain a relatively high transmission rate when packet loss is the result of other factors.

Fast TCP and Compound TCP are typical delay-based TCP optimization algorithms.

Delay-based TCP optimization algorithms are sensitive to the round-trip time (RTT) and are unaware of packet loss. As such, they have the following limitations:

a. Poor fairness, as TCP flows cannot compete for bandwidth resources or hand over bandwidth resources to other TCP flows

b. Stringent requirements on RTT accuracy call for high-precision clocks

c. No fine-grained optimization in scenarios with a small RTT

Type 3: TCP optimization algorithms that more accurately measure bandwidth and delay

We know that measuring the congestion window size based on packet loss or delay has their respective limitations. We can get a more accurate value by measuring the maximum bandwidth and the minimum delay. BBR is a typical TCP optimization algorithm that applies this method. The next section will further discuss this algorithm.

2. Dual-end optimization

Dual-end optimization for TCP introduces multiple feedback mechanisms to achieve high bandwidth utilization and allows for fair bandwidth utilization. This greatly improves transmission efficiency in scenarios with high delay and high packet loss rate.

6.5.3.2 BBR: Optimization at the Transmit End

The BBR algorithm solves the problems in packet loss-based congestion control algorithms as follows:

1. As packet loss does not necessarily mean network congestion, the BBR algorithm does not calculate the link bandwidth based on packet loss. If packet loss occurs, only the lost packets are retransmitted, which does not affect the link bandwidth calculation process. Therefore, when packet loss occurs occasionally, the BBR algorithm ensures that the TCP transmission rate can be continuously increased.

This is completely different from traditional TCP algorithms that only maintain a low transmission rate.

2. The BBR algorithm measures the bandwidth and delay separately. It also uses the maximum bandwidth and minimum delay to more accurately estimate the network capacity. Transmitting packets based on a more accurate network capacity almost eliminates the possibility of occupying the buffer of the device with a network bottleneck. Ultimately, the delay decreases.

The implementation process of the BBR algorithm is as follows:

1. After a TCP connection is set up, the BBR algorithm uses a mechanism similar to the slow-start mechanism of TCP to exponentially increase the transmission rate. According to the received acknowledgment packets, when it finds that the effective bandwidth does not increase any more, the BBR algorithm enters the drain phase. If the sending rate continues to increase and the buffer starts to be occupied, the effective bandwidth does not increase any more. The BBR algorithm terminates the exponential growth in the transmission rate. This prevents packet loss due to buffer exhaustion.

   During the early stage of the startup, almost no buffer is occupied. Therefore, the minimum delay is the initial estimated delay. When the startup is complete, the maximum effective bandwidth is the initial estimated bandwidth.

2. After the startup phase ends, in order to release the occupied buffer, the BBR algorithm enters the drain phase, and packets are gradually drained out. Theoretically, when the buffer is drained, the bandwidth of the bottleneck device is still fully occupied. At this stage, the BBR algorithm enters the steady state.

3. After the drain phase ends, the BBR algorithm enters the steady state and switches between detecting the bandwidth and the delay. The network bandwidth changes more frequently than the delay. Therefore, the BBR algorithm detects the bandwidth most of the time. The Probe Bandwidth phase uses a positive feedback system. The BBR algorithm periodically attempts to increase the packet transmission rate. If the rate of packets received also increases, the system further increases the packet transmission rate.

4. The BBR algorithm uses a very low packet transmission rate to measure the delay over a very short time, without occupying the buffer of the bottleneck device. Therefore, the delay detection is accurate. If an application happens to send packets at a low rate in a certain period of time, the BBR algorithm takes the opportunity to calculate the delay without the need to reduce the rate.

   The BBR algorithm is used to control congestion at the TCP sender and optimizes only unidirectional traffic sent by the sender. Therefore, this technology can be deployed only at the transmit end. Typically, the BBR algorithm is used for optimization on servers.

   For example, on the network shown in Figure 6.43, branch users and traveling users need to access servers deployed in the DC, but WAN optimization capabilities are unavailable on the user side. To enable users to obtain materials on the servers more quickly, we deploy the BBR algorithm on the CPE.

6.5.3.3 FillP: Dual-End Optimization

If we need a higher sending rate than the BBR algorithm can support, FillP is recommended. For this, WAN optimization technologies can be deployed at both the transmit end and the receive end.

FillP is a dual-end optimization algorithm designed to reliably transmit a large amount of data. Leveraging proxy TCP connections, FillP uses UDP to transmit data that is originally carried by TCP. Based on the UDP transmission mechanism, FillP introduces a feedback mechanism to deliver a sending rate that is nearly 100 times higher than that of traditional TCP in scenarios with high latency and high packet loss rate.

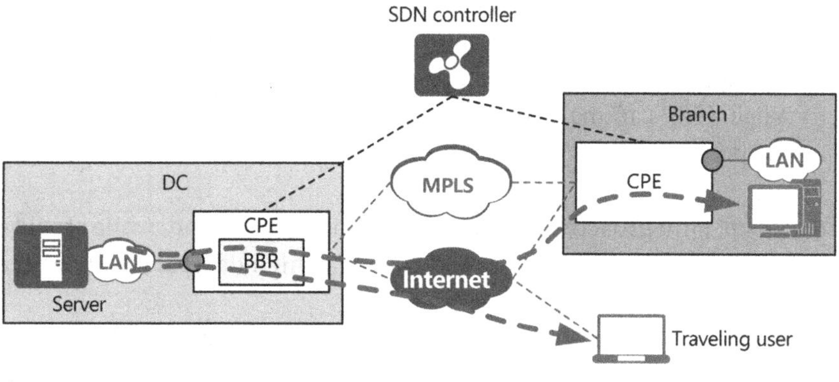

FIGURE 6.43 Optimization on the server.

FillP has the following advantages over traditional TCP algorithms:

1. **Insensitive to packet loss and delay:** FillP uses UDP to carry TCP data and abandons the TCP window. Therefore, FillP does not face the issue inherent to traditional TCP algorithms — that data cannot be sent out. In addition, even if packets are lost on a link, FillP can retransmit packets cached in the protocol stack to restore the lost packets. In summary, FillP ensures efficient utilization of link bandwidth.

2. **Precise flow control:** The transmit end performs comprehensive evaluation based on the sending capability of itself, packet loss information provided by the receive end, receiving rate, and other factors. It also maintains the sending rate in case packet loss occurs on a link. Both ends calculate the sending rate by estimating the bandwidth and delay. This prevents congestion from worsening when the link is congested, ensuring a stable sending rate.

3. **Double-confirmation mechanism:** FillP uses the Negative-Acknowledge mechanism to quickly provide packet loss information from the receive end to the transmit end. With this mechanism, the receive end can quickly request the transmit end to re-transmit lost packets and release received packets. Also, FillP periodically sends Periodic-Acknowledge (PACK) packets. This prevents the transmit end from stopping packet transmission if ACK packets are lost, as it would in the traditional mode.

4. **Dual-SEQ mechanism:** Traditional TCP algorithms use the sequence number mechanism to transmit bytes. If packet loss occurs, the receive end only knows the number of lost bytes but not the specific packets that are lost. With the dual-SEQ mechanism, the receive end can quickly identify the lost packets and retransmit these packets. As such, FillP can maintain a high sending rate, without being affected by packet loss or RTT growth.

To sum up, FillP proactively pushes data at the transmit end, while providing feedback at the receive end, and transmits high data rates even when packet loss rate and delay are high.

FillP applies when a large amount of data needs to be migrated. Let's use VM migration as an example. Site A and site B are two DCs. An employee

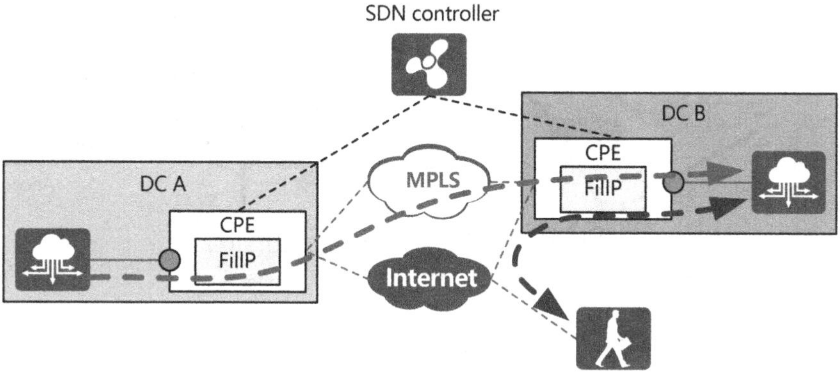

FIGURE 6.44 Application data migration scenario.

works near DC A and uses a VM that runs on its server before going on a business trip. After going on a business trip, the employee needs to access the VM in a location near DC B. To achieve this, the VM needs to migrate from DC A to DC B, as shown in Figure 6.44.

Migrating a VM across DCs involves a large amount of data, and if multiple employees need to migrate VMs, the entire process will slow down. In this situation, we can deploy FillP on the egress gateways (CPEs) of both DC A and DC B to optimize the VM migration traffic passing through the CPEs.

6.5.4 Third-Party WAN Optimization Technologies

Typically, data compression and deduplication, and application optimization are implemented by third-party WAN optimization devices. This section only describes their implementation.

6.5.4.1 Data Deduplication and Compression

Transmission optimization algorithms such as BBR and FillP can accelerate data transmission. However, the bandwidth purchased by an enterprise or allocated to TCP applications may still limit the sending rate. If performance is limited due to bandwidth, TCP optimization algorithms cannot significantly improve the application acceleration performance. Instead, WAN traffic needs to be reduced.

Data deduplication deployed at both ends of the process can significantly reduce the amount of transmitted data, as shown in Figure 6.45.

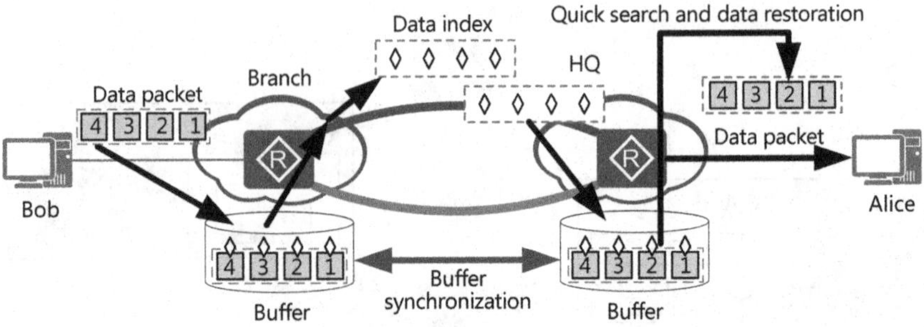

FIGURE 6.45 Data deduplication process.

The process of data deduplication is as follows:

Step 1: When a data packet is transmitted from Bob's client to Alice's client for the first time, the CPE buffers data, creates a block, creates an index, and sends the original data packet. The buffer module caches data and synchronizes the index.

Step 2: When the same data is transmitted again, the CPE at the transmit end first searches the local buffer. If a data block is found, only the corresponding index is transmitted, not the entire data block.

Step 3: The CPE at the receive end quickly searches for the data block in the local buffer based on the received index, restores the original data, and sends the data to Alice's client.

Besides data deduplication, data compression technology can significantly reduce the amount of data to be transmitted. That is, the transmit end compresses data using a lossless compression algorithm to minimize the redundancy of the original data. For example, the four original packets can be compressed into two data packets, which are then sent to the CPE at the receive end. This CPE then restores the original data packets based on the two compressed ones it has received and sends the restored data to the receive end, as shown in Figure 6.46.

Overall, data deduplication and compression technologies can minimize the volume of network traffic to be transmitted and accelerate data transmission.

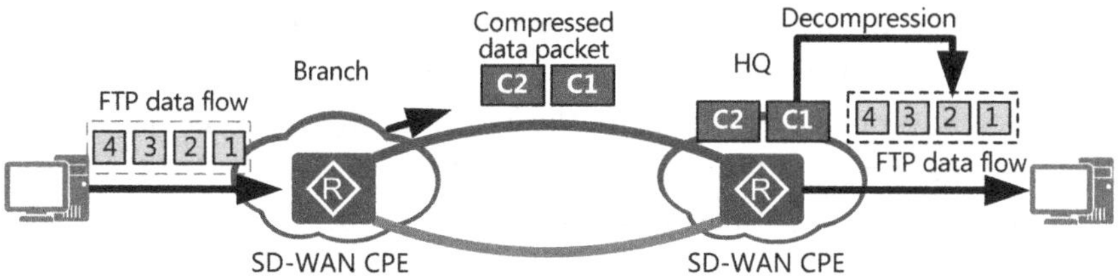

FIGURE 6.46 Data compression process.

6.5.4.2 Optimization Technologies for Web Pages and Web-Based Applications

An enterprise may have many web-based software systems. In addition, to adapt to multiple systems, more and more applications are being developed for use on the web. Although these software systems are provided as independent applications, we need to access web pages when using them. Therefore, if we can access the web pages faster, we will also accelerate access to most applications.

Figure 6.47 shows the optimization technologies that can be used to accelerate web page access.

1. **Buffering:** Web page access through a browser generally requires multiple TCP connections. After many text and image files are

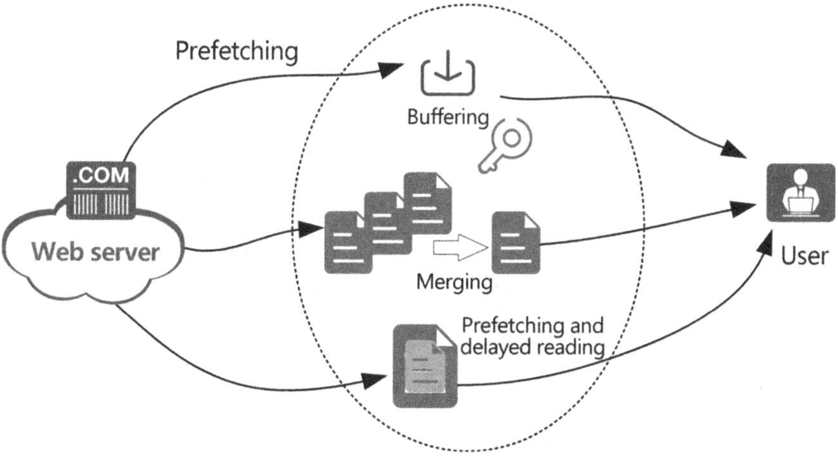

FIGURE 6.47 Optimization technologies for accelerating web page access.

obtained through requests, a web page will be displayed. A WAN optimization device can cache these files locally during the first access. When accessed again, the WAN optimization device locally searches for the files and sends them back to the client without the need to obtain the files from the remote server.

2. **Merging:** Files on static web pages can be obtained in batches. We can merge these files and obtain all of them with only one HTTP request. This minimizes the delay in HTTP interaction and accelerates web page access.

3. **Prefetching and delayed reading:** If a web page has multiple links, we can prefetch the files corresponding to the links and store them on the local host. When a user clicks a link, the system directly sends the file back to the user. In addition, on an excessively large web page, we can load and display only a part of the web page first, and load the remaining content later.

6.5.4.3 Optimization Technologies for File Transfer and Database Access Protocols

Some file transfer protocols and database access protocols are designed only for the LAN. If they are used on the Internet, many interactions are involved due to high latency. A WAN acceleration device can combine these requests and centrally respond to them as a proxy. This eliminates frequent interactions and accelerates file access, thereby enhancing file transfer performance, as shown in Figure 6.48.

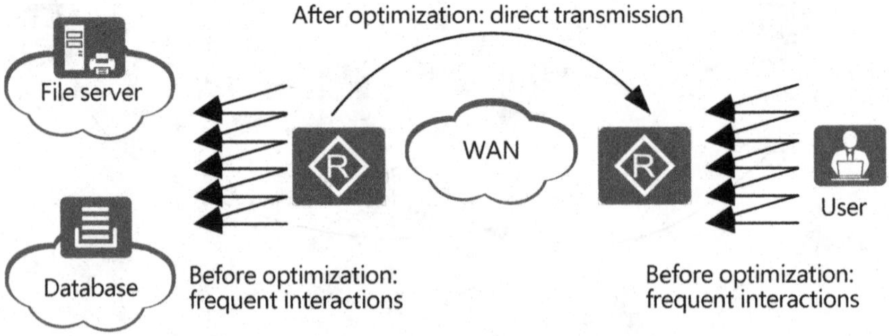

FIGURE 6.48 File transfer and database access.

6.5.4.4 Video Optimization Technologies

Video applications fall into three categories: IM, live streaming, and VoD. Characteristics and optimization methods vary according to the type of video application, as shown in Figure 6.49.

IM applications, such as bidirectional video calls and unidirectional video surveillance, use the packet loss mitigation technology. Images captured by cameras have stringent requirements on real-time transmission performance and are generally transmitted using UDP. Video applications are sensitive to packet loss. Therefore, we use packet loss mitigation technologies such as FEC and packet replication.

Live streaming applications use video stream aggregation technologies. A typical example is live streaming for large events, where many users access the server concurrently. Live streaming applications allow for a delay in seconds and use TCP for transmission. With data buffering and the TCP retransmission mechanism, they are insensitive to packet loss. However, such applications heavily consume the link bandwidth due to a large number of concurrent users, leading to network congestion. For optimization, a centralized proxy is connected to the video source and caches videos in the buffer through one link.

VoD applications use data buffering technologies, for example when videos are embedded on a web page or an application user interface (UI).

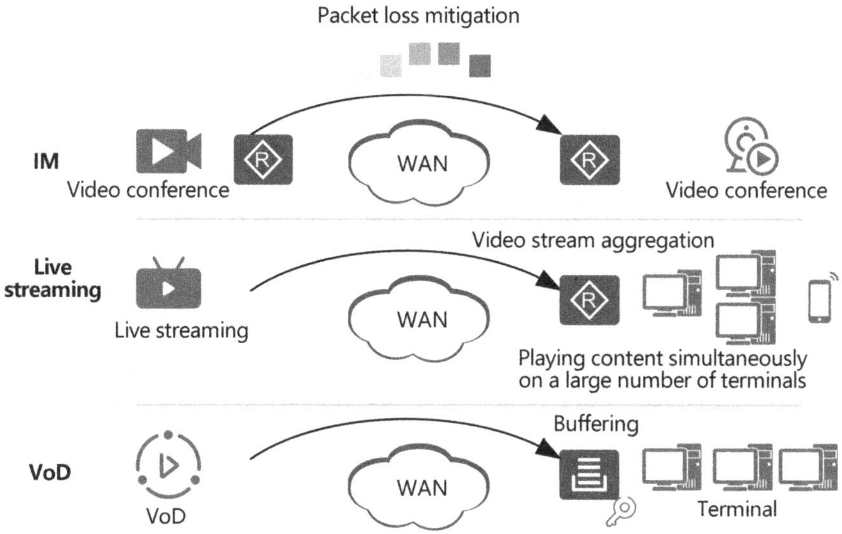

FIGURE 6.49 Video optimization.

It is rare that many users access these videos at the same time, but the videos do occupy the WAN-side bandwidth when accessed multiple times. With data buffering technologies, videos are segmented and stored on hard drives. For security purposes, data blocks are encrypted before being stored on hard drives.

6.5.4.5 SaaS Application Optimization Technologies

SaaS applications are web-based applications deployed on the cloud where optimization technologies for web page applications also apply. Traffic steering can also be optimized for common SaaS applications (such as Office 365, cloud disk, and ERP), as shown in Figure 6.50.

1. Traffic steering on the SD-WAN external network

   To access SaaS applications, enterprises can use a WAN egress with good link quality, which is near the DC where SaaS applications are deployed.

   Doing this greatly improves the experience of SaaS applications. For example, both site B and site C can access SaaS applications, but site C can achieve a better experience. Therefore, site C is selected as the egress site.

2. Traffic steering on the SD-WAN internal network

   By comprehensively measuring the quality of the entire SD-WAN network, applications can select the optimal path over the SD-WAN network to reach the enterprise egress. In our example, traffic can

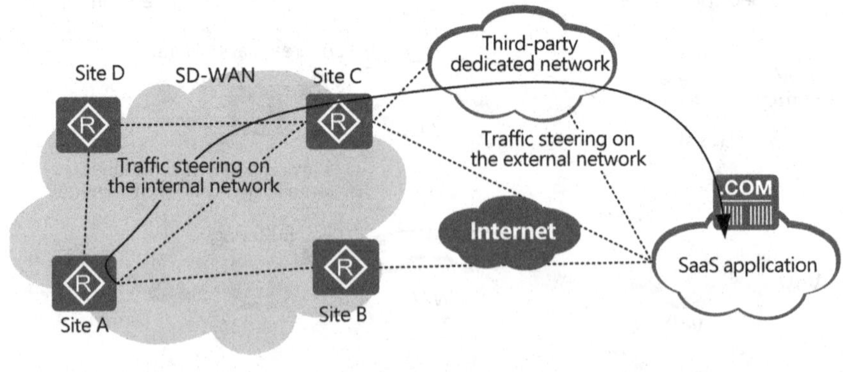

FIGURE 6.50 Access to SaaS applications.

be transmitted from site A to site C along two paths: site A → site D → site C; site A → site C. An optimal link needs to be selected using traffic steering technology on the internal network.

3. Third-party dedicated network

   Enterprises with a small number of sites may fail to achieve the optimal SaaS application experience regardless of any egress site in use. In this case, a WAN optimization vendor can collaborate with the SaaS provider to deliver a third-party dedicated network for enterprises to access SaaS applications.

   Typically, the abovementioned traffic steering technologies are combined to obtain an optimal E2E solution.

6.6 INTENT-DRIVEN APPLICATION EXPERIENCE OPTIMIZATION

As described in the preceding sections, the application experience assurance sub-solution includes traffic steering, QoS, and WAN optimization technologies. How can we employ these independent technologies to deliver an overall desirable application experience?

There are two methods:

Method 1: Independently configure these technologies by specifying conditions and parameters in fine-grained templates.

Method 2: Effectively combine these technologies and provide intent-driven assurance experience in SD-WAN.

In method 1, the assurance technologies seem to be configured intuitively but face the following issues:

1. The assurance technologies are configured separately. Each one involves complex parameters and configurations, which are difficult to understand and use.

2. The separate assurance technologies cannot associate with each other, failing to deliver an integrated application experience and even causing configuration conflicts.

   a. If traffic steering is configured separately, there may be no link available (none of the links meet quality requirements). This will compromise the application experience. When congestion occurs on a link, frequent link switchovers will further deteriorate user experience.

b. If packet loss mitigation is configured separately, enabling the FEC function causes unnecessary performance loss because the FEC function is not associated with the link status. In addition, if the bandwidth is insufficient, congestion will worsen due to the FEC function.

c. If QoS is configured separately, idle links cannot be fully used. When a link is disconnected or the link quality deteriorates, traffic cannot be dynamically switched to the link with good quality, and QoS cannot mitigate packet loss on the link.

The three assurance technologies mentioned above use comprehensive approaches to ensure user experience. In terms of service processes, the three are executed separately and entries are searched multiple times during service processing, affecting performance.

Method 2 simplifies configurations and enhances service operational efficiency, thereby bringing a better user experience.

The intent-driven experience optimization sub-solution is more suited to most enterprises, as shown in Figure 6.51.

The following sections describe the intent-driven application experience optimization sub-solution in detail.

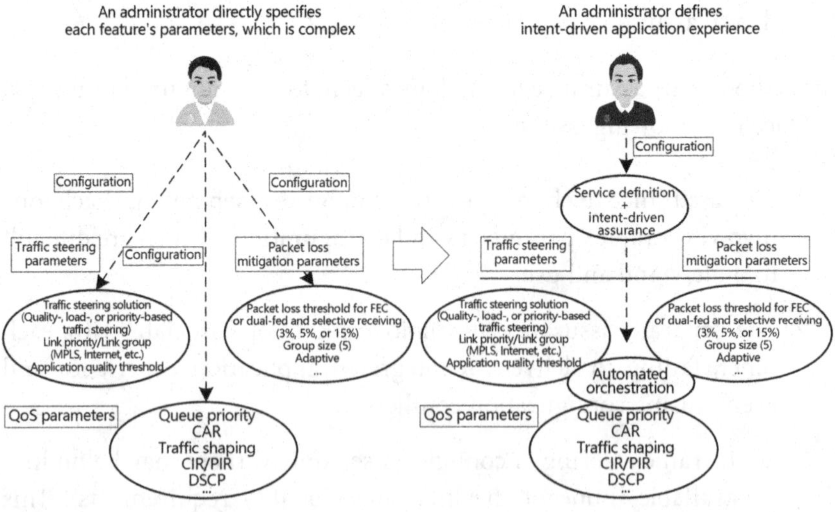

FIGURE 6.51 Application experience optimization solution comparison.

6.6.1 Experience Optimization

Figure 6.52 shows the improved intent-driven application experience optimization sub-solution, which resolves the above-mentioned issues of complex configurations, the lack of association between functions, and low operational efficiency.

The following intent-driven application experience optimization practices are available:

- **Centralized configuration approach:** Services can be classified based on service characteristics and requirements without considering specific configuration parameters, reducing the configuration workload.
- **Converged scheduling approach:** The CPE offers the following functions:
  - Matches and classifies traffic only once, reducing the calculation workload.
  - Improves application quality by dynamically combining traffic steering, packet loss mitigation, and QoS based on application characteristics and real-time network quality.

6.6.2 Centralized Configuration Approach

The centralized configuration approach provides default experience classification templates. With these templates, you simply need to classify services into corresponding categories based on service characteristics and requirements. During internal processing, the SDN controller automatically orchestrates and associates related policies and parameters.

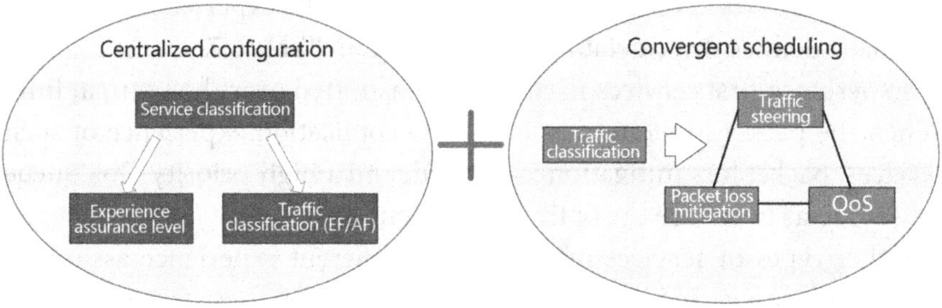

FIGURE 6.52 Improved intent-driven application experience optimization sub-solution.

The centralized configuration process is as follows:

Step 1: An administrator creates an intent-driven service template.

Step 2: The administrator selects applications and creates traffic classifiers based on services. The configuration of a traffic classifier is the same as that for other services such as QoS.

Step 3: The administrator configures the service experience assurance level. Based on this, the SDN controller references the corresponding quality parameters. The service experience assurance level can be configured as follows:

- Default service level (high, medium, or low), which reflects the network quality requirements of a service
- Customized service level

Step 4: The administrator delivers and executes services. The SDN controller delivers the orchestrated template and parameters to network devices, where corresponding service processes are executed based on policy configurations.

Figure 6.53 shows the SDN controller configuration page.

As shown in Figure 6.53, the service experience assurance level is critical to the centralized configuration approach.

In RFC 4594, we can define multiple service experience assurance levels based on the link quality requirements of services. Then, we can use the corresponding default QoS queues according to the service experience assurance level. Different QoS priorities can be specified for customized services (QoS scheduling is based on the HQoS scheduling model of SD-WAN). Based on the service distribution, the experience assurance approach defines three default levels, as listed in Table 6.7.

Experience-first services need to be transmitted over the optimal link. When the packet loss rate may affect the application experience of such services, packet loss mitigation is enabled, and a high-priority QoS queue is used, so as to ensure the optimal experience.

Other types of services may require a different experience assurance level and have a varied level of importance. Based on these factors, we can implement other related quality assurance technologies:

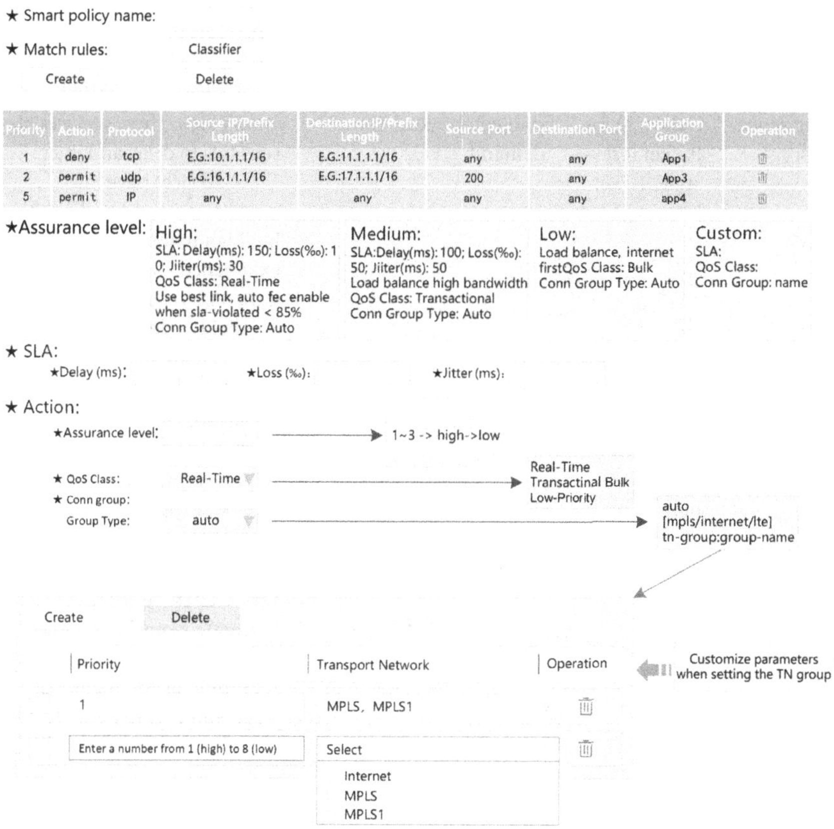

FIGURE 6.53 SDN controller configuration page.

- **Real-time:** high-level QoS guarantee for enterprises' voice and video services
- **Transactional:** medium-level QoS guarantee for enterprises' major interactive services
- **Bulk:** low-level QoS guarantee for enterprises' transmission services

6.6.3 Convergent Scheduling Approach

When multiple services are processed at the same time, applications need to be matched only once to implement experience assurance for these applications based on the SFC. Figure 6.54 shows a CPE performing application assurance technologies in associated mode based on the configuration on the SDN controller.

TABLE 6.7 Service Experience Assurance Levels

Category	Link Quality	Packet Loss Mitigation	QoS	Service Description
Experience first	Delay: 30 ms Packet loss rate: 10‰ Jitter: 150 ms	Enabled on demand	Real-time	Voice/Video: These services are sensitive to delay and jitter. Voice services have low requirements on bandwidth but high requirements on user experience. Video conferences have high requirements on bandwidth, delay, and packet loss rate and need to be preferentially guaranteed
Interaction/ Transaction	Delay: 100 ms Packet loss rate: 50‰ Jitter: 50 ms	N/A	Transactional	IM/Media: IM/Media applications have high requirements on transmission rates and application experience
Data	N/A	N/A	Bulk	Email, file sharing, and backup services: These services consume many bandwidth resources and have no requirements on the delay, jitter, and packet loss rate

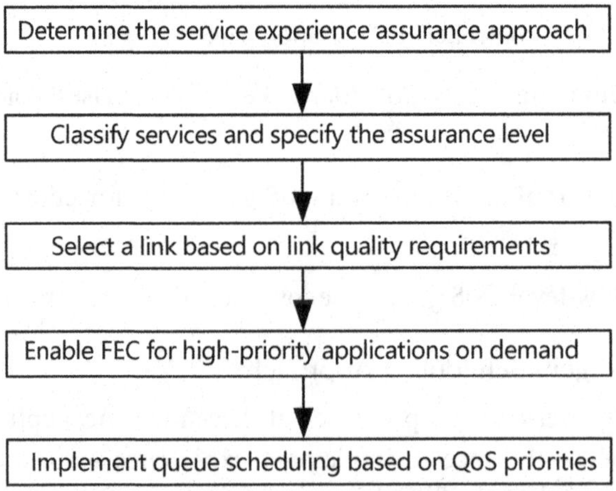

FIGURE 6.54 Associated application assurance process.

The process in which a CPE performs associated application assurance is described as follows:

1. Determine the optimal service experience assurance sub-solution based on the service traffic matching rule.
2. Classify services based on the configuration of the assurance approach, and specify the assurance level and QoS priority based on the classification result.
3. Select a link that meets the link quality requirements based on the service assurance level. For an experience-first service, the optimal link is selected.
4. Automatically enable packet loss mitigation technologies, such as FEC, when the link quality falls below the threshold.
5. Implement QoS bandwidth limiting and queue scheduling based on service types. QoS is implemented based on the three-level scheduling in the SD-WAN service model.

CHAPTER 7

Security: Top Priority

In recent years, enterprises have suffered full-scale security breaches, including frequent distributed denial of service (DDoS) attacks, large-scale user data breaches, and ransomware spreads. This has severely affected their day-to-day service operations.

As such, security is the top priority for any information system, and the SD-WAN solution is no exception. It is impossible to overemphasize the key role that security plays. To cope with security challenges, we need comprehensive security protection approaches. Therefore, this chapter discusses the security approaches and requirements for the SD-WAN solution.

7.1 NEW SECURITY CHALLENGES

The relatively closed architecture of a traditional enterprise WAN ensures some extent of network security. Each branch on a conventional enterprise WAN connects to the headquarters, perhaps through MPLS private lines. They do so to access the headquarters or DC for service operations or access the Internet through the headquarters. Therefore, an enterprise can implement security policies at the headquarters (e.g., by deploying firewalls) to control branch users' access to the DC and the Internet.

The SD-WAN solution fundamentally changes enterprise WANs by adding new components and optimizing service models. However, these changes pose new security challenges. For example, the SDN controller — the brain of the entire solution — is vitally important, which increases the possible impact of potential risks. The solution also introduces hybrid WAN links, which include Internet links that are less secure.

Specifically, an enterprise may face the following security risks when deploying an SD-WAN solution:

- **Unauthorized access:** An unauthorized device, for example, a forged CPE, registers with the SDN controller and accesses the network, posing serious security risks.
- **Intrusion behavior:** Solution components provide interfaces for interacting with external systems. This means that the components are vulnerable to various intrusion attacks on the Internet, compromising the stability and availability of the system.
- **Data breach:** Communication data between solution components and service data between sites are transmitted over the Internet. Therefore, the data may be stolen or tampered with.
- **Service damage:** Branch sites directly access the Internet for service operations. This facilitates the spread of viruses, malicious files, and ransomware on the Internet and may cause service damage.

As we can see, the SD-WAN solution comes with undeniable benefits (better enterprise branch interconnection and cloud access) but also poses new security challenges. To ensure the security, reliability, and stability of the operating environment, we must mitigate these new risks.

Addressing security challenges requires a systematic approach. To better analyze the security risks faced by the SD-WAN solution and work out risk mitigation strategies, we can divide the SD-WAN solution security into two levels: system security and service security, as shown in Figure 7.1.

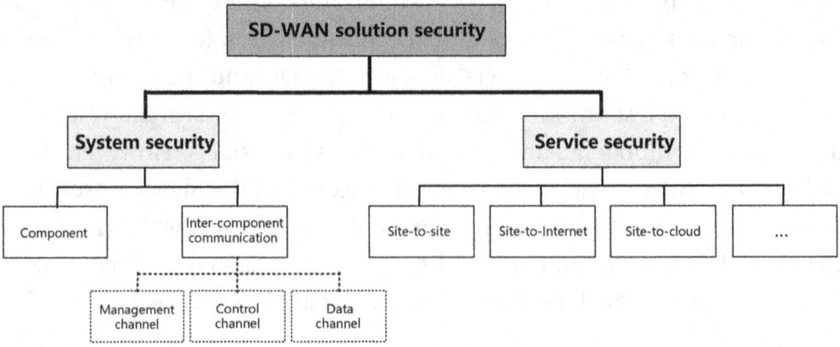

FIGURE 7.1 Security levels in the SD-WAN solution.

System security protection is an essential security capability necessary for the SD-WAN solution. Therefore, it is built into the system and is available as soon as the system starts working. This ensures that the system runs securely and reliably. However, enterprises must independently deploy service security protection approaches that are flexible and adaptable to varying service requirements.

7.2 SYSTEM SECURITY

The SD-WAN solution is a complex system that consists of multiple components. Each component and the communication between them are subject to security threats. To safeguard the system, we must implement risk mitigation strategies.

1. Component security

   To ensure their security, components require a robust system architecture and comprehensive security hardening policies. It is necessary to adopt preventive measures, including permission control, account and password management, data protection, and security audit. It is also important to protect the physical and network environments where components reside. To prevent attacks, components, especially the SDN controller, need to be deployed in areas protected by security devices, such as firewalls and anti-DDoS devices, when necessary.

2. Communication security between components

   In the SD-WAN solution, multiple security technologies work together to resolve security issues related to communication between components. Communication security between components includes mutual trust, secure access, and data encryption. For this, we must implement authentication, encryption, and verification.

We will detail component security and inter-component communication security in the following sections.

7.2.1 Component Security

The SDN controller and CPEs are basic components of the SD-WAN solution and will inevitably face various security threats. The security and stability of these components will ensure that the SD-WAN system operates normally.

7.2.1.1 SDN Controller

In the SD-WAN solution, the SDN controller is the brain of the entire network and its security is especially important. The SDN controller is more likely to be attacked because it is logically centralized. Its open APIs and southbound protocols are also vulnerable to security threats and attacks.

The security of the SDN controller is directly related to the reliability and availability of the entire network. For example, if someone successfully attacks the SDN controller by launching a denial of service (DoS) attack and stealing the SDN controller permissions, they will affect the entire network, leading to serious security problems. Considering its importance, we must implement dedicated security hardening policies for the SDN controller, as described in Table 7.1.

7.2.1.2 CPE

Regardless of its model, a CPE must have a secure system architecture and support multiple security protection measures to defend it against various security threats.

The CPE system architecture must comply with the X.805 standard formulated by the International Telecommunication Union-Telecommunication Standardization Sector (ITU-T). The CPE uses a three-layer and three-plane security isolation mechanism, meaning that its control plane, management plane, and forwarding plane are isolated from each other. Therefore, an attack on any plane does not affect the other two. In addition, the CPE needs multiple security protection capabilities on each plane, including but not limited to:

1. Physical security protection

   The CPE must be able to disable service ports, serial ports, and services that are not in use, to prevent them from attacks.

2. Data security protection

   The CPE must be able to encrypt sensitive information, such as service data, user names, and passwords, to prevent sensitive information leakage. The CPE must also control data access permissions to prevent unauthorized access to data.

3. Authentication and authorization

   The CPE must possess system permission control and account permission management functions. These implement strict identity

TABLE 7.1 Security Hardening Policies for the SDN Controller

Category		Measure	Security Risk to Eliminate
Authentication and permission control	Identity authentication	Local and remote authentication	Faking an administrator
	Permission control	Role-based permission control; tenant- and domain-based control	Tampering with or escalating user permissions
Data protection	Key management	Secure encryption algorithms; hierarchical key management	Key data breaches
Data protection	Data storage	Data access control; encrypted data storage	Key data breaches
	Data transmission	Secure communication protocols; encrypted data transmission	Key data breaches
Security detection and response	Attack detection	Attack detection in multiple dimensions, including ports, web pages, and OSs	External intrusion
	Intrusion prevention	Defense against various intrusion behaviors	External intrusion
	Anti-DoS/Anti-DDoS	Defense against common traffic attacks	DoS attacks
Privacy protection		Strict access permission control	Privacy data breaches
Security management	Security audit	Comprehensive logging and event recording functions	Repudiation
	Secure upgrade and patch installation	Administrator authentication (only authenticated administrators can perform upgrades or patch installation operations)	Data tampering during the upgrade or patch installation process
System protection		Integrity protection and signature verification for software and patches	Data tampering
Secure deployment		Zone planning, hierarchical deployment, and firewall deployment for isolation	Intrusion and important data breaches

authentication and permission control on login behaviors, and support account/password protection, password complexity check, and anti-brute force password cracking.

4. Attack defense

   The CPE must be capable of defending against various network attacks, such as IP flood attacks, ICMP flood attacks, Address

Resolution Protocol (ARP) flood attacks, ARP spoofing, Smurf attacks, malformed packet attacks, and invalid packet attacks.

The CPE must support Control Plane Committed Access Rate (CPCAR), which limits the rate of packets of each protocol (or a typical message of a protocol) sent to the control plane, preventing the CPU from being attacked.

5. Security audit

   The CPE must have a comprehensive logging system to log all configuration operations and abnormal status during system running, which can be used for future audit.

7.2.2 Inter-component Communication Security

7.2.2.1 Basic Principles of Authentication, Encryption, and Verification

In the network world, authentication, encryption, and verification are the fundamentals for secure interaction between communication parties. The SD-WAN solution uses these technologies to ensure secure and reliable communication between components. Let's first take a look at the theoretical bases for these processes: the symmetric key mechanism, the public key mechanism (also called asymmetric key mechanism), and hash calculation.

In the symmetric key mechanism, both communication parties use the same algorithm and key to encrypt and decrypt data. Both know and share the same key during encryption and decryption. On the other hand, the public key mechanism uses two different keys, one public and one private. The information encrypted with a public key can only be decrypted using the paired private key, and the information encrypted with a private key can only be decrypted using the paired public key.

Advanced Encryption Standard (AES) is a typical algorithm based on the symmetric key mechanism. Algorithms based on the public key mechanism include Rivest–Shamir–Adleman (RSA) and Diffie–Hellman (DH) algorithms. RSA can implement data encryption and decryption, authenticity verification, and integrity verification. The DH algorithm is a key exchange algorithm, based on which communication parties calculate the key through a series of data exchanges.

Hash calculation is another important step to implement data integrity verification. The process maps a string of a variable length to a string of a fixed length. It is unidirectional and not reversible. Most importantly, changes in any character in the original string result in different calculation results. The values returned after a hash calculation are called digests of the original strings. You can determine whether the data is modified by comparing these digests. The Secure Hash Algorithm (SHA) is a typical hash algorithm.

After reviewing the theoretical bases, let's look at the basic principles of authentication, encryption, and verification.

1. Authentication

   Authentication is a process in which two communication parties authenticate each other's identity. This is a widely used security protection measure in the network security field. Common identity authentication modes include user name/password authentication and the more prevailing certificate authentication. A certificate, or digital certificate, is a secure and trusted carrier. It's the network-world equivalent of an identity card. The certificate associates the holder's identity information with the public key, ensuring that the key belongs to the certificate holder. By verifying the certificate, the holder's identity can be confirmed. Figure 7.2 shows the typical certificate structure.

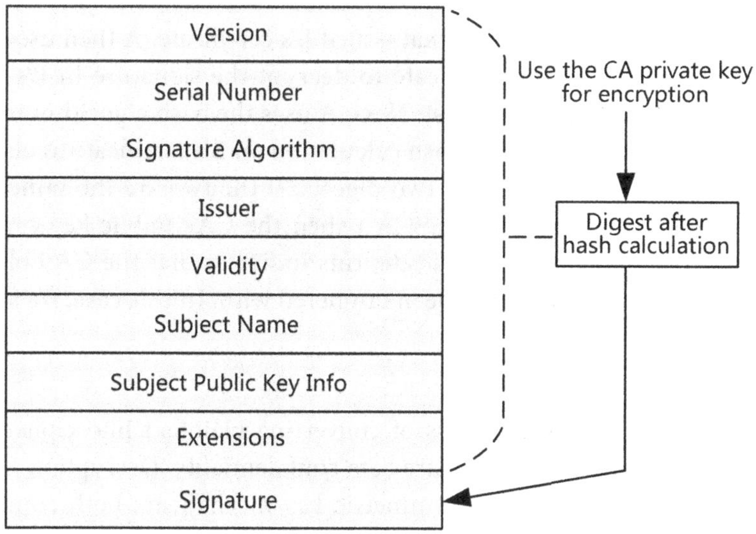

FIGURE 7.2 Certificate structure.

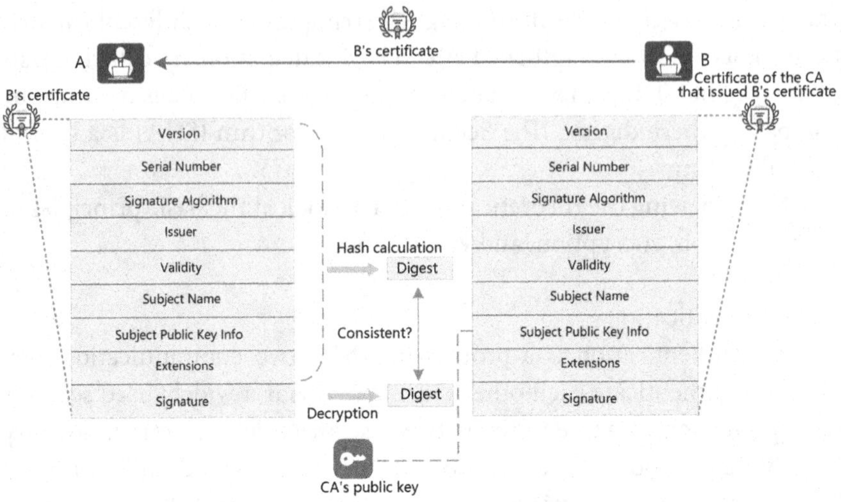

FIGURE 7.3 Certificate authentication process.

Certificate authentication is based on the public key infrastructure (PKI) architecture, where trusted CAs issue the certificates. When communication parties use certificates to authenticate each other, they need a CA certificate in addition to their own certificates.

Figure 7.3 illustrates how A authenticates B. A needs to obtain the certificate of the CA that issued B's certificate. A then uses the public key in CA's certificate to decrypt the signature in B's certificate to obtain the digest. Next, A uses the hash algorithm in B's certificate to perform a hash calculation on B's certificate to obtain a digest. A compares the two digests. If the two are the same, B's certificate is issued by the CA (when the CA's public key can be used to decrypt the certificate, this indicates that the CA holds a private key) and has not been tampered with. In this case, B's identity is authenticated.

2. Encryption

Encryption is a process of converting plaintext into ciphertext using an algorithm to ensure data confidentiality. Decryption is the reverse process. In the symmetric key mechanism, both communication parties use the same key to encrypt and decrypt data, as shown in Figure 7.4.

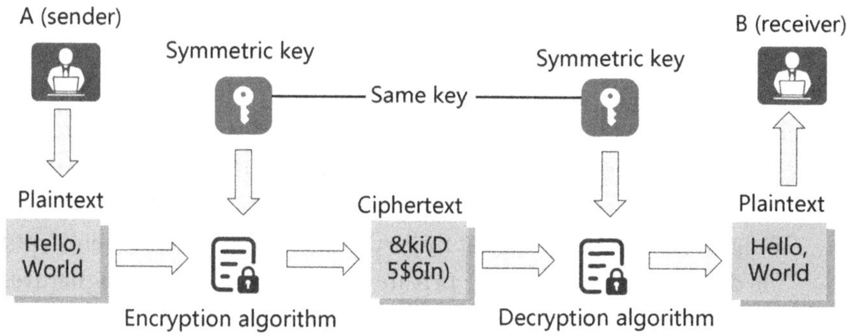

FIGURE 7.4 Encryption and decryption process of the symmetric key mechanism.

In the public key mechanism, two communication parties use different keys to encrypt and decrypt data, as shown in Figure 7.5.

The public key mechanism is more complex and time-consuming than the symmetric key mechanism, due to significant overhead in data encryption. Therefore, we cannot use the public key mechanism to encrypt a large amount of data. In applications, two communication parties normally use a public key mechanism-based algorithm (such as DH) for key negotiation. This makes sure that the key exchange materials are secure. After they calculate the key, the two parties use an algorithm (such as AES) based on the symmetric key mechanism to encrypt actual data. The symmetric key mechanism and the public key mechanism are used separately but cooperatively, balancing efficiency and security.

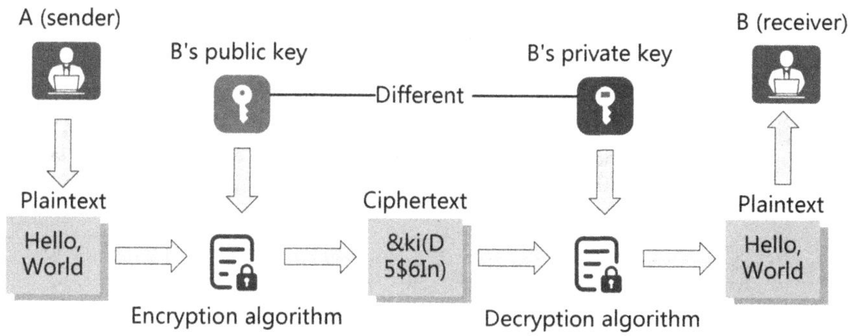

FIGURE 7.5 Encryption and decryption process in the public key mechanism.

3. Verification

During the verification process, two communication parties check whether data has been tampered with during transmission, by checking its integrity. During integrity verification, a private key is needed for encryption and a public one for decryption. First, the sender processes the original data using a hash algorithm to obtain a digest (also called a digital fingerprint). Then, the sender encrypts the digest using the sender's private key to create a digital signature. Finally, it sends the data and the digital signature together.

After receiving the data and the digital signature, the receiver uses the same hash algorithm to perform a hash calculation on the data to obtain the sender's digest. The receiver then decrypts the signature using the sender's public key to obtain another digest. The receiver compares the two digests. If they are the same, data integrity is verified, meaning the data came from the sender and has not been tampered with.

Figure 7.6 shows the encryption and decryption process for a digital signature.

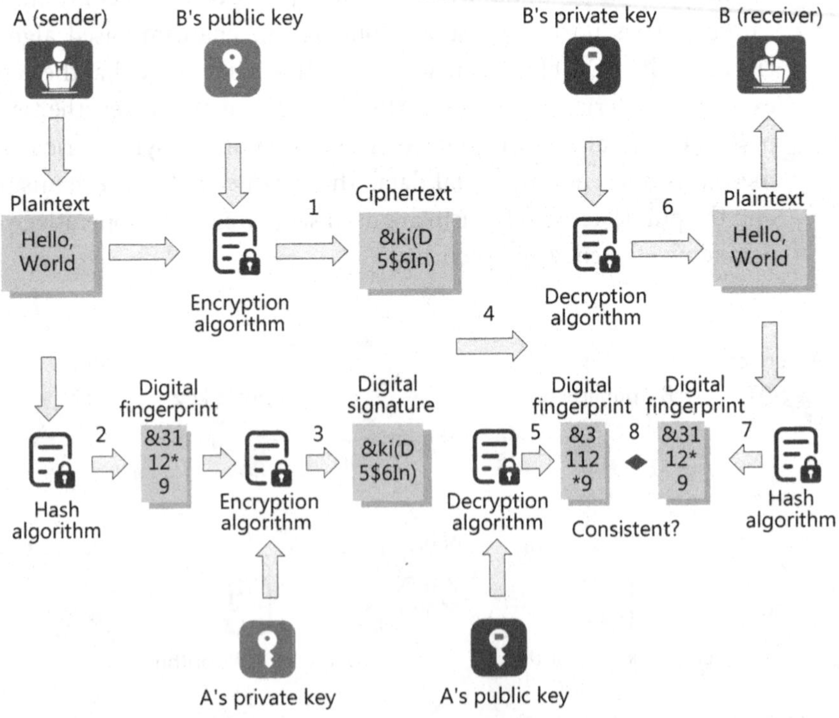

FIGURE 7.6 Encryption and decryption of a digital signature.

A and B obtain each other's public key in advance and perform the following steps to encrypt and decrypt the digital signature:

Step 1: A uses B's public key to encrypt plaintext, generating ciphertext.

Step 2: A uses the hash algorithm to perform a hash calculation on the plaintext, generating a digital fingerprint.

Step 3: A uses its own private key to encrypt the digital fingerprint, generating the digital signature.

Step 4: A sends both the ciphertext and the digital signature to B.

Step 5: B uses A's public key to decrypt the digital signature, obtaining the digital fingerprint.

Step 6: After receiving the ciphertext from A, B uses its own private key to decrypt the ciphertext, obtaining the original plaintext.

Step 7: B uses the hash algorithm to perform a hash calculation on the plaintext, generating a digital fingerprint.

Step 8: B compares the generated fingerprint with the received one. If the two fingerprints are the same, B accepts the plaintext; otherwise, B discards it.

7.2.2.2 Access Authentication Based on Zero Trust

In the SD-WAN solution, CPEs go online in plug-and-play mode. Although the deployment process is convenient, security issues, such as CPE spoofing and unauthorized access, may occur. When using a CPE, the SD-WAN solution applies the zero-trust security concept. It strictly authenticates CPE identities to ensure that only authorized CPEs can access the network.

Zero trust assumes that no entity inside or outside the network should be trusted by default. Instead, all entities should be authenticated before they are authorized to access the network. The SD-WAN solution leverages various authentication mechanisms based on zero trust to ensure that the SD-WAN system is accessed only by the right people using the correct devices.

1. Deployment

   We described CPE plug-and-play deployment in Chapter 4. In this phase, we must take the necessary security measures to ensure CPE deployment security. For personnel, only trusted site deployment personnel should receive the deployment email and the Universal Serial Bus (USB) flash drive. The URL in the deployment email must be encrypted, and the deployment personnel must be notified of the decryption password in a secure manner. This ensures that the right people will perform deployment operations.

If ESN-to-site binding is used during deployment, the SDN controller checks the CPEs based on their ESNs and determines which sites the CPEs belong to. If the ESNs are decoupled from sites, the SDN controller uses tokens to determine the sites. Since a token has a validity period, the SDN controller verifies the validity period of the token to ensure time-sensitive security of CPE deployment.

A CPE and the SDN controller use a certificate to authenticate each other when a management channel and a control channel are established between them. A CA issues a certificate, which represents the identity of the holder. Certificate authentication is mutual. That is, the SDN controller verifies the CPE's certificate to prevent access of a forged CPE. The CPE also authenticates the SDN controller by verifying the SDN controller's certificate, preventing a forged SDN controller from controlling the CPE.

Mutual certificate authentication ensures trust between the CPE and the SDN controller, indicating that the correct devices are used, as shown in Figure 7.7.

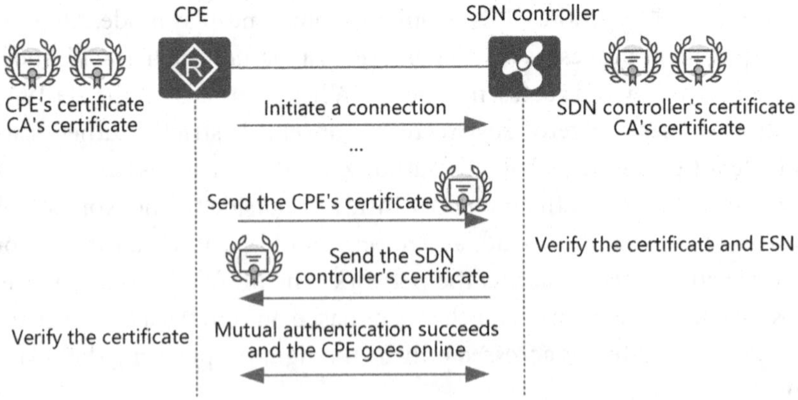

FIGURE 7.7 Mutual certificate authentication between a CPE and the SDN controller.

The following describes mutual certificate authentication between a CPE and the SDN controller:

Step 1: The CPE and the SDN controller are preconfigured with two certificates before factory delivery: a certificate issued by the same CA and the CA's certificate.

Step 2: After being powered on, the CPE sends a registration request that carries its certificate to the SDN controller.

Step 3: After receiving the CPE's certificate, the SDN controller uses the CA's certificate to verify the CPE's certificate. At the same time, it verifies the ESN of the CPE.

Step 4: The SDN controller sends its certificate to the CPE.

Step 5: After receiving the SDN controller's certificate, the CPE uses the CA's certificate to verify it.

Step 6: Mutual authentication is successful and the CPE goes online and registers with the SDN controller.

Generally, the preconfigured certificates are only used for CPE registration and login upon startup. An enterprise needs to replace these preconfigured certificates with its own certificates, for example, a certificate issued by the enterprise's CA or one purchased by the enterprise from a third-party CA.

2. Operations

Some enterprises have small-scale branch sites and do not have professional equipment rooms. Their CPEs are usually desktop devices, which are prone to theft. An attacker may use the existing configuration of a stolen CPE to connect to the enterprise intranet, causing security risks.

To address the CPE theft issue, the SDN controller must be able to detect a change in geographical location or access environment of a CPE. If one of these conditions changes, the SDN controller denies CPE access, or sends an alarm to the network administrator and only allows CPE access after the network administrator confirms that there is no problem. The SDN controller should also support the isolation function. This allows the network administrator to isolate suspicious CPEs from the network to reduce security risks caused by suspicious CPE access.

7.2.2.3 Data Protection Using Secure Communication Protocols

When the communication data between a CPE and the SDN controller is transmitted on the Internet, data breaches may occur. To ensure data confidentiality during transmission, an encryption mechanism is required. It is also essential to ensure communication data integrity, preventing data tampering.

Generally, the CPE and the SDN controller use SSH, TLS, or Datagram Transport Layer Security (DTLS) to establish a management channel and a control channel, as shown in Figure 7.8. SSH/TLS/DTLS provides encryption and authentication functions to ensure the confidentiality and integrity of the communication data between the CPE and the SDN controller.

When establishing an SSH/TLS/DTLS connection, the CPE and the SDN controller calculate a symmetric encryption key using the DH algorithm and negotiate an encryption algorithm. The subsequent interaction data between the CPE and the SDN controller is carried over the SSH/TLS/DTLS connection and is protected by encryption of the SSH/TLS/DTLS connection, such as NETCONF over SSH and HTTP/2 over TLS. This ensures data confidentiality during transmission. Similarly, data integrity also relies on SSH/TLS/DTLS.

The data channel established between CPEs represents the interconnection between sites. The data channel is an overlay tunnel established

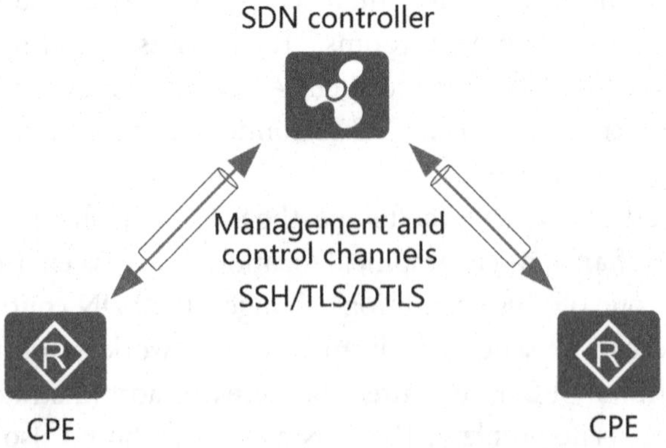

FIGURE 7.8 Management channel and control channel.

between CPEs on the underlay network to transmit service traffic. Data channels have relatively simple security requirements. The key is to ensure the confidentiality and integrity of the overlay tunnel between CPEs. In most cases, IPsec between CPEs can ensure data security, as shown in Figure 7.9.

In the traditional mode, the IKE protocol establishes IPsec tunnels between CPEs. The CPEs negotiate to establish an IKE connection, based on which they establish IPsec SAs. If a CPE needs to establish IPsec tunnels with multiple other CPEs, it needs to negotiate with each of the CPEs to establish IKE connections and IPsec SAs. This consumes more CPE system resources and makes it difficult to manage IPsec keys.

The SD-WAN solution addresses the above issues by allowing the SDN controller to distribute information required for establishing an IPsec SA to each CPE without first establishing an IPsec tunnel using the IKE protocol, as shown in Figure 7.10.

The SDN controller receives and distributes IPsec SA information through the control channel. This reduces CPE system resource consumption and improves the flexibility and scalability of establishing IPsec connections between CPEs. In addition, IPsec SAs are updated periodically to improve security.

Note that there are different ways to generate the keys used by CPEs to establish IPsec SAs. For example, keys can be preconfigured and

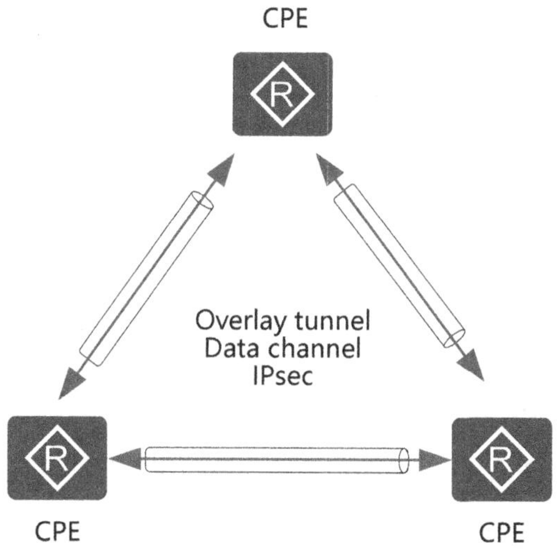

FIGURE 7.9 Data channel.

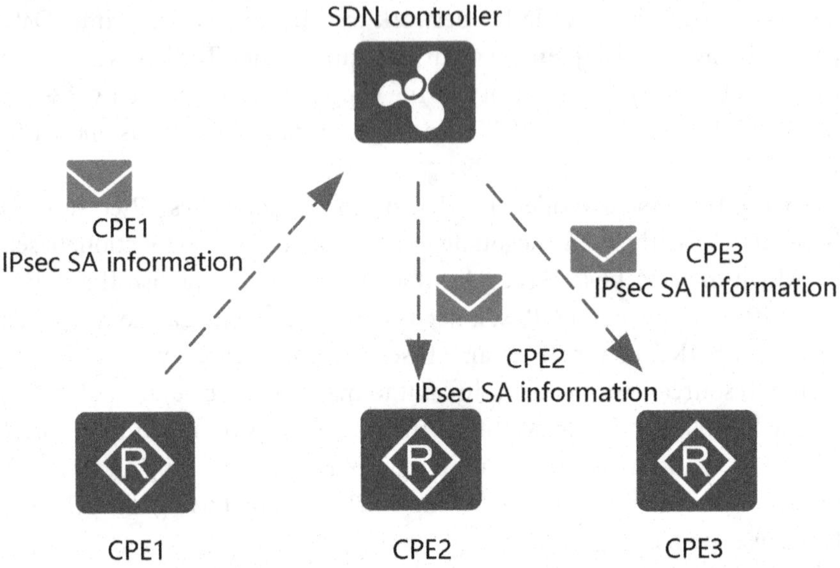

FIGURE 7.10 SDN controller distributing IPsec SA information.

distributed directly by the SDN controller to each CPE; or keys may be obtained through negotiation and calculation between CPEs.

7.2.2.4 Security Isolation in a Multi-tenant Scenario

In a carrier/MSP resale scenario, the carrier/MSP needs to provide SD-WAN services for a large number of enterprise users. An enterprise is usually regarded as a tenant, so the network provided by the carrier or MSP must support concurrent management of multiple tenants. To achieve this, the SDN controller must support multi-tenant management and implement security isolation between tenants.

A multi-tenant scenario may face two main security threats:

- **Unauthorized CPE access:** A tenant's CPE establishes a management channel and a control channel with another tenant's SDN controller, and then establishes a data channel with yet this tenant's CPE.
- **Data breaches:** One tenant steals other tenants' service data.

The SDN controller should provide strict security isolation functions to enable different tenants to have their own management and control channels.

These restrictions will ensure that only the same tenant's CPEs establish data channels with each other. For example, if tenant A's CPE releases information, the information cannot be spread to tenant B's CPE. Neither can tenant B's CPE access tenant A's network to establish a management channel, a control channel, or a data channel.

Figure 7.11 shows security isolation in a multi-tenant scenario where the SDN controller manages both tenant A and tenant B.

In this scenario, multiple tenants share the IWG in the SD-WAN solution. If the traditional mode of directly distributing IPsec SA keys is used, the same key may be used to encrypt and decrypt traffic sent from different tenants' CPEs to the IWG. As a result, a tenant can decrypt other tenants' traffic, causing data security problems.

To avoid such problems, the DH algorithm is used for negotiation between CPEs and between each CPE and the IWG. The SDN controller distributes key materials, and devices calculate keys by exchanging key materials to establish IPsec SAs. Using encryption key negotiation

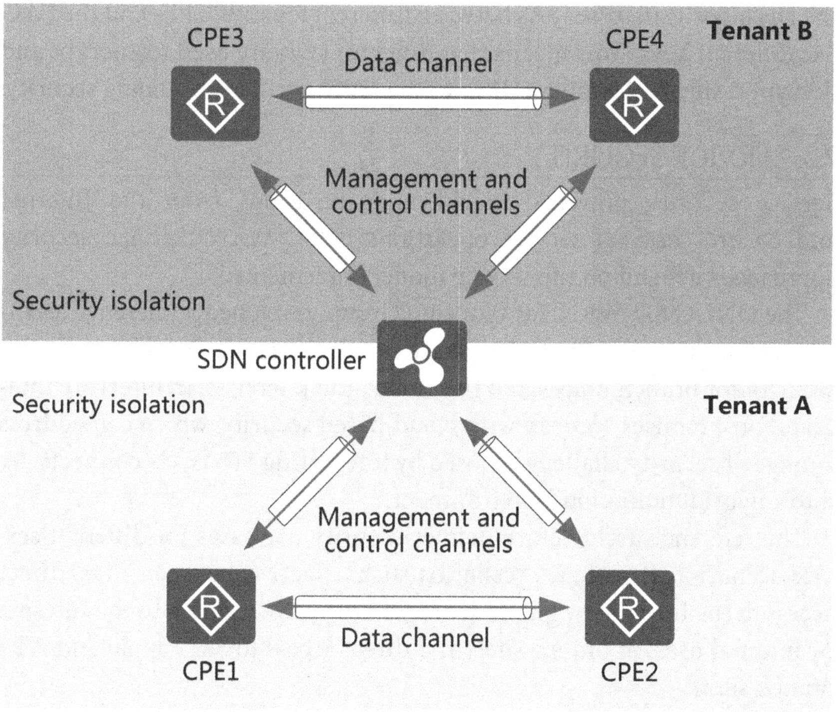

FIGURE 7.11 Security isolation in a multi-tenant scenario.

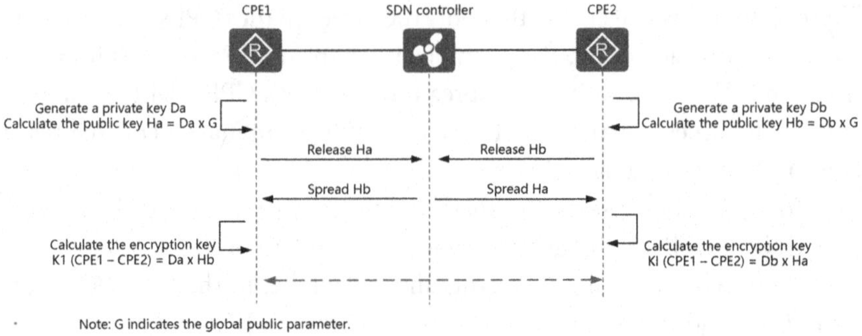

FIGURE 7.12 Generation of an encryption key.

between CPEs as an example, Figure 7.12 shows how an encryption key is generated.

The DH algorithm negotiation mode is more secure than the direct key distribution mode. Even if attackers obtain key materials exchanged between devices, they cannot calculate the keys. Additionally, when using the DH algorithm, IPsec SAs between different tenants' CPEs and the IWG use different keys. This means that different keys are used to encrypt and decrypt traffic sent by these CPEs to the IWG, further enhancing security.

7.3 SERVICE SECURITY

Service security prevents attacks and intrusions from the Internet and ensures normal service operations of SD-WAN. Service security approaches depend on the service model implemented.

The ONUG SD-WAN 2.0 Working Group developed a reference architecture for enterprise SD-WAN multi-cloud integration, which focuses on security for branch offices and the cloud. The reference architecture integrates on-premises security with cloud-based security, which can address a range of security challenges posed by integrating SD-WAN connectivity into a hybrid multi-cloud environment.

The reference architecture defines security use cases for different service scenarios. It includes scenarios such as access between sites, direct access to the Internet by guests at branch sites, direct access to the Internet by internal users at branch sites, and direct access to SaaS applications by branch sites.

Before analyzing service security requirements, we need to understand SD-WAN service scenarios. For this, we will refer to the security use cases

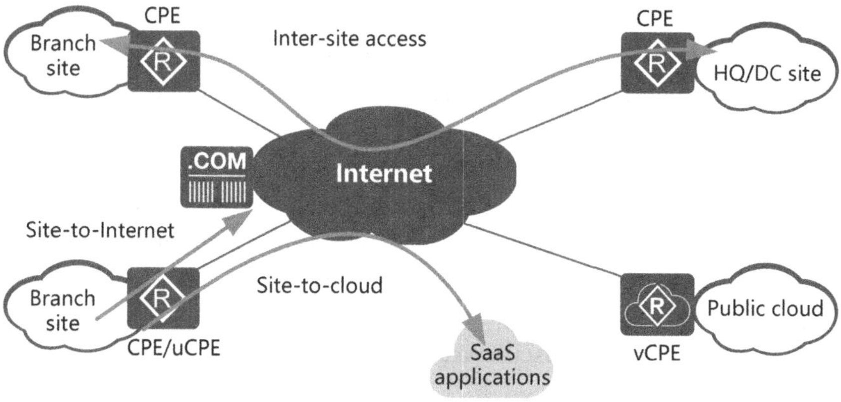

FIGURE 7.13 SD-WAN service scenarios.

defined by the SD-WAN 2.0 Working Group. Besides legacy enterprise WAN links, SD-WAN introduces the Internet to transmit service traffic, including service traffic between sites. Branch sites can directly access the Internet through local Internet access. In addition, as more enterprise applications migrate to the cloud, branch sites may use the Internet to directly access SaaS applications and operate services. Figure 7.13 shows these service scenarios.

Without exception, these service scenarios face security challenges and therefore require adequate security approaches:

1. Site-to-site

   The connection between CPEs transmits inter-site service traffic. For inter-site services, we need to focus on the security of service traffic transmitted on the Internet. Generally, CPEs use IPsec to set up connections and use mechanisms such as encryption and integrity verification to prevent data from being stolen or tampered with.

2. Site-to-Internet

   After Internet links are introduced in SD-WAN, enterprise branch sites can directly access the Internet. This change in the service model makes Internet access more convenient but simultaneously exposes branch sites to Internet-related security risks. In this service scenario, sites themselves usually deploy security protection measures where the CPE/uCPE provides security protection.

3. Site-to-cloud

   As traditional enterprise applications gradually migrate to the cloud, enterprise branch sites can directly access applications on the public cloud through the Internet, generating yet another security risk. In this service scenario, security protection can be deployed on the cloud. For example, we can use third-party cloud security protection services to safeguard the service traffic from branches to the cloud in SD-WAN.

The following sections discuss security protection for the three service scenarios.

7.3.1 Site-to-Site

Site-to-site communication relies on the connection between CPEs. Based on the underlay network, CPEs establish overlay tunnels, which carry the inter-site service traffic. Inter-site service traffic is transmitted on the Internet. Therefore, overlay tunnels use IPsec to protect service traffic, ensuring confidentiality and integrity during transmission, as illustrated in Figure 7.14.

When establishing IPsec tunnels between CPEs, the SDN controller distributes information required to establish IPsec SAs, without using the IKE protocol for setting up a connection. As mentioned earlier, it is possible to preconfigure the keys used by IPsec SAs and have the SDN controller distribute them directly to each CPE. Alternatively, CPEs can negotiate with each other and calculate the keys using the DH algorithm.

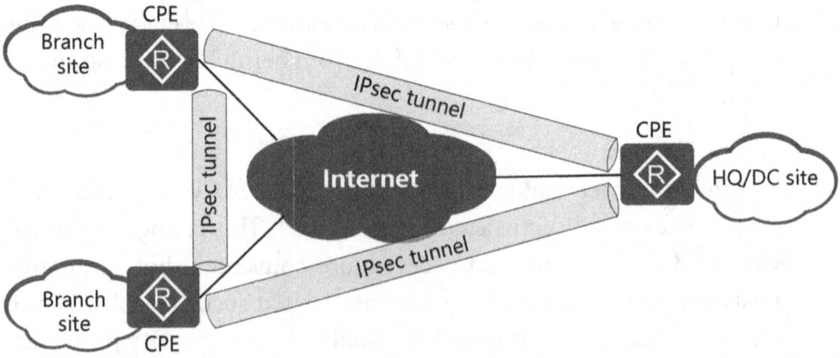

FIGURE 7.14 Site-to-site service security.

7.3.2 Site-to-Internet

The introduction of Internet links allows sites to have direct access to the Internet through CPEs, where CPEs become the front line of defense. Considering this, CPEs must have some service security protection capabilities, including but not limited to the following:

- Traffic filtering is based on ACLs. ACLs are the most basic function of network devices and are widely used for security protection. Traffic can be classified based on ACL rules, blocking specific types of traffic.
- Firewall, an isolation technology that logically isolates the internal network from the external network by dividing security zones. This protects the internal network from unauthorized access.
- An intrusion prevention system (IPS) is another security mechanism. An IPS analyzes network traffic characteristics to detect intrusion behaviors, such as buffer overflow attacks, Trojan horses, and worms. It then responds to and blocks detected intrusion behaviors to protect networks from being attacked.
- URL filtering controls the URLs that users can access at a site, prevents users from accessing certain web page resources, and regulates users' online behaviors, hence mitigating security risks.

These are some of the basic security functions. If advanced security functions are required, we can deploy VASs. In SD-WAN, VAS devices can be virtual NEs — for example, VNFs deployed on uCPEs — or physical firewalls connected to CPEs in off-path mode. We can deploy virtualized security devices, such as virtual firewalls (vFWs), on uCPEs to provide advanced security functions, such as content filtering, file blocking, and advanced threat defense. This is a common solution used in SD-WAN to provide security protection for services.

To sum up, security protection for site-to-Internet access services can be implemented using CPEs' built-in security functions or the VAS-based advanced security protection functions provided by uCPEs, as shown in Figure 7.15. We can flexibly select either or both of them based on service requirements.

The following describes site-to-Internet access security in more detail, including the built-in security functions of CPEs and VAS-based advanced security protection functions provided by uCPEs.

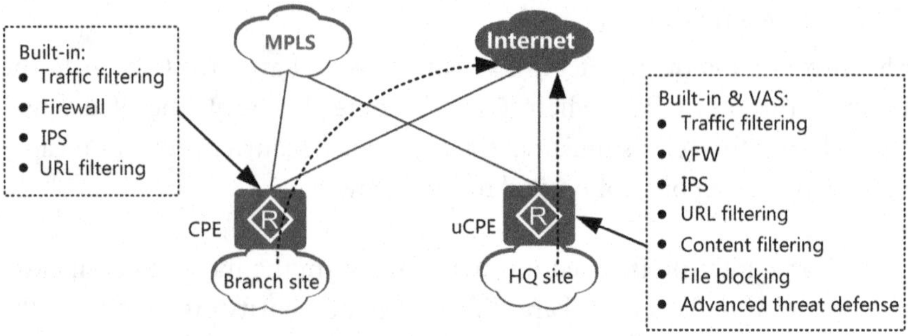

FIGURE 7.15 Site-to-Internet service security.

7.3.2.1 Built-in Security Functions of CPEs

1. Traffic filtering

   If we need to control specific types of traffic exchanged between the enterprise intranet and the Internet, we can configure the traffic filtering function. For this, we need to configure ACL rules on a CPE to classify packets based on the source IP address, destination IP address, source port number, and destination port number, and then filter the classified packets. We can deploy the traffic filtering function on the WAN- or the LAN-side interface of a CPE to filter traffic passing through the CPE.

   A CPE can filter traffic based on policies. After the network administrator configures a policy on the SDN controller, the SDN controller applies the ACL rules to CPEs. Figure 7.16 shows the traffic filtering configuration page on the SDN controller.

   Figure 7.17 shows a typical traffic filtering application scenario. Applying an ACL to the WAN-side interface of a CPE can prevent specific traffic from the external network from entering the branch intranet. Applying an ACL to the LAN-side interface of a CPE can block specific traffic sent from the branch intranet to the external network.

2. Firewall

   The firewall function used for isolating internal networks from external networks involves two concepts:

   a. **Security zone (zone for short):** Consists of one or more interfaces. The networks connected to these interfaces have the same security attributes.

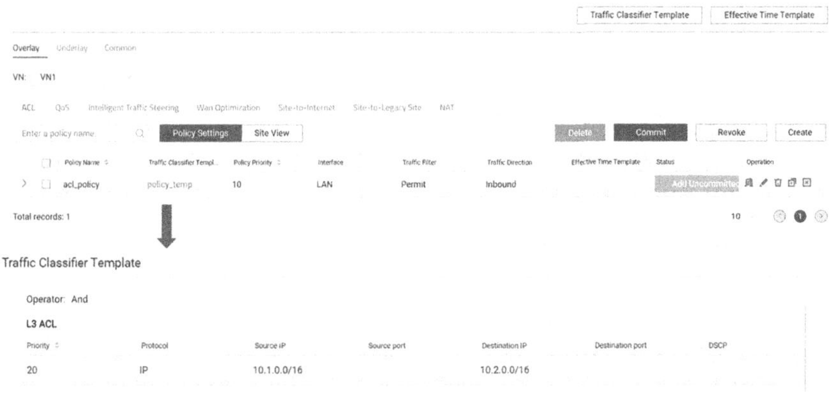

FIGURE 7.16 Traffic filtering configuration page.

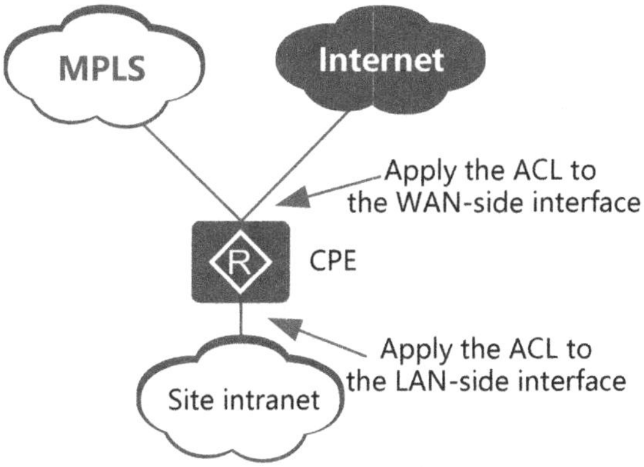

FIGURE 7.17 Traffic filtering application scenario.

b. **Interzone:** Consists of any two security zones. The transmission of interzone packets triggers the firewall function.

Generally, we can enable the firewall function on a CPE through policies applied to interzones. For this, we need to consider security zone planning and the scope of the interzone to which the policies are applied. In SD-WAN, simple operations on the SDN controller will automatically set up security zones based on the actual conditions of the CPEs and apply policies to the corresponding interzones. In this scenario, the network administrator does not need to perform complex configuration on CPEs through the CLI, which simplifies the configuration process. Figure 7.18 shows the firewall function configuration page on the SDN controller.

Figure 7.19 uses centralized Internet access and local Internet access as examples to describe the firewall function application scenarios. The firewall function is enabled on CPEs to protect the service traffic sent from the enterprise branch and headquarters sites to the Internet.

In the centralized Internet access scenario, Internet traffic of all branch sites is diverted to the headquarters site for Internet access. The firewall function is enabled on the CPE at the headquarters site to protect Internet traffic.

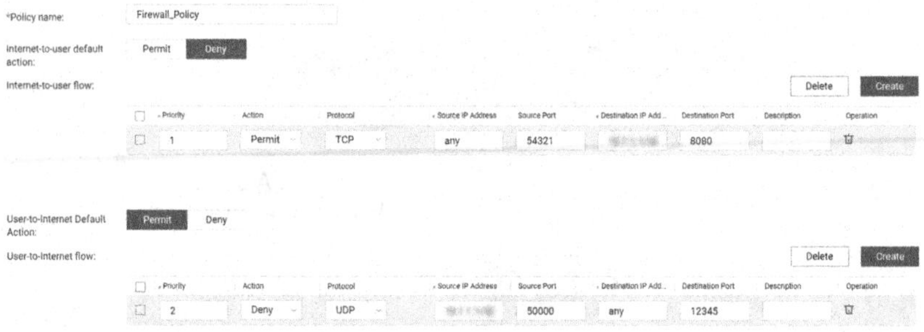

FIGURE 7.18 Firewall function configuration page.

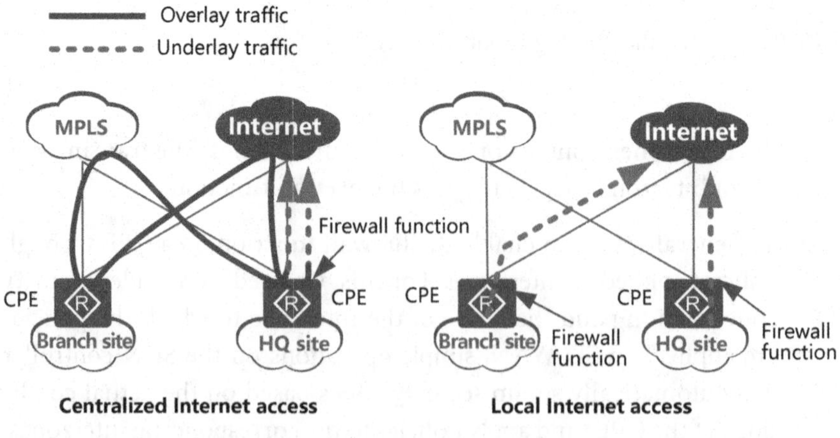

FIGURE 7.19 Firewall function application scenarios.

In the local Internet access scenario, the branch and headquarters sites directly access the Internet through local CPEs. The firewall function is enabled on the CPEs at the branch and headquarters sites to protect Internet traffic.

3. IPS

The IPS signature database contains common intrusion behavior signatures and is preconfigured on a CPE. The IPS compares packet characteristics with signatures in the database. If it finds a match, the IPS detects an intrusion and takes corresponding defense measures. It is essential to periodically update the IPS signature database of a CPE so that the CPE can identify new attacks and intrusion behaviors to better defend against network threats.

A CPE can also implement the IPS function through policies that are applied to interzones. In SD-WAN, simple operations on the SDN controller will automatically set up security zones based on the actual conditions of the CPEs and apply policies to the corresponding interzones. In this scenario, the network administrator does not need to perform complex configuration on CPEs through the CLI, which simplifies the configuration process. Figure 7.20 shows the IPS function configuration page on the SDN controller.

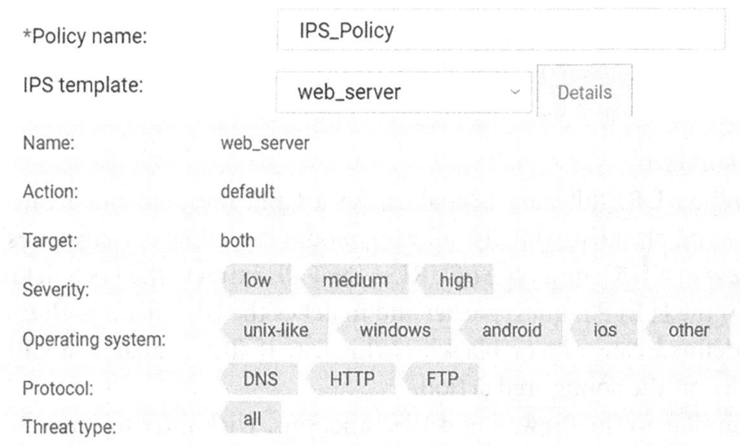

FIGURE 7.20 IPS function configuration page.

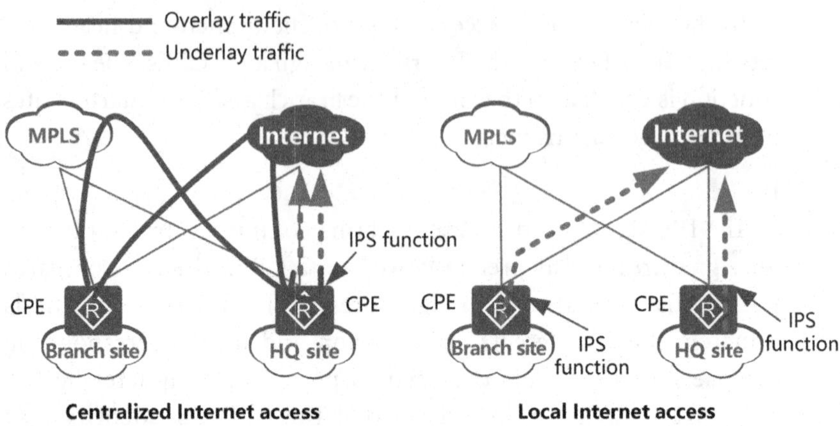

FIGURE 7.21 IPS function application scenarios.

Figure 7.21 shows the IPS function application scenarios, including centralized Internet access and local Internet access. The IPS function is enabled on CPEs to protect the service traffic sent from the enterprise branch and headquarters sites to the Internet.

In the centralized Internet access scenario, Internet traffic of all branch sites is diverted to the headquarters site for Internet access. The IPS function is enabled on the CPE at the headquarters site to block various intrusion behaviors from the Internet.

In the local Internet access scenario, Internet traffic of the branch sites is directly transmitted from local CPEs to the Internet. The IPS function is enabled on the CPEs at the branch and headquarters sites to block intrusion behaviors from the Internet.

4. URL filtering

When URL filtering is enabled on a CPE, the predefined category and blacklist/whitelist filtering modes can allow or deny users' access to a URL or a type of URLs. The CPE extracts the URL fields from the HTTP request packet and matches the URL fields with the predefined category or blacklist/whitelist. If they match, the CPE performs the configured action.

Similar to the firewall and IPS functions, URL filtering can also be implemented through policies that are applied to interzones. The network administrator only performs simple operations on the SDN controller, which will automatically set up security zones based on

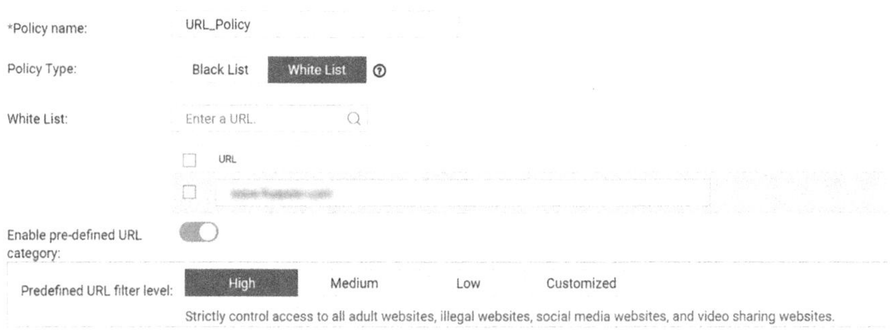

FIGURE 7.22 URL filtering configuration page.

the actual conditions of the CPEs and apply policies to the corresponding interzones. Figure 7.22 shows the URL filtering configuration page on the SDN controller.

Figure 7.23 shows a typical URL filtering application scenario, where a user at a branch site accesses the web servers at the headquarters site and on the Internet. To control the URLs the user accesses, URL filtering is enabled on the CPE at the branch site. As such, when the user accesses the web server on the Internet or at the headquarters site, the URL filtering function regulates the user's online behavior.

7.3.2.2 VAS-Based Advanced Security Protection Functions of uCPEs

Virtualized security devices exist logically on a uCPE as independent devices. The SDN controller orchestrates SFCs to control how service traffic entering the uCPE is forwarded. An SFC defines and instantiates a group of ordered SFs, and then enables traffic to pass through these SFs in sequence.

As shown in Figure 7.24, service traffic from the enterprise intranet (LAN side) is first diverted to the vFW by the virtual switch, processed by the vFW, returned to the virtual switch, and finally sent by the router. The path of the return service traffic is similar.

A vFW — in the form of a VNF — is deployed on the uCPE to provide advanced security functions such as antivirus, content filtering, and file blocking to defend against advanced threats. Enterprises can obtain on-demand VAS-based advanced security protection services without additional hardware investment. This also simplifies network management and reduces network O&M costs.

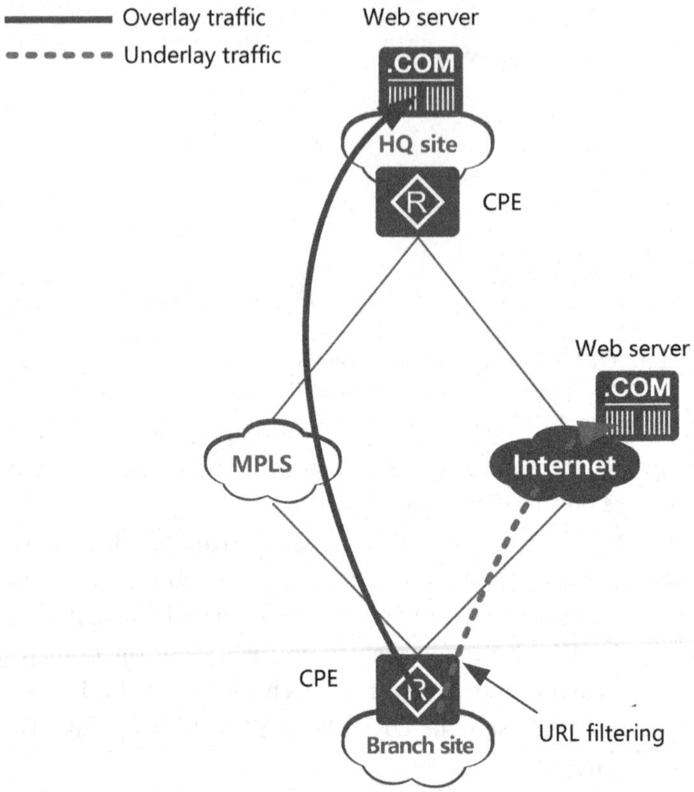

FIGURE 7.23 URL filtering application scenario.

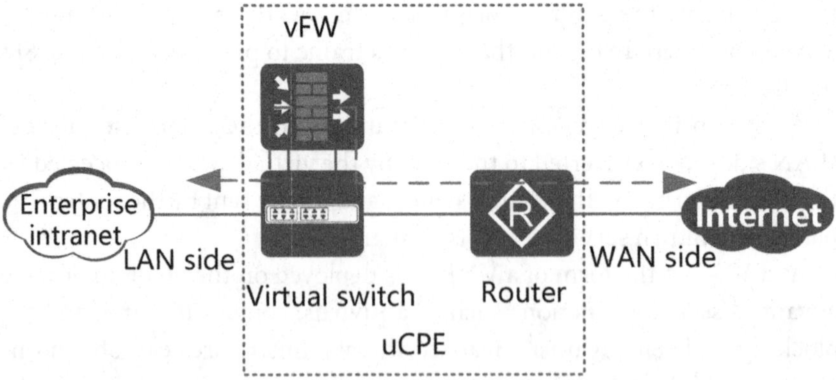

FIGURE 7.24 vFWs deployed on uCPEs.

7.3.3 Site-to-Cloud

As enterprise applications migrate to the cloud, enterprise branch sites can directly access SaaS applications, such as office software and databases, through the Internet. In addition to the CPE/uCPE security functions at branch sites, we can also use security functions on the cloud to protect related service traffic.

Some cloud security service providers can offer services such as access control, threat detection, attack defense, and data protection for enterprise access to SaaS applications. Cloud-based security protection services are widely available and scalable, and can be easily integrated with other vendors' services in the network ecosystem.

As shown in Figure 7.25, CPEs at the enterprise branch sites connect to the cloud security gateway provided by a third-party cloud security service provider. Typically, the CPEs establish tunnels with the third-party cloud security gateway and send the branch sites' SaaS application access traffic to the third-party cloud security gateway for security detection. After the

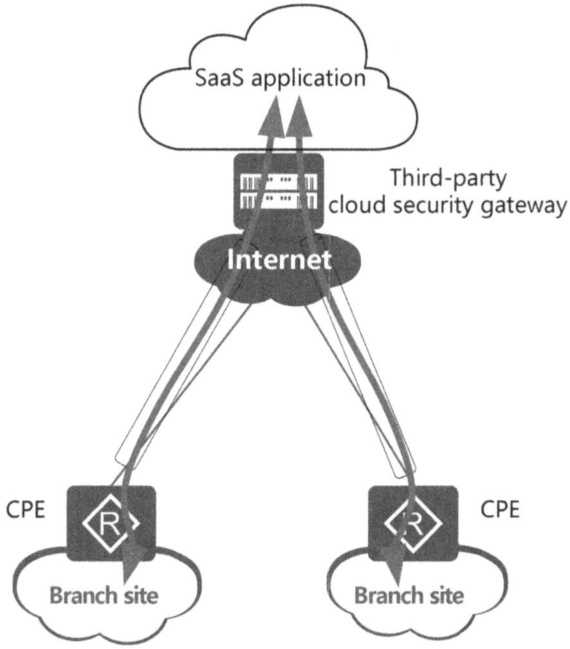

FIGURE 7.25 Site-to-cloud service security.

detection, the third-party cloud security gateway sends legitimate traffic to the corresponding SaaS applications.

The SD-WAN solution integrates the cloud security gateway provided by a third party and uses these third-party security services to protect site-to-cloud traffic. This highly effective security protection approach can reduce costs, simplify management, and improve the access experience and the security of cloud applications.

CHAPTER 8

Easy O&M

TRADITIONAL NETWORK O&M IS device-centric, with network administrators using commands to manage a single device or the NMS to manage multiple devices in a centralized manner. In order to locate faults and perform routine maintenance, administrators must analyze the data provided by devices. As such, traditional O&M places high demands on the expertise of network administrators and can incur significant costs as a result.

The SD-WAN solution offers simplified O&M through the use of the SDN controller, which provides a network-wide monitoring function capable of obtaining site and link status information in real time and displaying this data in a visualized manner. In addition, the SDN controller can analyze network operational data, proactively identify possible fault points, and quickly notify network administrators, enabling them to efficiently locate and rectify network faults. Equipped with such intelligent and automatic O&M functions, the SD-WAN solution reduces management costs and boosts O&M efficiency.

8.1 O&M MODE TRANSFORMATION

With the diversification, globalization, and cloudification of services, enterprises are increasingly dependent on networks and their sites continue to grow rapidly in number. Against this evolving landscape, traditional network O&M modes can no longer meet the requirements of modern enterprises. Consequently, a transformation of O&M modes has begun.

Network administrators face the following pain points relating to network O&M:

- The O&M platform lacks permission control, and the network can be managed by multiple administrators, leading to operational errors and adversely impacting services on the network.
- Network administrators are unable to detect the status of traditional networks in real time, and potential threats cannot be predicted, necessitating round-the-clock standby. Simple and efficient proactive O&M methods are urgently required to solve these issues.
- The network structure is complex, and fault locating is difficult as a result. Network administrators must focus on how to quickly locate and rectify faults (typically underlay and overlay network faults and service faults) in order to ensure normal service operations.
- Device migration is challenging, and services cannot be switched smoothly. To minimize the impact on the live network, administrators must perform such migration during off-peak hours.

The SD-WAN solution provides comprehensive O&M functions capable of addressing each of these pain points and achieving the goal of intelligent O&M. It provides such O&M functions at four layers based on a brand-new O&M architecture.

- Role-based permission control: Specific O&M modes are provided for each customer to ensure most fine-grained permission control, while various access permissions can be set for different roles to implement rights- and domain-based management.
- Multidimensional visualized monitoring: The SDN controller provides visualized data, such as the historical trends and current health of networks and services, and displays the real-time status of tenant networks.
- Various fault diagnosis functions: The SDN controller's integrated fault locating tool can quickly locate complex faults, lowering the skill requirements on O&M personnel and increasing O&M efficiency.

- Migration from traditional networks to SD-WAN. The SD-WAN solution can incorporate devices on legacy sites or reconstruct them as SD-WAN sites, thereby transforming traditional networks to SD-WAN.

8.2 O&M ARCHITECTURE

Enterprise networks can be maintained in a proactive or passive mode.

Proactive O&M employs proactive monitoring and management measures to detect potential faults and eliminate potential risks in advance. Such techniques include monitoring device alarms and logs, and the quality of links between sites.

Passive O&M locates and rectifies a fault that has already occurred. For example, we can use the network quality diagnosis tool to locate a fault, remotely log in to the device where the fault occurs through SSH, and perform such operations as version upgrade or patch installation.

The following describes the O&M functions of the SD-WAN solution based on the overall O&M architecture, as shown in Figure 8.1.

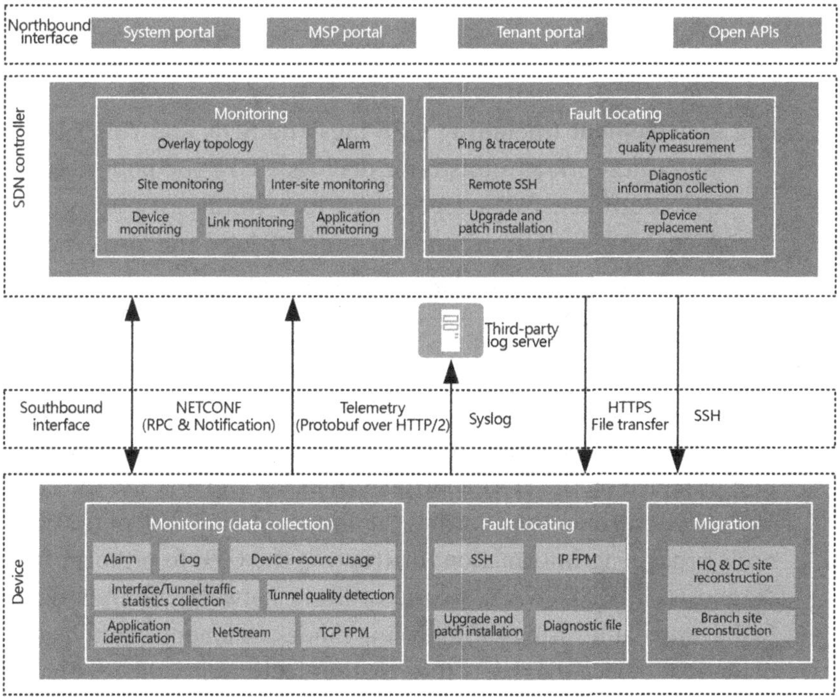

FIGURE 8.1 O&M architecture.

1. Role management

   The SD-WAN solution supports three administrator roles: system administrator, MSP administrator, and tenant administrator. The SDN controller provides O&M management portals from which all three roles can perform O&M operations.

2. Monitoring

   Alarm monitoring: The SDN controller provides functions for receiving, storing, centrally monitoring, and querying real-time and historical alarms, enabling network administrators to quickly locate faults based on alarm information. CPEs can report alarm information to the SDN controller.

   Link, device, site, topology, and application monitoring: CPEs collect raw data using technologies such as NetStream, and then report the collected raw data to the SDN controller which classifies and displays the data to administrators in different dimensions. This approach allows the traffic or quality of the network topology, sites, devices, links, and applications to be effectively monitored.

   Log management (including CPE logs and SDN controller logs): After a network administrator deploys a CPE log policy on the SDN controller, CPEs report logs to a third-party log server, which can collect, store, and analyze CPE logs, and also generate reports. SDN controller logs are stored and analyzed by the SDN controller itself.

3. Fault locating

   If a network exception is detected using the alarm and monitoring functions, network administrators can use the fault diagnosis function provided by the SDN controller to locate and analyze the fault. The SDN controller then notifies the corresponding CPE of the diagnosis processing information, instructing the CPE to perform fault diagnosis. The SD-WAN solution allows network administrators to diagnose faults using the following modes: link/application quality diagnosis, device diagnosis information collection, and SSH-based remote login. After fault diagnosis and locating, network administrators can replace devices or install patches on the SDN controller.

4. Migration

   To reuse or even expect more of legacy sites in the SD-WAN solution, we must first reconstruct them into new sites equipped with SD-WAN attributes. The migration of legacy sites includes migration of the headquarters, DC, and branch sites.

The SD-WAN solution provides the following five types of interfaces between O&M components:

- **NETCONF:** The SDN controller delivers service configurations to devices through NETCONF interfaces, whereby devices report alarms to the SDN controller through NETCONF notification, and the SDN controller remotely maintains devices through NETCONF RPC messages.
- **Telemetry:** Devices use telemetry technology to periodically report collected performance data to the SDN controller, and to provide high-precision data sources for the SDN controller to monitor device performance and analyze performance trends based on historical data. Devices connect to the SDN controller through HTTP/2, and the SDN controller encodes the reported data through Protocol Buffers.
- **Syslog:** Device logs can be reported to a third-party log server using the Syslog protocol.
- **HTTPS file transfer:** Files, including software packages and patch files, are transferred using HTTPS between devices and the SDN controller.
- **SSH:** Network administrators can log in to devices through SSH provided by the SDN controller and perform remote O&M and diagnosis. Based on SSH port forwarding, this mode allows device login after NAT traversal.

8.3 O&M PERMISSION CONTROL

8.3.1 O&M Mode and Management Roles

The SD-WAN solution provides two O&M modes for different service scenarios: enterprise O&M and carrier/MSP O&M.

The enterprise O&M mode is applicable to scenarios where large enterprises build their own SD-WAN services. In this scenario, enterprises maintain their own networks.

The carrier/MSP O&M mode is applicable to scenarios where an MSP provides SD-WAN management services for small and midsize enterprises. In this scenario, enterprise tenants can maintain networks themselves or authorize the MSP to manage tenant networks. This O&M mode is also used in cloud service scenarios.

In each O&M mode, the SD-WAN solution assigns management permissions to users based on their user levels. The solution supports various administrator roles (system administrator, MSP administrator, and tenant administrator), each of which performs network O&M management through corresponding portals.

Table 8.1 describes the relationships between O&M modes and administrator roles.

Table 8.2 describes the relationships between O&M modes and visualized O&M portals.

As the executor of O&M, a network administrator must ensure that the network provides 24/7 services for customers. On complex and large-scale networks, rights- and domain-based management is required to guarantee stable and reliable network operations. To achieve this, various network administrators with specific management permissions are assigned to networks in different areas or at different layers.

The following describes various types of administrator roles and their management rights.

TABLE 8.1 Relationships between O&M Modes and Administrator Roles

O&M Mode	System Administrator	MSP Administrator	Tenant Administrator
Enterprise O&M mode	Supported	Not supported	Supported
Carrier/MSP O&M mode	Supported	Supported	Supported

TABLE 8.2 Relationships between O&M Modes and Visualized O&M Portals

O&M Mode	System Administrator Portal	MSP Administrator Portal	Tenant Administrator Portal
Enterprise O&M mode	Supported	Not supported	Supported
Carrier/MSP O&M mode	Supported	Supported	Supported

8.3.1.1 System Administrator

A system administrator possesses the highest level of permissions and is responsible for managing and maintaining the SDN controller, including managing administrator accounts at sublevels, such as MSP administrator accounts or tenant administrator accounts.

8.3.1.2 MSP Administrator

An MSP administrator has the permission to manage tenant networks and assist tenants in network management. When a tenant subscribes to the SD-WAN service from the MSP, the MSP administrator manages the tenant administrator account. If a tenant does not have O&M capabilities, they can authorize the MSP to oversee their network O&M, in which case the MSP administrator manages the tenant network.

8.3.1.3 Tenant Administrator

Tenant administrators manage and maintain their own networks, which includes managing their own accounts, devices, end user information, and network services.

Figure 8.2 shows the administrator hierarchy. A system administrator can create multiple MSP administrators or tenant administrators, an MSP

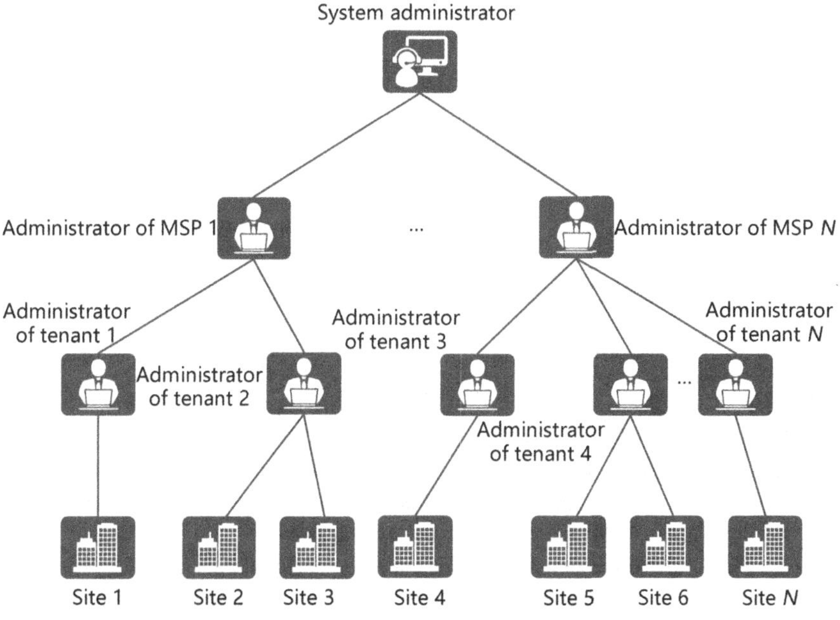

FIGURE 8.2 Administrator hierarchy.

administrator can create multiple tenant administrators, and a tenant administrator manages one or more sites.

8.3.2 Role-Based Permission Control

In the SD-WAN solution, different management permissions can be assigned to system, MSP, and tenant administrators to ensure most fine-grained permission control, facilitating enterprise network O&M. Table 8.3 lists the management permissions of these types of administrators.

In addition, multiple administrator accounts can be configured for each type of administrator, and different roles can be assigned to each administrator account, thereby implementing role-based permission control. Table 8.4 lists the available administrator roles.

The following example provides a better understanding of how to utilize administrator roles:

1. Create a tenant administrator account **admin** for a tenant.
2. Use the **admin** account to create multiple tenant administrator sub-accounts, such as tenant administrator A and tenant administrator B.

TABLE 8.3 Administrator Management Permissions

Administrator Type	Management Permission
System administrator	Management of SDN controllers, devices, alarms, clusters, logs, systems, system accounts, certificates, servers, files, faults, and others
MSP administrator	Management of multi-tenant accounts, multi-tenant devices, multi-tenant networks, MSP accounts, logs, files, email servers, and others
Tenant administrator	Authorizing the MSP to manage tenant networks; management of tenant devices, tenant networks, logs, local users, files, and others

TABLE 8.4 Available Administrator Roles

Role	Permission
Management	Global administrator, possessing all permissions
Monitoring	Global monitoring personnel, possessing all monitoring permissions
Configuration	Network configuration personnel, with the permission to configure the network, traffic policies, and security policies
Maintenance	O&M personnel, with the permission to manage devices, files, and logs
Query	Nonoperation personnel, with the permission to only query each module

3. Use the **admin** account to assign the monitoring role to tenant administrator A and the configuration role to tenant administrator B.

Tenant administrator A and tenant administrator B now possess the corresponding management capabilities on the tenant network.

8.4 NETWORK-WIDE MONITORING CAPABILITIES

Recent years have witnessed a surge in enterprise branches, increasing amounts of complex applications, and the emergence of ever more bandwidth-hungry services. Such rapid evolution presents significant challenges for existing network fault locating techniques and greatly increases O&M costs. The SD-WAN solution empowers network administrators with excellent real-time network visibility and delivers statistical views based on sites, links, and applications, enabling network administrators to quickly locate faults. The solution also enables network administrators to more accurately predict future network usage and to implement proactive O&M based on changes to quality and traffic of links or applications.

8.4.1 Dashboard View

The SD-WAN solution provides multidimensional monitoring functions and displays key indicator data on the GUI, enabling network administrators to efficiently obtain global network information.

The dashboard centrally displays key information relating to operational services, such as alarm information, site health status, and application traffic rankings. The availability of such data allows network administrators to stay informed of the overall service status, and to quickly identify potential problems while also providing strong support for subsequent troubleshooting.

- **Alarm information:** Network administrators are notified of device alarm information, which they use to quickly rectify network faults and ensure proper service operations. As shown in Figure 8.3, the SD-WAN solution supports four types of alarm information. Each color (a specific gray scale in the figure) indicates a different emergency level.
- **Site health:** Network administrators are notified of sites with poor health, encouraging prompt analysis of these sites so that appropriate

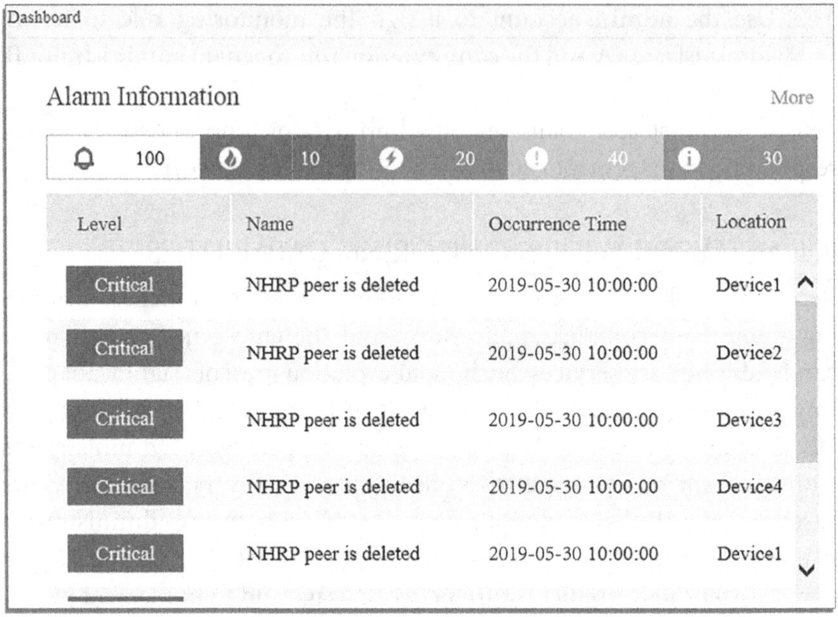

FIGURE 8.3 Alarm information.

measures can be taken. Figure 8.4 shows an example of health status displayed for different sites. A health score with a specific background color (a specific gray scale in the figure) indicates the health status of a site. For example, green (1 in the figure) indicates that the site is in a healthy state, while red (2 in the figure) indicates that the site is in a poor health state.

- **Application traffic ranking:** Top applications with the highest traffic are prominently displayed, enabling network administrators to efficiently adjust policies based on the traffic usage of those applications. As shown in Figure 8.5, the traffic from Skype currently exceeds the limit. In this case, you can configure a traffic policy to increase the bandwidth allocated to Skype.

8.4.2 Network-Wide Monitoring

The SD-WAN solution provides multidimensional monitoring functions, including alarm, overlay network topology, site, and inter-site monitoring. The monitored objects include NEs, sites, inter-site links, and service applications. The comprehensive monitoring information facilitates quick and efficient fault detection and rectification.

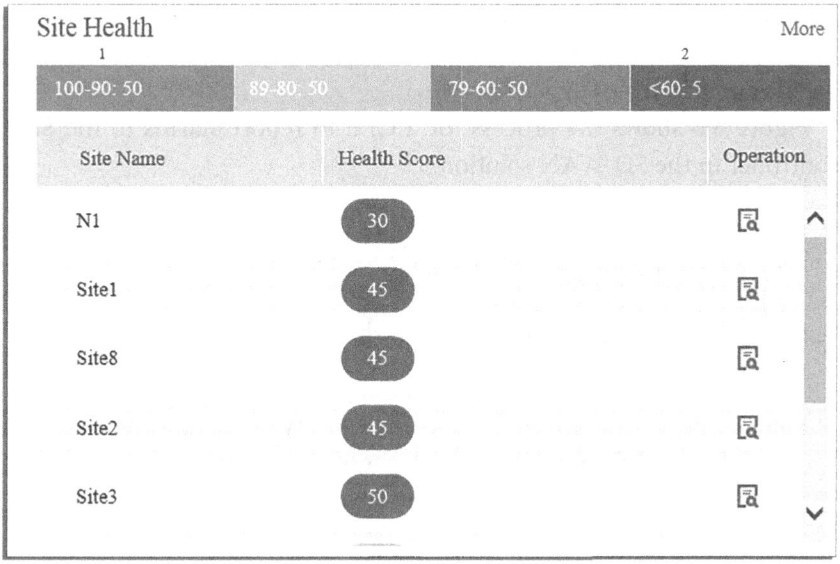

FIGURE 8.4 Site health.

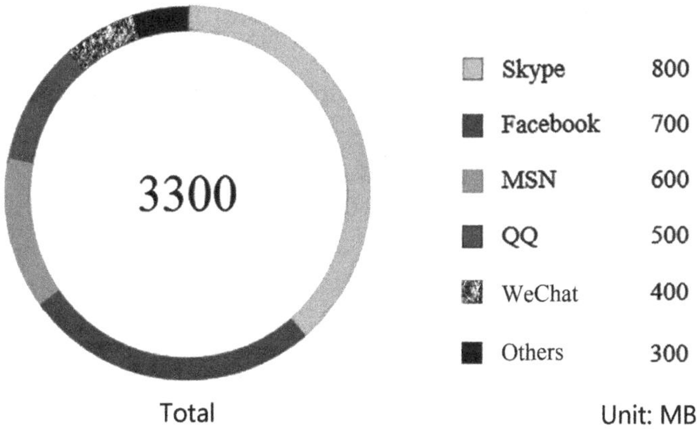

FIGURE 8.5 Application traffic ranking.

8.4.2.1 Alarm Monitoring

Real-time, accurate, and intuitive alarms directly reflect system faults, and the solution provides handling suggestions for alarms to help network administrators effectively locate faults.

When a device or service in the SD-WAN solution becomes faulty or exhibits potential risks, an alarm is generated and reported to the SDN controller, which then displays the alarm information to network administrators.

Network administrators can then use the alarm management function to obtain alarm information in real time, and rectify faults based on alarm details and handling suggestions.

Figure 8.6 shows the process for a CPE to report alarms to the SDN controller in the SD-WAN solution.

Step 1: The SDN controller instructs the CPE to enable the alarm reporting function.

Step 2: The CPE generates an alarm which is then reported to the SDN controller through NETCONF.

Step 3: The SDN controller checks the validity of the reported alarm. If valid, the SDN controller records the alarm in the database and notifies the network administrator by email or SMS.

Based on the received alarm information, the network administrator queries the alarm clearance operation guide released by the solution provider and rectifies the fault. Alarms reported by CPEs include device alarms (such as device restart and user login and logout), network alarms (e.g., the

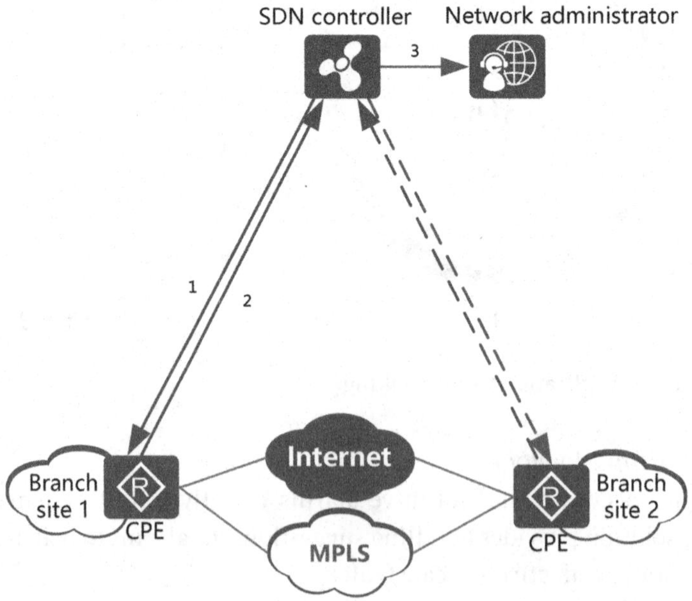

FIGURE 8.6 Alarm reporting process.

link status of an interface goes down or an OSPF interface goes down), and service policy alarms (e.g., when the link status of a voice service changes).

8.4.2.2 Overlay Network Topology Monitoring

Network administrators can view the overlay network status through topology monitoring, allowing them to quickly detect abnormal sites and inter-site links, and to quickly rectify faults and restore services.

Overlay network topology monitoring primarily displays the site status and inter-site link status. The site status can be normal, abnormal, or offline, and only the latter two require troubleshooting. Inter-site link status includes normal (green), poor quality (yellow), and faulty (red), allowing network administrators to filter links by color and focus only on those links that require troubleshooting.

8.4.2.3 Site Monitoring

In traditional network O&M, network administrators intervene only after a fault has already occurred. Even if the network administrator engages in proactive inspection, it is not feasible to cover thousands of branch sites considering the massive workload involved.

The SD-WAN solution leverages the SDN controller to intuitively display the health scores of all sites and sort those with poor health on the GUI, enabling network administrators to perform proactive O&M. Figure 8.7 is an example.

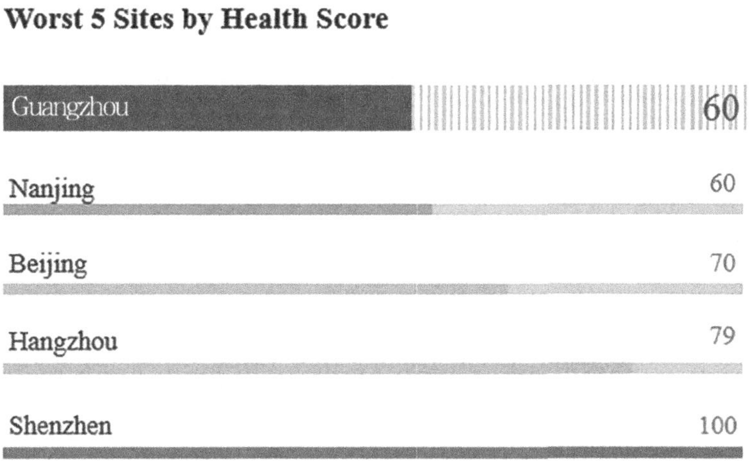

FIGURE 8.7 Worst five sites by health score.

If the health status of a site is poor, network administrators can query the site data, including the link quality, uplink and downlink bandwidth; link status, device ESN, and device model in the site topology; and device hardware running status, in order to quickly locate faults.

Site link quality and uplink and downlink bandwidth: The SDN controller analyzes the quality of the links connected to a site and the uplink and downlink bandwidth to evaluate the health status of the site, as shown in Figure 8.8.

Link status, device ESN, and device model in the site topology: Network administrators can view the location of the network fault node based on the link status in the topology. If a device is faulty, network administrators can view the device details in the site topology view for troubleshooting, as shown in Figure 8.9.

Device hardware running status: After identifying a faulty device on a site, network administrators can view site information to determine

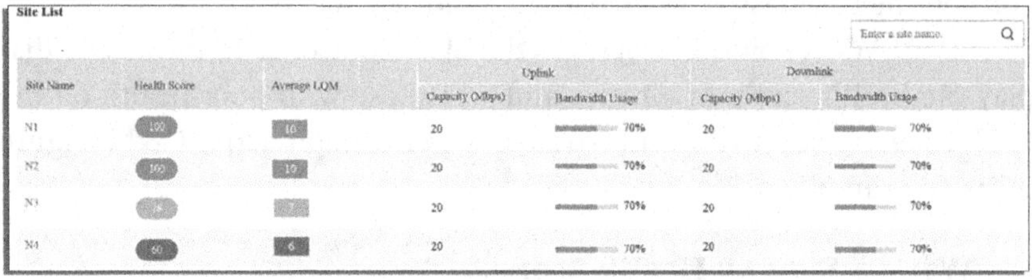
Site List

Enter a site name.

Site Name	Health Score	Average LQM	Uplink		Downlink	
			Capacity (Mbps)	Bandwidth Usage	Capacity (Mbps)	Bandwidth Usage
N1	100	10	20	70%	20	70%
N2	100	10	20	70%	20	70%
N3	[illegible]	7	20	70%	20	70%
N4	60	6	20	70%	20	70%

FIGURE 8.8 Site link quality and uplink and downlink bandwidth.

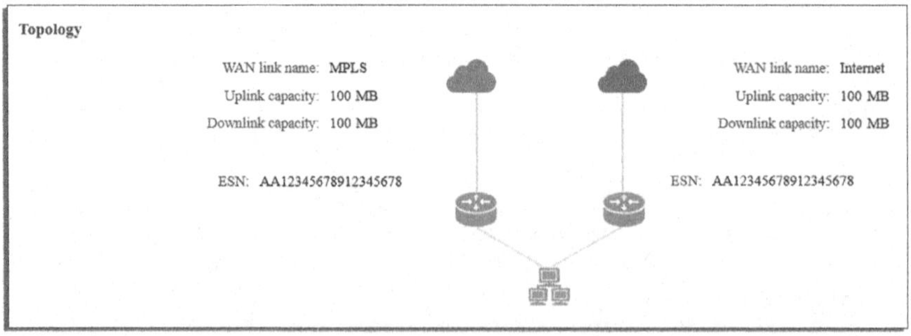

FIGURE 8.9 Link status, device ESN, and device model in the site topology.

the hardware status of the device and further locate the fault, as shown in Figure 8.10.

8.4.2.4 Inter-site Monitoring

Network administrators can view the link quality, communication traffic, and service application quality between sites through inter-site monitoring, thereby determining the working status of inter-site links and services on the GUI.

Worst five inter-site links by quality: Based on the displayed worst five inter-site links on the tenant network (as shown in Figure 8.11), network administrators can quickly and efficiently detect and rectify the area where service quality is most impacted.

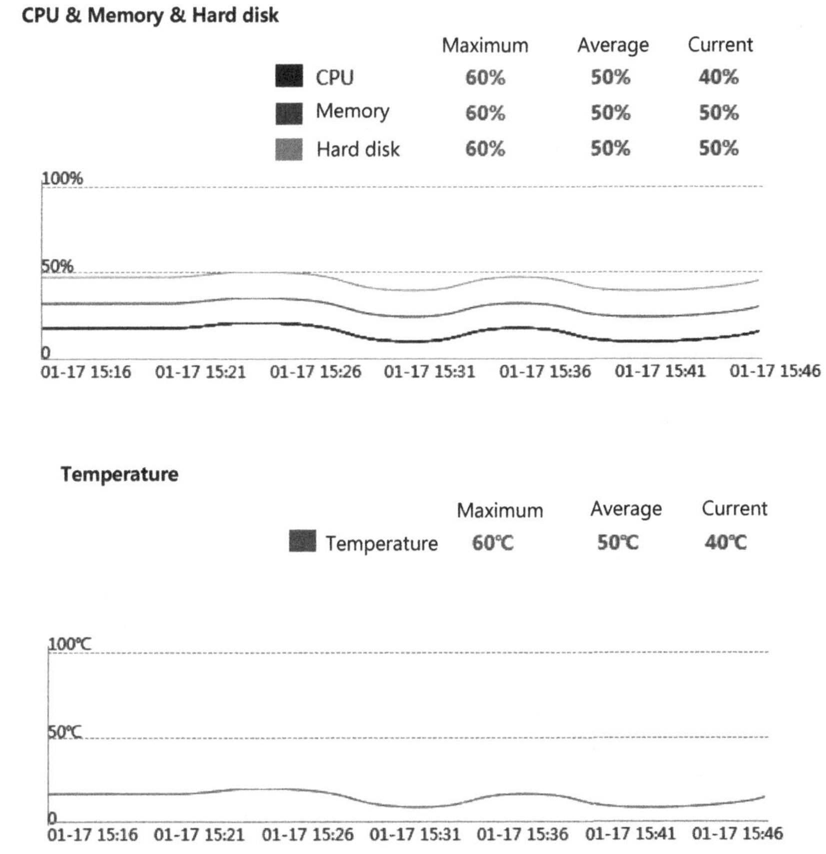

FIGURE 8.10 Site information.

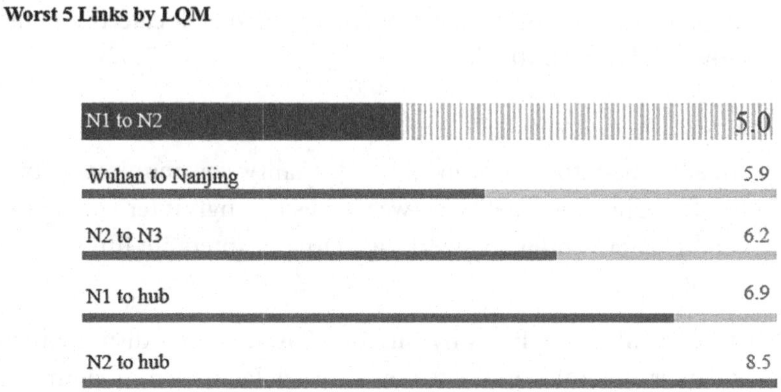

FIGURE 8.11 Worst five inter-site links by quality.

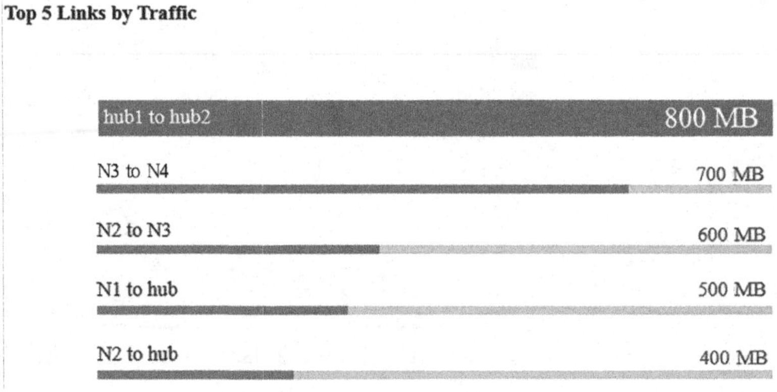

FIGURE 8.12 Top five inter-site links by traffic.

Top five inter-site links by traffic: Based on the displayed status of top five inter-site links ranked by traffic (as shown in Figure 8.12), when a fault such as network congestion occurs, network administrators can quickly locate the impacted area and rectify the fault.

8.4.3 Quality Prediction

With traditional enterprise WANs, network administrators can intervene only after a network fault has already occurred. This type of passive O&M overextends network administrators and, in worst-case scenarios, can even lead to customer complaints. The SD-WAN solution proposes a "trend prediction and proactive O&M" approach, whereby network issues are proactively detected in the early stages, and network administrators are instructed to perform proactive O&M in order to effectively prevent network faults.

Network instability is usually caused by severe delay and packet loss. The SD-WAN solution provides link- and application-based network quality prediction to effectively prevent such problems.

8.4.3.1 Link-Based Network Quality Prediction

By analyzing the trends of link throughput and link quality, the SD-WAN solution can identify poorer quality inter-site links. Traffic can then be switched to another link offering higher quality based on the configured link quality-based traffic steering function. If the bandwidths of the MPLS network and the Internet are unbalanced, the bandwidth-based traffic steering function can be used to divert traffic to the link with higher bandwidth.

- **Link quality trend and throughput trend:** Network administrators can view the link quality and throughput trends across various time ranges in real time through the SDN controller, accurately gauging the network link quality, as shown in Figure 8.13.
- **Link bandwidth trend:** Network administrators can view the bandwidth trend of the transport network (including the MPLS network and the Internet) in real time through the SDN controller, as shown in Figure 8.14.

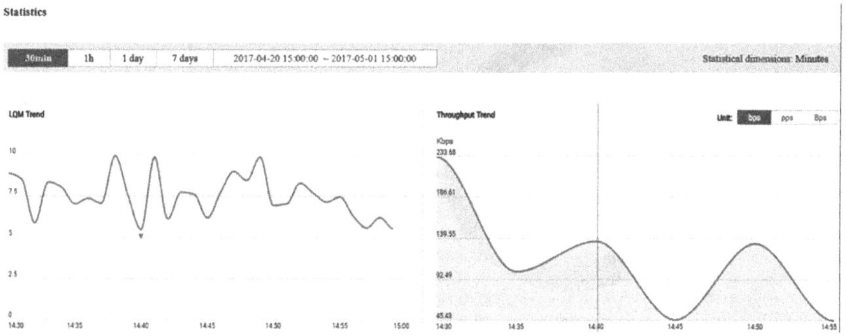

FIGURE 8.13 Link quality trend and throughput trend.

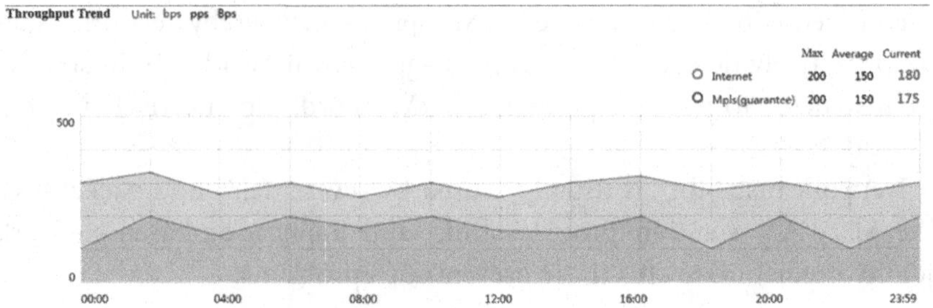

FIGURE 8.14 Link bandwidth trend.

8.4.3.2 Application-Based Network Quality Prediction

Applications are important carriers of critical network services such as video, instant messaging, and gaming. To monitor the quality of applications, network administrators must first identify application types on the network. Accurate identification of applications provides a basis for getting the most out of network services including intelligent traffic steering, QoS, application optimization, and security.

The SD-WAN solution can effectively identify applications and monitor the quality of various types. Network administrators can observe the trends of application quality, application traffic, and application access user quantity and adjust inappropriate application policies in real-time.

- **Application distribution by application quality measurement (AQM) score:** The SDN controller displays the distribution of applications by AQM score, as shown in Figure 8.15, allowing network

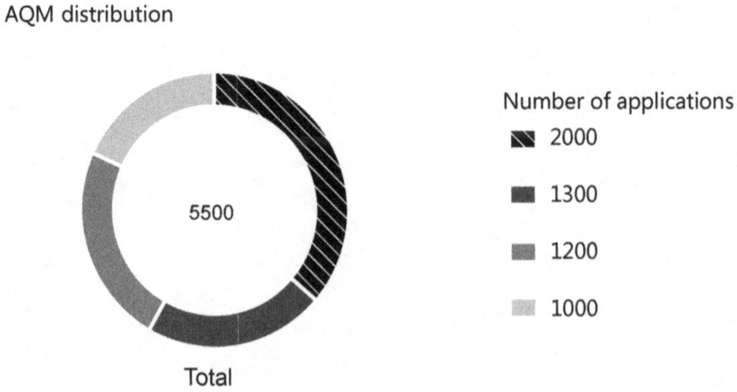

FIGURE 8.15 Application distribution by AQM score.

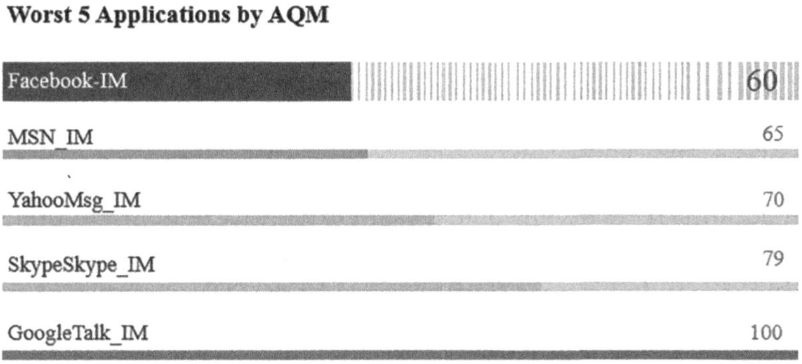

FIGURE 8.16 Worst five applications by AQM score.

administrators to determine the overall quality of applications on the network.

- **Worst five applications by quality (shown in Figure 8.16):** Upon discovering that the quality of a mission-critical application is poor, network administrators can adjust the bandwidth of other applications on the link or adjust the routing policy of the application to improve its quality.
- **Top five applications and traffic trend:** Network administrators can quickly and efficiently adjust QoS policies for applications in cases where traffic usage exceeds expectations. Based on the application trend monitored by the SDN controller, as shown in Figure 8.17, network administrators can predict key applications where future traffic may exceed the preset threshold and adjust QoS policies accordingly.

8.5 FAULT LOCATING ASSISTANT

Fault locating, routine device maintenance, and log management are the three pillars of routine maintenance. Intelligent O&M is emerging as the

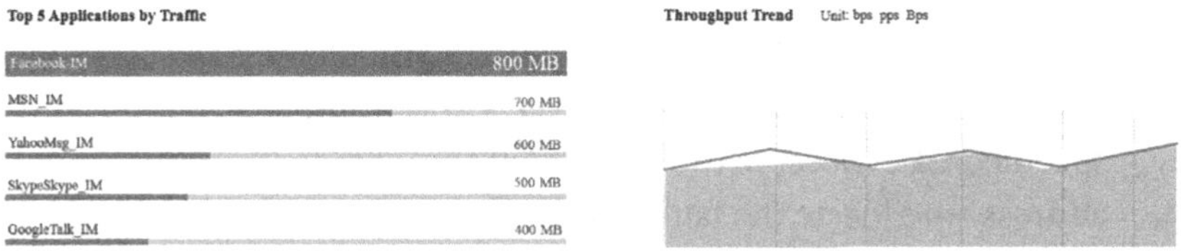

FIGURE 8.17 Top five applications by traffic and traffic trend.

new driving force behind routine maintenance, becoming an efficient and reliable assistant to network administrators and enabling even greater maintenance workloads to be automatically accomplished.

8.5.1 Diagnosis Tools

Fault locating relies heavily on the O&M experience of network administrators. However, the SD-WAN solution allows network administrators to leverage the fault diagnosis tools provided by the SDN controller to automatically locate faults, greatly simplifying O&M.

8.5.1.1 Application Quality Diagnosis

In the SD-WAN solution, CPEs collect raw data using technologies such as NetStream and report the collected data to the SDN controller. Network administrators can use AQM indicators to analyze network vulnerabilities and determine appropriate countermeasures.

8.5.1.2 Network Diagnosis

1. Network connectivity diagnosis

   The ping operation is available for checking the connectivity of both underlay and overlay networks. Here, we use an overlay network as an example to describe how to diagnose network connectivity.

   On the SDN controller, select the source site and the device to be pinged. Configure the SDN controller to deliver a ping command to the specified CPE, and observe the packet loss rate and the average delay to determine whether the overlay network is faulty. As shown in Figure 8.18, the packet loss rate is 0, indicating a proper connection between the overlay network and the destination CPE.

2. Network faulty node diagnosis

   Traceroute is applicable to both underlay and overlay networks, locating the node where a fault occurs. Through the SDN controller, network administrators can deliver a traceroute instruction to the specified CPE, observe the route from this CPE to the destination IP address, and locate the faulty node on the network.

   In addition, the SDN controller can display network paths in a visualized manner based on traceroute, displaying all sites and

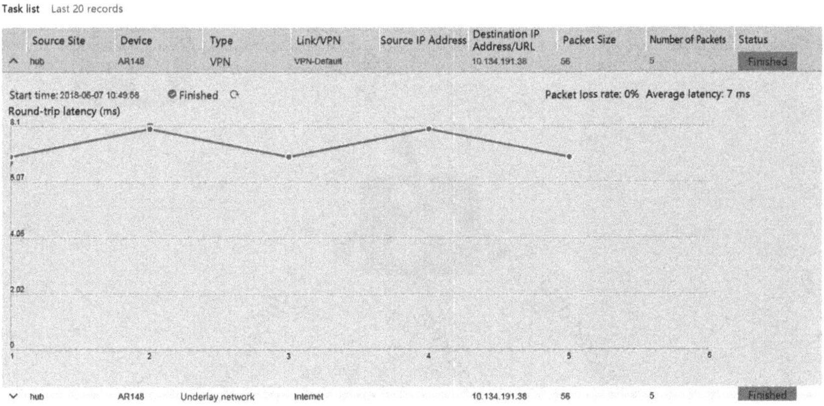

FIGURE 8.18 Overlay network connectivity.

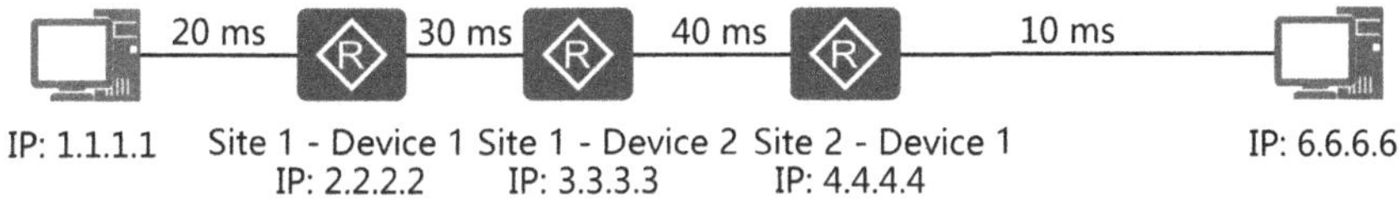

FIGURE 8.19 Path visualization.

devices between a specific device and the destination IP address, as well as information relating to the corresponding sites, devices, and interfaces. As a result, network administrators can more efficiently locate faults, as shown in Figure 8.19.

8.5.1.3 Device Diagnosis

Local and remote diagnosis methods are provided for network administrators to flexibly choose from.

1. Local diagnosis

   A network administrator obtains the device's diagnostic information file through one simple click on the SDN controller that provides the diagnostic information collection function and saves the file to the local PC for fault analysis. Figure 8.20 illustrates the collection process.

   Step 1: The network administrator delivers a diagnostic information collection instruction to a CPE through the SDN controller.

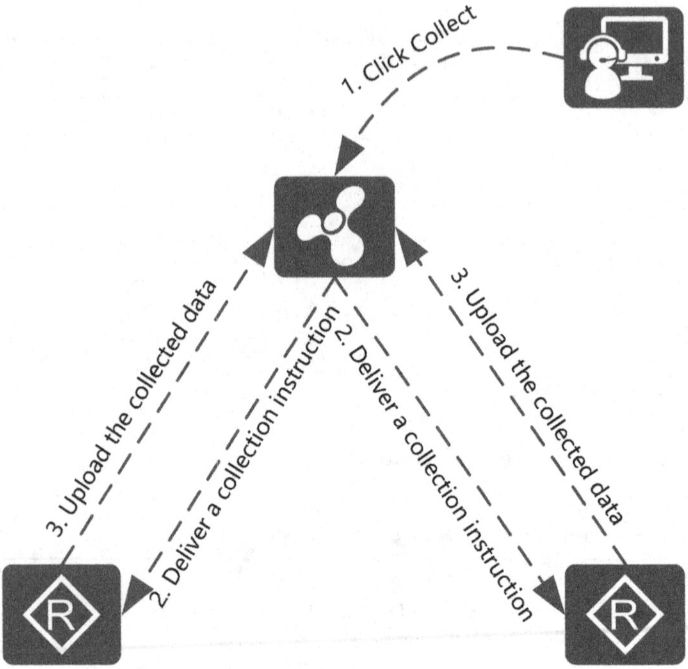

FIGURE 8.20 Collection process.

Step 2: After receiving the instruction, the CPE collects information and compresses the collected data.

Step 3: The CPE uploads the collected data file through HTTPS and deletes the file after the upload is complete.

2. Remote diagnosis

   On the SDN controller, the network administrator logs in to the CPE remotely in SSH mode, and then runs commands on the CPE to diagnose faults.

   Commands that can be executed on a CPE mainly include diagnosis and query commands, which include **display**, **debugging**, and **diagnose**.

8.5.2 Device Maintenance

Device upgrade and replacement are frequent but error-prone routine maintenance tasks. The following describes how to efficiently complete routine maintenance without errors in the SD-WAN solution.

8.5.2.1 Device Upgrade and Downgrade

Upgrade and downgrade operations are almost identical, apart from the relationship between the source and target versions: upgrade when the source version is earlier than the target version; downgrade when the source version is later than the target version. Here, we use the device upgrade as an example to describe the operation process.

1. Determine an upgrade policy.

   As the upgrade affects the current services, it is necessary to first evaluate the impacts and determine an appropriate upgrade policy before performing an upgrade. The SD-WAN solution provides the following configurable parameters for administrators to flexibly plan in advance:

   a. **Upgrade file download time:** immediate or within the specified time

   b. **Device restart time:** immediate or within the specified time

   c. **Upgrade mode:** batch upgrade by device type or device name

2. Upload the upgrade file.

   The SD-WAN solution allows administrators to upload the upgrade file (also called the software package) in either of the following ways:

   a. Through the built-in server of the SDN controller. You can upload the upgrade file to the SDN controller, which then delivers it to the corresponding device.

   b. Through a third-party file server. Before using this method, ensure that the SDN controller can properly communicate with a third-party file server. You can then upload the upgrade file to the SDN controller through the third-party file server, and deliver the upgrade file to the corresponding device through the SDN controller.

3. Upgrade the device and check the upgrade status.

   An upgrade cannot be manually interrupted, and no configurations will be lost following a successful upgrade. If the upgrade fails, check the failure cause, and perform the upgrade again once the fault has been rectified.

8.5.2.2 *Device Replacement*

When a device becomes faulty or obsolete, you must replace it in order to ensure functionality. The SD-WAN solution offers a replacement process far more convenient than traditional methods, requiring just a single site visit from deployment personnel.

1. Add the replacement device on the SDN controller.
2. Synchronize device information (including the site, location, and configuration information) from the legacy device to the new one, and complete device replacement on the SDN controller.
3. Replace the legacy device onsite. Specifically, site deployment personnel transport the new device to the site to complete device deployment and hardware replacement. After the new device registers with the SDN controller, the SDN controller automatically delivers information about the legacy device to the replacement, ensuring consistent services.

8.5.3 Log Management

The SD-WAN solution enables administrators to query and manage logs on the SDN controller or on third-party log servers, greatly facilitating routine maintenance. Figure 8.21 illustrates the process of managing logs through a third-party log server.

Step 1: The SDN controller is interconnected with the log server. To ensure successful communication between the two, the configurations on both sides must be the same, including the IP address,

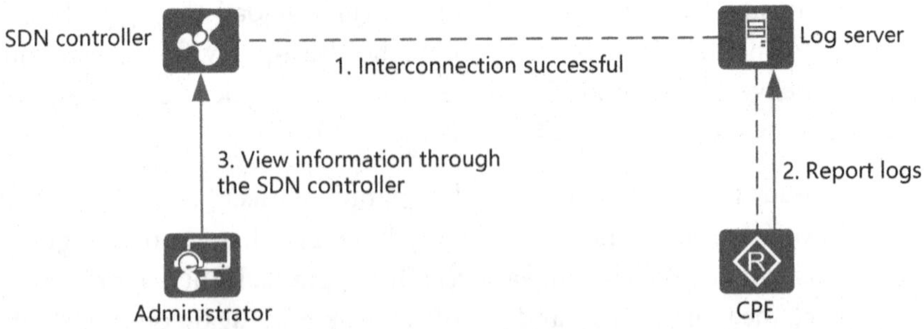

FIGURE 8.21 Log management through a third-party log server.

port number, log storage path, server certificate, and reporting mode of the log server.

Step 2: The CPE reports logs to the log server through TLS or UDP. TLS capable of encryption is recommended.

Step 3: The SDN controller receives and displays the CPE log information from the log server. On the SDN controller, the network administrator may then perform further filtering queries on the device logs, such as querying logs by time, tenant ID, and device ID.

8.5.4 Intelligent Troubleshooting

The ultimate goal of intelligent O&M is automatic fault analysis and recovery, completely relieving network administrators of tedious O&M tasks.

1. **Automatic analysis:** The SD-WAN solution leverages big data analytics to locate faults and quickly analyze root causes from a large amount of source data, providing network administrators with appropriate rectification suggestions.
2. **Automatic recovery:** Network administrators can customize execution policies based on their troubleshooting experience from common issues to achieve automatic network recovery.

The following uses an example to describe the process of proactive detection and identification of audio and video quality, automatic fault locating, and automatic recovery, all of which will occur through the SD-WAN solution in the future.

Video conferences are a staple of modern enterprise communication. However, traditional networks often encounter congestion, which deteriorates the link quality of video services and leads to pixelation or voice delay during video conferences. The SD-WAN solution can effectively resolve this issue. First, the site CPE detects the audio and video streams of the conference terminals and periodically reports the quality data to the SDN controller. The SDN controller then analyzes the reported quality data and identifies the audio and video streams with poor quality. Finally, the SDN controller obtains a complete path of the audio and video streams, and detects the link quality.

In the SD-WAN solution, the following two actions will be taken once a fault is detected on the path of audio and video streams:

- Intelligent traffic steering policies are dynamically adjusted and the traffic steering effect is visualized, facilitating subsequent tracing by administrators.
- The root cause of the fault is analyzed and effective optimization suggestions ensure that the original link can be restored as soon as possible.

8.6 MIGRATION FROM TRADITIONAL NETWORKS TO SD-WAN

Migration from legacy sites to SD-WAN can greatly benefit enterprises.

Figure 8.22 illustrates the process of migrating a legacy site to SD-WAN. After planning SD-WAN services, we must formulate a migration plan, determine the migration procedure, and gradually migrate sites based on the migration plan.

The sites to be migrated are usually classified into two types: headquarters/DC sites and branch sites. Figure 8.23 outlines their respective migration solutions.

For a headquarters/DC site, we choose the gradual migration mode, which is also referred to as hitless upgrade. In this mode, we must connect one or two SD-WAN CPEs (two are recommended to ensure device-level reliability) to legacy CEs in off-path mode, and add (or use existing) Internet links to gradually migrate services to the new SD-WAN CPEs.

For a branch site, if hitless upgrade is used, the operations match those for a headquarters/DC site. If lossy upgrade is used, however, we can

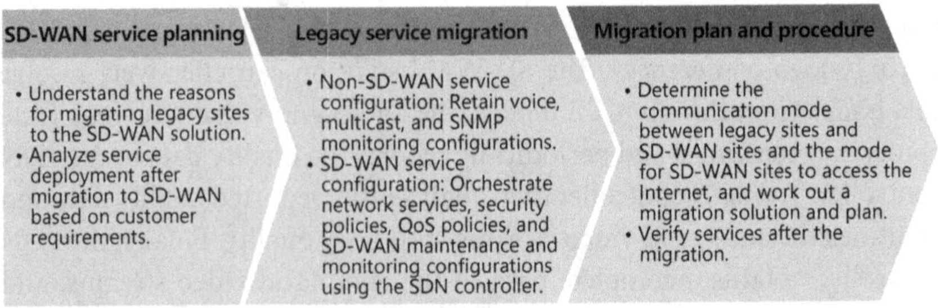

SD-WAN service planning
• Understand the reasons for migrating legacy sites to the SD-WAN solution.
• Analyze service deployment after migration to SD-WAN based on customer requirements.
Legacy service migration
• Non-SD-WAN service configuration: Retain voice, multicast, and SNMP monitoring configurations.
• SD-WAN service configuration: Orchestrate network services, security policies, QoS policies, and SD-WAN maintenance and monitoring configurations using the SDN controller.
Migration plan and procedure
• Determine the communication mode between legacy sites and SD-WAN sites and the mode for SD-WAN sites to access the Internet, and work out a migration solution and plan.
• Verify services after the migration.

FIGURE 8.22 Migration process.

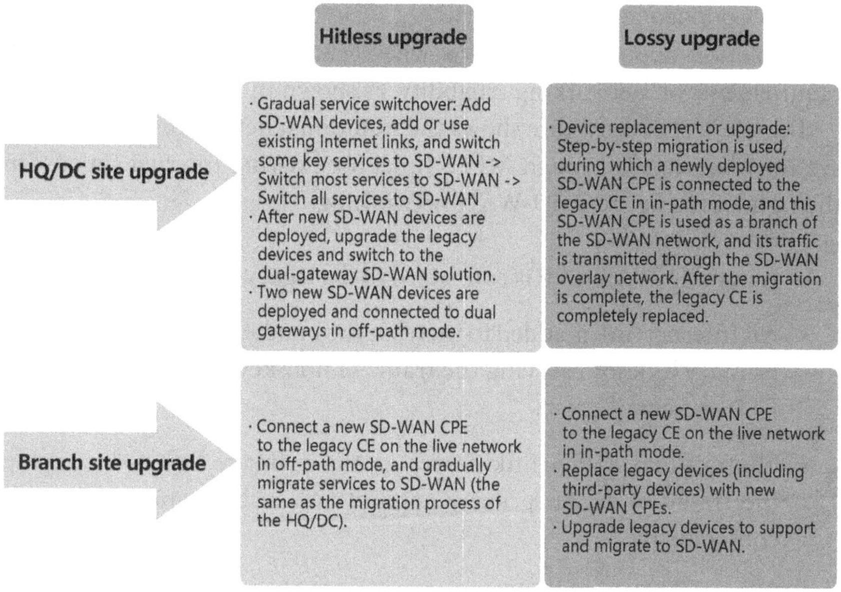

Hitless upgrade
Lossy upgrade
HQ/DC site upgrade
· Gradual service switchover: Add SD-WAN devices, add or use existing Internet links, and switch some key services to SD-WAN -> Switch most services to SD-WAN -> Switch all services to SD-WAN
· After new SD-WAN devices are deployed, upgrade the legacy devices and switch to the dual-gateway SD-WAN solution.
· Two new SD-WAN devices are deployed and connected to dual gateways in off-path mode.
· Device replacement or upgrade: Step-by-step migration is used, during which a newly deployed SD-WAN CPE is connected to the legacy CE in in-path mode, and this SD-WAN CPE is used as a branch of the SD-WAN network, and its traffic is transmitted through the SD-WAN overlay network. After the migration is complete, the legacy CE is completely replaced.
Branch site upgrade
· Connect a new SD-WAN CPE to the legacy CE on the live network in off-path mode, and gradually migrate services to SD-WAN (the same as the migration process of the HQ/DC).
· Connect a new SD-WAN CPE to the legacy CE on the live network in in-path mode.
· Replace legacy devices (including third-party devices) with new SD-WAN CPEs.
· Upgrade legacy devices to support and migrate to SD-WAN.

FIGURE 8.23 Migration solution.

directly replace a legacy CPE with a new SD-WAN CPE, or upgrade the legacy CPE to a version that supports SD-WAN.

8.6.1 Headquarters/DC Site Migration

Table 8.5 describes two migration solutions for a headquarters/DC site.

TABLE 8.5 Migration Solutions for a Headquarters/DC Site

Scenario	Networking before the Migration	Networking after the Migration	Deployment Description
Connecting a single SD-WAN CPE in off-path mode	Legacy CEs in the headquarters/DC site are connected to other branches through MPLS (or MPLS+Internet) links	An SD-WAN CPE is added and connected to legacy CEs in off-path mode, and an Internet link is added to expand bandwidth	The reliability of the headquarters or DC site is low. The new SD-WAN CPE carries SD-WAN service traffic
Connecting two SD-WAN CPEs in off-path mode	Legacy CEs in the headquarters/DC site are connected to other branches through MPLS (or MPLS+Internet) links	Two SD-WAN CPEs are added and connected to legacy CEs in off-path mode, and Internet links are added to expand bandwidth	The reliability of the headquarters or DC site is high. The new SD-WAN CPEs carry SD-WAN service traffic

8.6.1.1 Connecting a Single SD-WAN CPE in Off-Path Mode

This solution applies to scenarios where enterprises do not have high requirements on networking reliability, as shown in Figure 8.24.

The left part of the figure shows the traditional networking, where sites are interconnected through MPLS links (or existing Internet links). On the right of the figure is SD-WAN networking.

- A new SD-WAN CPE is connected to the legacy CE in off-path mode.
- An Internet link is added to increase the bandwidth and serves as the primary link for carrying the traffic of non-key applications such as FTP and email.
- The original MPLS link serves as the primary link for carrying the traffic of higher-priority applications such as voice and video conferencing.
- The new SD-WAN CPE carries SD-WAN service traffic, and the legacy CE carries service traffic of the legacy site (which can be used as a centralized gateway for other SD-WAN sites to access the legacy site).

8.6.1.2 Connecting Two SD-WAN CPEs in Off-Path Mode

This highly reliable solution is recommended for most enterprise headquarters or DC sites, as shown in Figure 8.25.

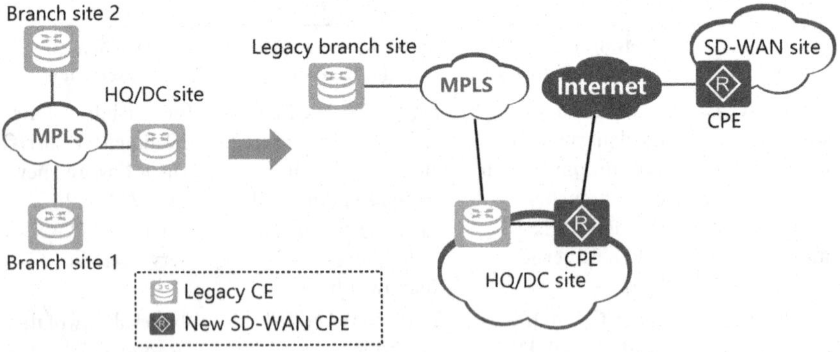

FIGURE 8.24 Connecting a single SD-WAN CPE in off-path mode for migrating a headquarters/DC site to SD-WAN.

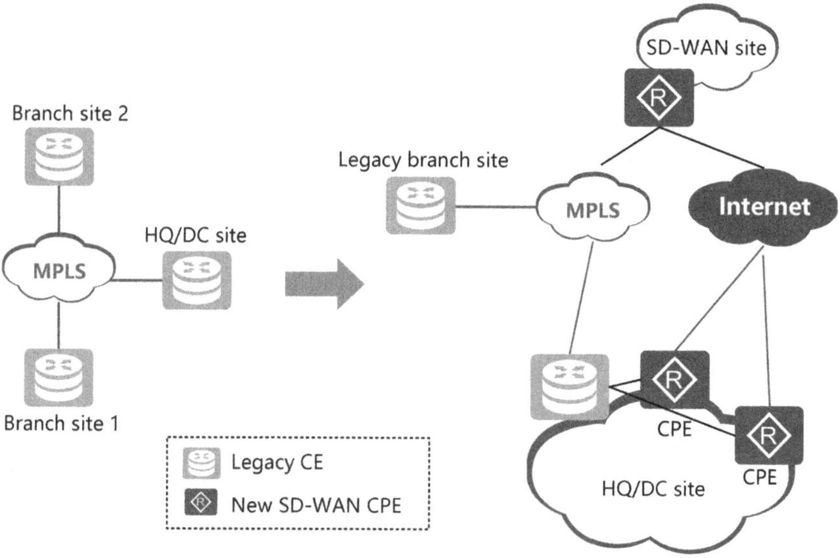

FIGURE 8.25 Connecting two SD-WAN CPEs in off-path mode for migrating a headquarters/DC site to SD-WAN.

The left part of the figure shows the traditional networking, where sites are interconnected through MPLS links (or existing Internet links). On the right of the figure is SD-WAN networking.

- Two new SD-WAN CPEs are connected to the legacy CE in off-path mode.
- An Internet link is added for each SD-WAN CPE to increase the bandwidth and serves as the primary link for carrying the traffic of non-key applications such as FTP and email.
- The original MPLS link serves as the primary link for carrying the traffic of higher-priority applications such as voice and video conferencing.
- The new SD-WAN CPEs carry SD-WAN service traffic, and the legacy CE carries service traffic of the legacy site (which can be used as a centralized gateway for other SD-WAN sites to access the legacy site).

8.6.2 Branch Site Migration

Table 8.6 describes two migration solutions for branch sites.

8.6.2.1 Directly Replacing or Upgrading a Legacy CE

This solution applies to scenarios where enterprises do not have high requirements on reliability, or where enterprises have a small number of branch sites, as shown in Figure 8.26.

The left part of the figure shows the traditional networking, where branch sites are connected to the network through MPLS links (or existing Internet links). On the right is SD-WAN networking.

TABLE 8.6 Migration Solution for Branch Sites

Scenario	Networking before the Migration	Networking after the Migration	Deployment Description
Directly replacing or upgrading a legacy CE	Legacy CEs at branch sites are connected to the network through MPLS (or MPLS+Internet or only Internet) links	A new SD-WAN CPE replaces a legacy CE or the software version of a legacy CE is upgraded, and Internet links are added to expand the bandwidth	Lossy upgrade
Connecting an SD-WAN CPE in off-path mode	Legacy CEs at branch sites are connected to the network through MPLS (or MPLS+Internet or only Internet) links	New SD-WAN CPEs are connected to legacy CEs in off-path mode, and Internet links are added to expand the bandwidth	Hitless upgrade

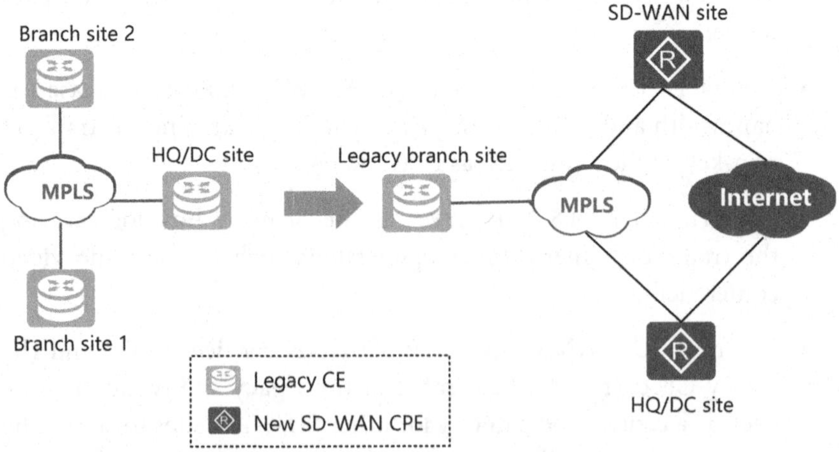

FIGURE 8.26 Device replacement/upgrade for migrating branch sites.

- An Internet link is added for each SD-WAN CPE to increase the bandwidth and serves as the primary link for carrying traffic of non-key applications such as FTP and email.
- The original MPLS link serves as the primary link for carrying the traffic of higher-priority applications such as voice and video conferencing.
- Branch SD-WAN sites and legacy sites can communicate with each other directly or through the headquarters or DC site.
- Through intelligent traffic steering, available paths can be dynamically selected to transmit the traffic of a specified application.

8.6.2.2 Connecting SD-WAN CPEs in Off-Path Mode

An SD-WAN CPE is added to a branch site and connected to the legacy CE in off-path mode, and some services are migrated to SD-WAN. Enterprises typically have high reliability requirements and require high-priority services to be ensured during SD-WAN migration. In this case, some low-priority services can be switched to SD-WAN first. Figure 8.27 gives an example.

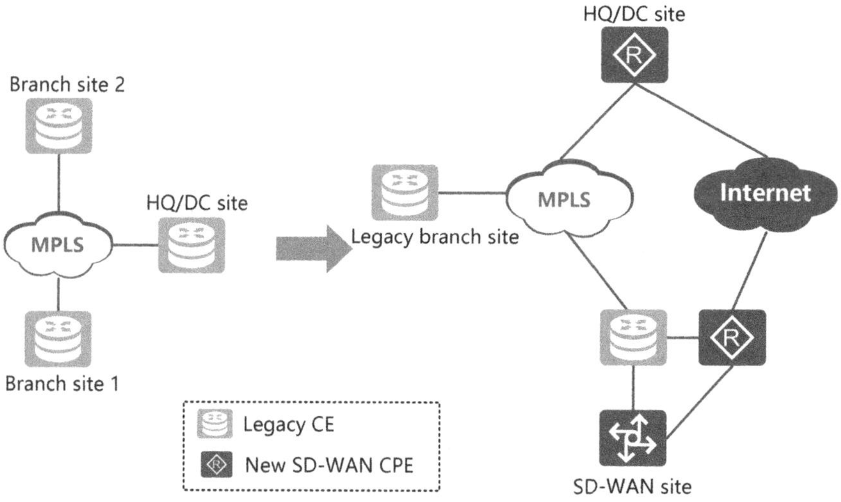

FIGURE 8.27 Connecting SD-WAN CPEs in off-path mode for migrating a branch site to SD-WAN.

The left part of the figure shows the traditional networking, where sites are interconnected through MPLS links (or existing Internet links). On the right of the figure is SD-WAN networking.

- An Internet link is added to increase the bandwidth and serves as the primary link for carrying the traffic of non-key applications such as FTP and email.
- MPLS links carry the traffic of key services such as voice and video services.
- Branch sites use the new SD-WAN CPEs to carry SD-WAN traffic.
- The legacy CPE carries key service traffic (through the underlay network, as before).
- Branch SD-WAN sites and legacy sites can communicate with each other directly or through the headquarters or DC site.
- The new SD-WAN CPE dynamically selects available paths to transmit the traffic of a specified application using intelligent traffic steering.

CHAPTER 9

SD-WAN Best Practices

PRACTICE MAKES PERFECT. SD-WAN has been tested, tried, and proven across industries. In this chapter, we take a close look at SD-WAN practices across several typical industries. To elaborate, we analyze the service characteristics and core requirements on SD-WAN in each industry, and then explain detailed SD-WAN design and deployment solutions, so that you can better understand the principles of SD-WAN solutions and their field-proven value across a wide range of industries.

9.1 SD-WAN APPLICATION SCENARIOS

Currently, there are two major SD-WAN business models: enterprise-built and carrier/MSP-resale. Both business models provide a range of application scenarios, as illustrated in Figure 9.1.

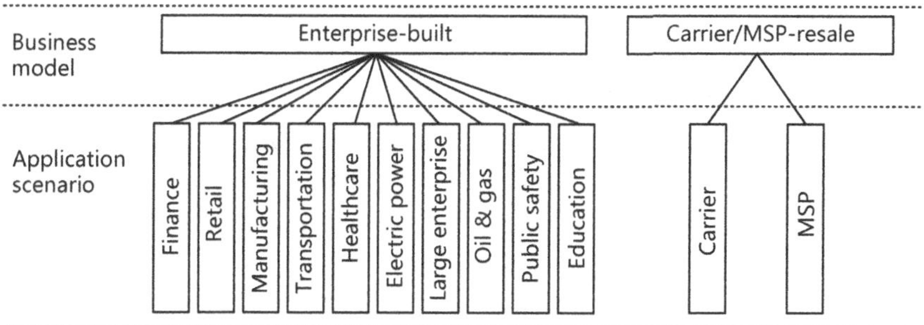

FIGURE 9.1 SD-WAN application scenarios.

SD-WAN solution providers should provide differentiated SD-WAN solutions tailored to specific business models and customer characteristics.

1. Business model 1: enterprise-built SD-WAN

   In this business model, SD-WAN solution providers directly sell SD-WAN solutions to end customers. Typical customers come from industries such as finance, retail, manufacturing, transportation, healthcare, electric power, oil and gas, public safety, and education.

   Customers vary greatly across industries, but their core requirements for SD-WAN are largely the same. Generally speaking, they all require a unified visualized O&M mode to improve O&M efficiency. They also look for hybrid WANs to reduce private line costs. Intelligent traffic steering is another priority, as it is needed to improve service experience. In addition to these, customers pursue other key features, including ZTP, local Internet access, multi-VPN, WAN optimization, security, connection to public clouds, and access to SaaS applications.

2. Business model 2: carrier/MSP-resale SD-WAN

   In this business model, SD-WAN solution providers sell SD-WAN solutions to carriers and MSPs who then resell SD-WAN solutions as a new type of service to end customers. This scenario adds to the requirements in the enterprise-built scenario with features suitable for business operations, including multi-tenant management, multi-tenant IWG, cloud gateway, POP networking, and open northbound APIs provided by the SDN controller.

9.2 TYPICAL SD-WAN CASES

The following sections describe the enterprise-built and carrier/MSP-resale business models in detail. To facilitate understanding of the enterprise-built scenario, we have selected two typical industries (namely, finance and large enterprise) as examples.

9.2.1 SD-WAN in Finance

The financial services industry is an important pillar of the economy. It includes subsectors such as banking, insurance, and securities.

With the development of social media and mobile technologies, Internet financial services are developing rapidly. Driven by this, the conventional financial business models are undergoing changes, including the following:

1. Service transformation at financial branches

   While conventional transaction services are still in place, new financial services are also growing rapidly and becoming ubiquitous. One example is the introduction of intelligent, self-service remote video devices to financial services for publicity and training. This improves customer experience while reducing the cost of manpower. Another example lies in the rise of mobile branches, breaking down geographical constraints for widespread coverage of financial services.

   Due to these changes, conventional transaction services generate a relatively small amount of data traffic, but new non-transaction services such as video, voice, and social services produce ever-larger quantities of data.

2. Increasingly flattened services and networks

   The financial services industry prefers to use a hierarchical network structure. Typically, service traffic from the lowest levels of branches must be first aggregated at intermediate branches before it is sent to the headquarters. This increases latency and affects the SLA.

   Performance bottlenecks on the aggregation devices at the upper-level intermediate branches affect stability of services at lower-level intermediate branches. To prevent such bottlenecks, we tend to expand the capacity of aggregation devices at the upper-level intermediate branches; however, this results in high capacity expansion and IT construction costs.

   Production services are flattened. Meanwhile, intermediate branches are also flattened into the lowest level of branches. This flattened management avoids bottlenecks in traffic aggregation, improves operational efficiency and outcomes, and reduces management costs.

3. Emergence of 5G smart branches

   With the advent of the 5G era, the financial services industry is exploring and fostering new service models. This is where smart unstaffed financial branches come in.

   By delivering higher bandwidth and lower latency, 5G makes it possible for video-assisted contact center agents to handle service requests remotely. It also realizes real-time HD video surveillance backhaul and security response, as well as IoT use, which is capable of connecting nearly everything. IoT endpoints (such as door, curtain, lighting, and fresh air system devices) at smart financial branches can be controlled automatically and intelligently for truly unmanned operations.

4. Service cloudification

   The rise of Internet finance accelerates the progress of financial technology (FinTech). FinTech advances are critical to enterprise transactions in scenario-specific financial services. Driven by the fast-growing Internet finance and scenario-specific financial services, the entire industry is stepping toward the cloud. Cloud will undoubtedly become a trend in financial branches at all levels, and third parties will be given convenient access to the financial services cloud.

   To meet these changes, a well-designed SD-WAN solution is a must. Suppose that Bank J is a nationwide commercial bank. We will use this case study to illustrate the SD-WAN scenario analysis process and solution design.

9.2.1.1 Scenario Analysis

Bank J has branches all over the country. The rapid development of banking services strains the bank's legacy WAN network architecture with a series of new challenges, for example:

1. Increased operational costs because of expensive private lines

   This bank deploys all applications in its headquarters/DC and connects its branches to the headquarters/DC through private lines. To improve the access reliability for large branches, two different private line networks are leased from carriers.

   With the rise of ubiquitous financial services, this bank requires larger WAN link bandwidth to ensure its branch services are stable, especially when bandwidth-hungry services such as facial recognition and video conferencing are introduced. As a result, the operational costs of private line services keep increasing.

2. Impossible to quickly roll out branches due to time-consuming private line service provisioning and poor service flexibility

   Conventional private line networks are not easy to obtain, as optical fibers and cables need to be deployed independently, which is time-consuming. When private lines traverse multiple networks or carriers, service provisioning is prolonged and as a result, services cannot be flexibly subscribed to.

In the Internet era, Bank J plans to quickly open temporary branches to handle banking services in order to attract more customers. Legacy private line approaches, however, cannot make this happen.

3. Limited network device functions and poor service scalability

   A large number of network devices make heavy use of inflexible hardware, which must be replaced if new functionality needs to be added. This new hardware can take a long time to deploy.

4. Unable to adapt to the development of cloud services

   IT infrastructure and software in the financial services industry have started to migrate to the cloud, requiring software-based network functions and automatic network adjustment according to service changes. This bank's legacy network, however, is heavily reliant on its hardware devices and cannot adapt to the development of cloud services.

5. Low network management and O&M efficiency

   Bank J has hundreds of branches across the country but lacks a unified and centralized O&M system to manage these branches. As a result, branches must be staffed with many network O&M professionals.

   With the pervasive use of cloud computing and mobile Internet, Bank J sees dramatic changes to internal applications and data traffic distribution. Problems on legacy networks become more unpredictable and complex than ever. However, conventional O&M and provisioning practices, which rely heavily on the CLI, cannot efficiently and quickly handle these problems.

9.2.1.2 Solution Design

Bank J decides to introduce an SD-WAN solution. According to the bank, SD-WAN should be open, intelligent, and multilayered, and meet the following expectations and goals:

- The networks for branches within the same province are flattened, improving service experience and network reliability.

- Internet links replace MSTP private lines between branches and the DC at the headquarters, reducing the operational costs of WAN lines and sites.
- Numerous branches are centrally managed using the SDN controller, achieving visualized O&M, policy delivery, and fault diagnosis and therefore simplifying O&M.

Figure 9.2 illustrates the SD-WAN architecture for Bank J.

1. Overall planning

   Bank J plans to deploy an SDN controller in the DC at the headquarters. Two geographically-redundant DCs are deployed for

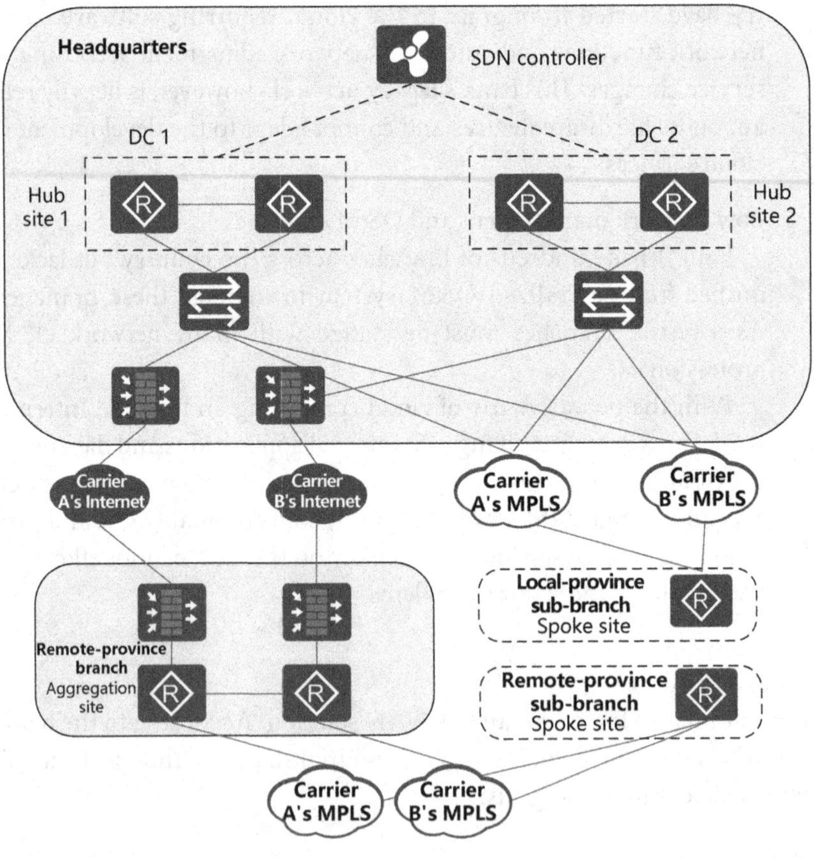

FIGURE 9.2 Bank J's SD-WAN architecture.

reliability purposes. The bank also hopes to reconstruct the networks for the two DCs (hub sites), branches (aggregation sites), and subbranches (spoke sites).

a. The SDN controller in the DC centrally manages all SD-WAN sites and orchestrates services accordingly. It provides two southbound IP addresses that correspond to MPLS and Internet links, respectively, so that SD-WAN sites can register with the SDN controller through MPLS or Internet links during deployment.

b. The legacy networks for subbranches in the same province as the DC are flattened. Hub sites are deployed at the active and standby DCs at the headquarters, with two CPEs deployed at each site. Spoke sites are deployed at branches within the same province, with one or two CPEs deployed at each site.

c. Branches and subbranches in remote provinces use hierarchical networking. Branches introduce Internet links to replace inter-province private lines. Aggregation sites are deployed at branches, with two CPEs deployed at each site. Spoke sites are deployed at subbranches, with one or two CPEs deployed at each site.

d. Two RDs and corresponding transport networks are planned for the overlay network, which match MPLS VPN and Internet links on the underlay network.

e. The deployment sequence is as follows: SDN controller, hub sites, aggregation sites, and spoke sites.

2. Site design

Local-province sites

Hub sites are deployed in the active and standby DCs at the headquarters. At each hub site, two gateways (namely, CPEs) are deployed and connected to legacy routers in off-path mode.

LAN- and WAN-side routes are designed as follows:

a. LAN-side interfaces establish OSPF neighbor relationships with the core switches at the headquarters.

b. Two WAN-side interfaces connect to the routers on the live network at the headquarters. On the underlay network, static routes are configured and destined for the routers on the live network at the headquarters. On the overlay network, BGP peer relationships are established with devices at branches and subbranches through tunnels. Each CPE at the hub site has two WAN links, one of which connects to the MPLS VPN and the other connects to the Internet. That is, each hub site with two CPEs would have a total of four WAN links (two MPLS links and two Internet links).

Subbranches in the same province as the DC are deployed with spoke sites. Each spoke site uses one or two gateways.

LAN- and WAN-side routes are designed as follows:

a. LAN-side interfaces establish OSPF neighbor relationships with the switches at subbranches.

b. WAN-side interfaces connect to the routers on the subbranch's live network. On the underlay network, static routes are configured and destined for the routers on the subbranch's live network. On the overlay network, BGP peer relationships are established with devices at the headquarters through tunnels. Each subbranch site has two WAN links to connect to MPLS VPNs from different carriers.

Remote-province sites

Aggregation sites are deployed at branches in remote provinces (which are outside the province where the DC is located). Each aggregation site uses two gateways.

LAN- and WAN-side routes are designed as follows:

a. LAN-side interfaces establish OSPF neighbor relationships with the routers at the branch's live network.

b. Two interfaces are configured on the WAN side, one of which connects to the branch's egress firewall and the other connects to the router on the branch's live network. On the underlay network, static routes are configured and destined for the egress firewall. On the overlay network, BGP peer relationships are established with devices at the hub sites of the headquarters through tunnels. Each CPE at a branch site has two WAN links, one of which connects to the MPLS

VPN and the other connects to the Internet. That is, each aggregation site with two CPEs would have a total of four WAN links (two MPLS links and two Internet links).

Subbranches in remote provinces are deployed with spoke sites. Each spoke site uses one or two gateways.

LAN- and WAN-side routes are designed as follows:

a. LAN-side interfaces establish OSPF neighbor relationships with the switches at sub-branches.

b. Two interfaces are configured on the WAN side, both of which connect to the routers on the subbranch's live network. On the underlay network, static routes are configured and destined for the routers on the subbranch's live network. On the overlay network, BGP peer relationships are established with devices at the branches through subbranches. Each sub-branch site has two WAN links to connect to MPLS VPNs of branches.

3. Networking design

Control plane

A hub site at the headquarters operates as an RR site, meaning that the RR site and hub site are combined. Other sites in the local province establish control channels with the RR site (hub site at the headquarters).

Aggregation sites at branches in remote provinces also operate as RR sites. They establish control channels with subbranches in the same region. In addition, they set up full-mesh control channels with both hub sites at the headquarters and the aggregation sites at branches in other regions. BGP is used to exchange overlay routing information.

Data plane

A hub-spoke topology is used in the province where the headquarters/DC is located.

a. A flattened architecture is generally used, where hub sites connect directly to spoke sites, without aggregation sites in between. All spoke sites in the local province use hub sites as regional border sites. In this hub-spoke topology, hub sites can directly communicate with spoke sites, and spokes sites communicate with each other indirectly through hub sites.

b. Overlay tunnels are established between subbranches in the local province and hub sites by using MPLS VPNs.

A hierarchical topology is used in remote provinces (which are outside the province where the headquarters/DC is located).

c. A three-layer architecture — which consists of the headquarters (hub sites), branches (border sites), and subbranches (spoke sites) — is used. Aggregation sites must be included in this architecture. Aggregation sites at branches in remote provinces are used as regional border sites. Each border site forwards east-west and north-south traffic for the region. Spoke sites in a region are connected to hub sites outside the region through border sites. Subbranches in a region communicate with each other indirectly through border sites. Hub sites are used as transit sites for inter-region communication: Traffic is forwarded from border sites in a region, then to hub sites at the headquarters, and finally to other sites.

d. Overlay tunnels are established between aggregation sites and hub sites through Internet links, and between aggregation sites and spoke sites by using MPLS VPN links.

4. Service design

Inter-site communication

Overlay tunnels between SD-WAN sites are mainly used to carry office and production services for Bank J.

a. Application policies are customized, and FPI technology is used to identify applications based on the source/destination IP address. In this way, production and office applications are easy to identify, and it is also effortless to monitor and collect statistics on application quality.

b. WAN-side egresses at branches and subbranches use dual WAN links. The WAN link SLA is detected in real time, and intelligent traffic steering is used to divert production and office services to different WAN links for optimal outcomes.

The two WAN links work in active/standby mode and allow fast link switchover based on the packet loss rate, delay, and jitter thresholds.

c. Application identification technologies and application-based QoS policies are used to ensure voice and video conference experience.

Internet access

Bank J has stringent security requirements on Internet access. SD-WAN ensures service security by implementing centralized Internet access as follows:

a. The hub site at the headquarters implements centralized Internet access for branches in the same province. Traffic from such branches is aggregated at the hub site and is then routed out to the Internet through a LAN-side interface at the hub site.

b. Branches in a remote province implement centralized Internet access for subbranches in the same province. Traffic from such subbranches is aggregated at branches through overlay tunnels and is then routed out to the Internet through a LAN-side interface at branches.

Interworking with legacy MPLS networks

Bank J's SD-WAN sites should communicate with legacy sites. As such, Bank J adopts a centralized access approach, where the hub site at the headquarters is used as the centralized egress for traffic. Specifically, Bank J enables the centralized access function and selects the hub site as the centralized access site on the SDN controller.

5. Security design

Bank J has stringent requirements on data transmission security and adopts a series of security measures, including the following:

a. **Inter-firewall tunnels:** IPsec tunnels are established between interprovince firewalls to ensure that private network traffic can traverse the public network, without compromising secure data transmission.

b. **SD-WAN overlay tunnels:** All overlay tunnels in SD-WAN use GRE over IPsec to ensure secure data transmission. These tunnels include MPLS and Internet links between branches in the local province and the headquarters, between subbranches and branches in remote provinces, and between branches in remote provinces and the headquarters.

6. O&M design

Bank J uses an SDN controller — a centralized management, control, and O&M platform — to implement visibility, manageability, and maintainability. Using the SDN controller, Bank J's O&M staff can:

a. View network and service status through a global dashboard, receive anomaly alerts in real time, and monitor the topology, intra-site communication, inter-site communication, and application status.

b. Configure sites individually, deliver service policies in batches, and flexibly define multiple O&M roles.

c. Locate faults with visualized tools, log in remotely through SSH, and upgrade the systems/patches for devices in batches.

Primary functions delivered by the SDN controller include:

a. **Automated device deployment:** Site underlay network parameters can be configured using the SDN controller, whereby the network administrator specifies the URL in the deployment email, configures relevant deployment information, and sends the deployment email to a deployment professional's mailbox. After receiving the deployment email, the deployment professional uses a browser to access the URL in the email and initiates the deployment process. The device is then automatically deployed.

b. **Easy device configuration delivery:** Through the SDN controller, the network administrator can configure the VLANs, Layer 3 interfaces, application identification policies, QoS policies, ACLs, and traffic steering policies for site devices. Subsequently, once devices are successfully registered and go online, the SDN controller automatically delivers the previous configurations to these devices.

c. **Simple device replacement:** When a device can no longer be used due to a hardware fault, maintenance personnel can easily replace it with a new device and perform email-based deployment. After the new device goes online, it can still use the service configurations of the original device.

9.2.2 SD-WAN in Large Enterprises

Digital transformation is sweeping across enterprises of all sizes. Driven by this evolution, increasing numbers of enterprise applications, such as office services and local data storage applications, are migrating to the cloud. Consequently, cloud computing has become the underlying technology and foundation for many other emerging technologies. For example, the development and pervasive use of such promising technologies as VR, AI, self-driving vehicles, and blockchain are all closely related to cloud computing.

This ongoing cloudification trend leads to surges in enterprise egress traffic. According to statistics from Gartner, 30%–50% of large enterprises' traffic is shifting to the cloud. In addition, IDC predicts that by 2030, 80% of new enterprise applications will be deployed in the cloud and enterprise WAN egress traffic will double every three years.

Amid these dramatic changes, large enterprises face a number of challenges including expensive traditional private lines, high interconnection costs, inefficient network deployment, time-consuming service provisioning, weak application identification capabilities, and difficult service management. Effective new ICT is essential to transform, innovate, and grow businesses in today's digital transformation era.

Suppose that Company H is a large ICT company. We will use this case study to illustrate the SD-WAN scenario analysis process and solution design.

9.2.2.1 Scenario Analysis

Company H faces numerous WAN evolution challenges, including the following:

1. Inefficient O&M of many sites scattered across wide areas

   On Company H's network, capital expenditure (CAPEX), OPEX, maintenance costs, and IT support costs increase in tandem with the number of branches. Routing information is complex, and diverse service types coexist on the legacy network. As such, any network or service change requires site-by-site configuration.

2. Inability to properly utilize multiple types of WAN links

    Company H faces difficulties in monitoring and maintaining the WAN links provided by carriers. WAN egress bandwidth, reliability, and SLA vary with sites, as does application quality due to the varied transport networks, areas, and application types. As a result, multiple types of WAN links cannot be properly utilized, and bandwidth is sometimes preempted.

3. Difficult to ensure service experience

    When a branch accesses the DC or the public cloud, or when branches communicate with each other, traffic from bandwidth-hungry applications between branches must be routed to the DC for centralized forwarding in order to facilitate compliance, monitoring, and security.

    However, centralized access to the public cloud from the DC wastes bandwidth and increases costs. Direct access of local branches to the public cloud through the Internet can lead to security and reliability concerns. In addition, it is difficult to control application priorities at the "last mile" from enterprise branches, DCs, and public clouds to enterprise branches.

4. Impossible to grow enterprise departments

    Company H has multiple departments, all of which need to be isolated from one another. However, the existing isolation mechanism makes it difficult to expand departments, complicates the configuration operations, and fails to ensure the service experience of various departments.

5. Difficult to maintain cross-border services

    Company H is a multinational organization providing cross-border services. As network transmission quality and reliability, as well as OPEX, vary depending on the region, IT teams face great challenges in maintaining the networks worldwide. Compliance with the laws and regulations of each region is required, as well as ensuring global interconnection of regional networks.

9.2.2.2 Solution Design

Company H decides to introduce an SD-WAN solution from a leading provider. According to Company H, SD-WAN is designed to efficiently utilize Internet links, implement network awareness and intelligent service scheduling, and create a high-quality, cost-effective, and perceptible globally-connected WAN.

SD-WAN should deliver the following three major improvements for Company H:

- Improved SLA

  Application performance is monitored in real time. Intelligent multipath scheduling achieves fast link switchover when a network fault occurs, reducing the packet loss rate and latency of applications. Network resources dynamically adapt to application requirements. With these advantages, SLA is improved and user experience is enhanced.

- Reduced WAN costs

  More cost-effective Internet links are introduced, and traffic is easily scheduled to these links as a result of preset application and path priorities. Intelligent traffic steering of key applications is also supported. All of the above capabilities greatly reduce WAN costs.

- Simplified and more efficient O&M

  The SDN controller centrally controls and manages WAN resources to implement visualized monitoring and management as well as fast fault locating, greatly simplifying WAN O&M.

Figure 9.3 illustrates the SD-WAN architecture for Company H.

1. Overall planning

   Company H plans to deploy the SDN controller in the global DC site while also reconstructing the networks for the global DC site, regional DC sites, and representative office sites.

   a. The SDN controller in the global DC site centrally manages all SD-WAN devices and orchestrates services accordingly. It provides two southbound IP addresses that correspond to MPLS and Internet links, respectively, so that SD-WAN sites can register with the SDN controller through MPLS or Internet links during deployment.

   b. Two CPEs are deployed at the global DC site, at each regional DC site, and at each representative office site to become SD-WAN sites.

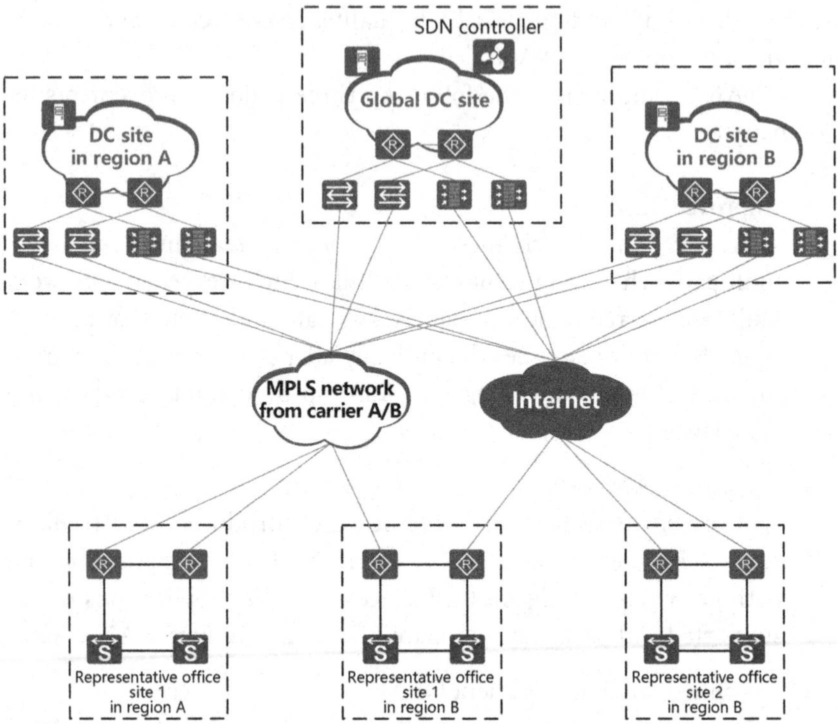

FIGURE 9.3 Company H's SD-WAN architecture.

c. Two RDs are planned for the overlay network, with each RD allocated with two transport networks. This results in four transport networks, MPLS 1 and MPLS 2 as well as Internet 1 and Internet 2, from different carriers.

d. The deployment sequence is as follows: SDN controller at the global DC site, RR-capable DC sites, and representative office sites.

2. Site design

DC sites

DC sites (including global and regional DCs) operate as both SD-WAN and RR sites, with two CPEs deployed as gateways at each site.

LAN- and WAN-side routes are designed as follows:

a. LAN-side interfaces and DC router interfaces learn each other's LAN-side routes through EBGP.

b. Two interfaces are configured on the WAN side. One interface exchanges underlay MPLS VPN routes with routers through EBGP, and the other connects to the Internet. Two gateways are deployed at each site, and there are four WAN links (two MPLS links and two Internet links). The two CPEs are directly connected through Eth-Trunk.

Representative office sites

Representative office sites also operate as SD-WAN sites, and two CPEs are deployed as gateways at each site.

LAN- and WAN-side routes are designed as follows:

a. LAN-side interfaces connect to LAN-side switches or terminals in wired or wireless mode. VRRP can be deployed on a network with two CPEs to improve reliability. If LAN-side switches work in Layer 2 mode, they can communicate with CPEs through VLANs. If LAN-side switches work in Layer 3 mode, they can communicate with CPEs through OSPF, BGP, direct routing protocols, or static routing protocols, in which case the split horizon feature is also enabled to prevent routing loops.

b. An interface is configured on the WAN side to connect to a WAN link. The two WAN links at a site can be dual MPLS links, dual Internet links, or one MPLS link plus one Internet link.

c. OSPF, EBGP, or static routing protocols are deployed on WAN-side interfaces, depending on the routing configuration of the upper-layer network.

3. Networking design

Control plane

DC sites, functioning as RR sites, establish BGP peer relationships with other RR sites through overlay control channels in order to synchronize overlay routes. As a result, a full-mesh overlay topology is created between sites.

Each DC site establishes control channels with representative office sites in the same region. In doing so, they can exchange overlay route information through BGP.

Data plane

Company H's sites around the world (including representative office sites, regional DC sites, and the global DC site) all function as SD-WAN sites on the overlay network. The overlay topology utilizes full-mesh networking, whereby sites are fully interconnected and services are directly accessed through overlay tunnels between sites. The global DC site serves as the redirect site for others in cases of inter-site communication failures.

4. Service design

Service access

Company H operates multiple types of services between sites on the live network. Typical examples include voice/video conference services between representative office sites; data synchronization and email services between representative office sites and DC sites; and data synchronization between DC sites. As a result, a full-mesh network is created to enable easy access of services between sites through directly connected overlay tunnels.

Each SD-WAN site has two WAN links: one MPLS VPN link and one Internet link. Service types are precisely identified using application identification (first packet identification or service awareness) technologies, and application and link quality is also efficiently detected. In addition, traffic steering based on SLA link quality indicators, such as the packet loss rate, delay, and jitter, ensures that different services leverage different primary and backup links, achieving load balancing of traffic on MPLS and Internet links.

a. Key applications, such as voice and video conference applications, are preferentially scheduled on MPLS links, with Internet links used as backups.

b. Traffic from applications that occupy large bandwidth but have low requirements on real-time performance (such as system/software upgrade and backup) is directed to

Internet links. When those links become unstable, the traffic is then intelligently redirected to MPLS links to ensure an uncompromised experience.

QoS policies are used to ensure the experience of key applications. In SD-WAN, applications are prioritized into six types based on their importance on the network, with key applications such as VoIP and video conferencing being preferentially guaranteed. Bandwidth for each of the six types of applications is guaranteed through specific QoS policies.

The six types of applications are listed in descending order of priority as follows:

a. Voice services (occupying 15% of the total bandwidth) are very sensitive to delay, jitter, and packet loss rates. They generate a relatively low traffic volume but have the highest priority. As such, voice services should be preferentially forwarded with guaranteed bandwidth. Packets that exceed the reserved bandwidth are discarded.

b. Streaming media services (occupying 10% of the total bandwidth) are moderately sensitive to delay, jitter, and packet loss rates. While their priority is medium, bandwidth should still be guaranteed for these services. Packets that exceed the reserved bandwidth compete with packets in other medium priority queues.

c. Video conferencing and desktop cloud services (consuming 50% of the total bandwidth) are also moderately sensitive to delay, jitter, and packet loss rates, and require fixed bandwidth. In particular, desktop cloud service traffic has an oversubscription ratio, whereby its priority is medium but bandwidth should still be guaranteed. Packets that exceed the reserved bandwidth compete with packets in other medium priority queues.

d. Key data and network management services (occupying 15% of the total bandwidth), including interactive applications and network infrastructure applications, are of high importance. They are relatively sensitive to delay and have low bandwidth requirements, and while their priority is medium, bandwidth should still be guaranteed for these services. Packets that exceed the reserved bandwidth compete with packets in other medium priority queues.

e. Basic office services (consuming 9% of the total bandwidth) do not have high requirements on real-time performance. However, they serve a large number of users, are in widespread use, and generate a large proportion of traffic. If bandwidth is insufficient, user experience will be negatively impacted. As such, the priority of these services is medium but bandwidth should still be guaranteed. Packets that exceed the reserved bandwidth compete with packets in other medium priority queues.

f. System/software upgrade and backup services (occupying 1% of the total bandwidth) generate significant burst traffic and occupy a large amount of bandwidth. However, they have low demands for real-time performance and require the lowest priority. To ensure these types of services, 1% of bandwidth is reserved, and packets that exceed the reserved bandwidth do not compete with packets in other queues.

Internet access

Company H adopts centralized Internet access plus local Internet access (for SaaS applications), with a detailed design as follows:

a. By default, traffic is centrally forwarded to the global DC site and then routed out to the Internet. Traffic from representative office sites is forwarded to DC sites and then routed out to the Internet through a LAN-side interface.

b. Internet access points can be flexibly selected for SaaS services. FPI technology is used to precisely identify SaaS applications. Representative office sites can access SaaS services in centralized Internet access mode or local Internet access mode to ensure the service experience of SaaS applications.

Interworking with legacy MPLS networks

Company H's SD-WAN sites should communicate with legacy sites. As such, Company H adopts a centralized access approach, where the global DC site is used as the centralized egress for traffic. Specifically, Company H enables the centralized access function and selects the global DC site as the centralized access site on the SDN controller.

5. Security design

   Company H designs a series of security measures, including the following:

   a. **ACL policy:** In compliance with corporate access regulations, Company H deploys ACL policies in the inbound direction on WAN-side interfaces at SD-WAN sites to filter out malicious packets and minimize high-risk interface access.

   b. **Overlay tunnel security:** Data transmitted over overlay tunnels based on Internet links is encrypted using IPsec to ensure secure data transmission.

6. O&M design

   Company H uses an SDN controller — a centralized management, control, and O&M platform — to implement visibility, manageability, and maintainability. Using the SDN controller, Company H's O&M staff can:

   a. View network and service status through a global dashboard, receive anomaly alerts in real time, and monitor the topology, intra-site communication, inter-site communication, and application status.

   b. Configure sites individually, deliver service policies in batches, and flexibly define multiple O&M roles.

   c. Locate faults with visualized tools, log in remotely through SSH, and upgrade the systems/patches for devices in batches.

   Primary functions delivered by the SDN controller include:

   d. **Automated device deployment:** Site underlay network parameters can be configured using the SDN controller, whereby the network administrator specifies the URL in the deployment email, configures relevant deployment information, and sends the deployment email to a deployment professional's mailbox. After receiving the deployment email, the deployment professional uses a browser to access the URL in the email and initiates the deployment process. The device is then automatically deployed.

e. **Easy device configuration delivery:** Through the SDN controller, the network administrator can configure the VLANs, Layer 3 interfaces, application identification policies, QoS policies, ACLs, and traffic steering policies for site devices. Subsequently, once devices are successfully registered and go online, the SDN controller automatically delivers the previous configurations to these devices.

f. **Simple device replacement:** When a device can no longer be used due to a hardware fault, maintenance personnel can easily replace it with a new device and perform email-based deployment. After the new device goes online, it can still use the service configurations of the original device.

9.2.3 SD-WAN in Carriers

The rise of SDN and cloud computing revolutionizes traditional closed IT architecture of enterprises. Currently, enterprises tend to access daily office applications on the cloud through Internet links.

Although carriers can provide ubiquitous network services, their approaches to traditional network construction cannot keep up with ever-growing Internet-based practices. In this regard, all carriers are urgently seeking new ways to transform deep-rooted conventional practices, explore new profit growth, and stay ahead of the competition amidst rising Internet enterprises.

Carrier and MSP resale is an important SD-WAN business model. Suppose that Carrier V is a leading carrier. We will use this case study to illustrate the SD-WAN scenario analysis process and solution design.

9.2.3.1 Scenario Analysis

Tackling fierce competition from Internet enterprises, Carrier V decided to make a change, and in doing so was faced with many challenges, including:

1. Expensive private lines increase the burden on enterprises.

   Enterprise private lines are expensive because they are complex to access and often used exclusively. However, with the rapid development of Ethernet and IP technologies, Internet performance and reliability are continuously being improved. As a result of the cost-effectiveness of enterprise-class Internet links compared to private lines, WANs consisting of both MPLS and Internet links have grown in popularity and become the preferred choice for enterprises. For these reasons, Carrier V has seen rapidly shrinking private line business, resulting in decreased revenue.

2. The provisioning of private lines is time- and labor-consuming, and is unable to support fast deployment.

   Services over long-established private lines are isolated using VPNs or physical isolation measures, which are inherently secure. The comprehensive QoS assurance features of private lines ensure reliable data transmission for delay-sensitive applications (video conferencing, VoIP services, etc.).

   However, private lines are provisioned slowly. Factors — such as the availability of local POPs, the reachability of physical lines to POPs, cross-carrier line leasing and collaboration, and device-by-device line and service deployment at branches — all affect the speed of supplying private lines.

   In addition, more services are migrating onto the cloud. Carriers and ISPs will compete to get ahead in the private line market by provisioning cloud-based private line services in minutes.

3. Conventional private line services use simple traffic steering policies, which cannot precisely detect services or efficiently steer traffic.

   When it comes to conventional private line services, traffic steering policies are formulated based on route information; for example, the destination (the headquarters or DC) of service flows. The traffic steering granularity is not specific to applications, and the requirements of applications on link quality are not considered.

   With the development of the Internet, a growing number of services can be provided. Enterprises have vastly varied requirements on the quality of both services and WAN links.

4. The O&M of typical private line networks is complex, making it difficult to quickly locate network faults.

   When a fault occurs on a typical private line network, it is difficult and time-consuming to locate it, and as a result, enterprise services cannot quickly recover, causing huge losses.

Rapid network progress drives enterprise office services to become cloud-based and remotely-operated. Intelligent networks, in particular, can help enterprises quickly locate issues and ensure connectivity between branches and headquarters.

5. Conventional network service configuration relies heavily on manual operations, which require highly trained network engineers and are prone to errors.

   Traditionally, enterprise WAN services are manually provisioned onsite by network engineers. In addition to being error-prone, this approach not only requires highly skilled network engineers but also results in time-consuming service provisioning and inefficient O&M. As enterprises expand their scope, more branches need to be built and services adjusted as required. Considering this, new enterprise services must be quickly rolled out to adapt to rapid enterprise growth.

Carrier V has meticulously analyzed the preceding pain points and therefore proposed the following network construction expectations and goals:

1. Improving the competitiveness of its private line business and enhancing customer loyalty

   Carrier V looks for a WAN construction solution that is simple and easy to deploy, thereby quickly winning more enterprise customers, especially high-value large enterprises, and expanding market shares. The solution must also be able to enhance Carrier V's competitiveness to outperform ISPs.

   In addition to building xDSL/MPLS/MSTP private line networks, Carrier V expects to create WANs based on Ethernet and IP technologies to reduce private line costs for its enterprise customers.

2. Transforming marketing practices and increasing the revenue from private lines

   Enterprises used to purchase private line services and VAS equipment from Carrier V. However, Carrier V now hopes to sell SD-WAN services and virtual VASs.

   In the past, Carrier V resold CPEs in a one-off manner and then charged the private line fees on a monthly or yearly basis. Carrier V now wishes to shift toward offering free CPEs and charging service fees in a pay-as-you-go or pay-as-per-use manner to gain recurring revenue.

Enterprises currently manage and maintain WANs independently. To free them from O&M burdens, Carrier V aims to provide managed services for enterprises, whereby it can manage and maintain WANs on their behalf.

3. Implementing intelligent network scheduling based on SDN

   Enterprise WAN connections have traditionally been difficult to manage and schedule. If network interruptions occur, they are difficult to locate and require network administrators to locate faults onsite. Further, it is difficult to switch over service traffic paths when faults occur, as improper network configuration, if any, may cause network breakdown.

   It is under these conditions that Carrier V will introduce SDN technology to make a difference. On the wizard-based GUIs, network administrators can effortlessly formulate fine-grained service path management and control policies based on customers' actual network service requirements. Big data analytics of real-time data, such as bandwidth quality and usage, helps proactively optimize traffic paths and intelligently schedule network resources. The resulting benefits include greatly reducing the workload of network administrators and helping enterprises reduce network maintenance costs.

4. Quickly provisioning private line services and implementing visualized O&M management for reduced OPEX

   Conventional private line services face issues such as slow provisioning, high costs, and complex configuration. Carrier V plans to deploy CPEs or uCPEs on the customer's premises and implement ZTP for faster deployment, greatly simplifying customers' network configuration operations. Carrier V also wishes to offer a visualized global network dashboard for enterprise customers to view network quality and security status in real time, thereby efficiently operating and maintaining their WANs.

9.2.3.2 Solution Design

In order to meet the analyzed requirements of Carrier V, an SD-WAN solution provider designs an open, intelligent SD-WAN solution that integrates NFV/SDN technology. With this solution, Carrier V can provide managed services and VASs for SMEs, as well as offer WAN construction and maintenance services, thereby reducing enterprises' investment in devices and O&M professionals.

The SD-WAN solution has the following three major advantages:

- Separate forwarding and control planes for enhanced network reliability

  An SDN controller is deployed to separate the control and data planes, preventing the impact of traffic surges on the control plane and therefore improving network reliability.

- Multi-tenant management and simplified O&M

  The SDN controller centrally manages SD-WAN sites and allows MSPs to provide managed network maintenance services for tenants. GUI-based monitoring, configuration, and diagnosis provided by the SDN controller improve network O&M efficiency.

- Open APIs for easy interoperability with other platforms and automated service provisioning

  The SDN controller provides open northbound RESTful APIs to easily interconnect with Carrier V's BSS and OSS. In this way, users' service subscription information can be automatically imported to the SDN controller, accelerating service provisioning. The SDN controller also supports automated provisioning of VASs, enhancing customer loyalty.

Figure 9.4 illustrates the SD-WAN architecture for Carrier V.

1. Overall planning

   Carrier V plans to deploy the SDN controller in the DC and introduce multi-tenant RR and IWG sites, while also reconstructing enterprise customers' legacy sites into SD-WAN sites and helping them deploy new ones.

   a. The SDN controller in the DC centrally manages all SD-WAN devices and orchestrates services accordingly. It provides two southbound IP addresses that correspond to MPLS and Internet links, respectively, so that SD-WAN sites can register with the SDN controller through MPLS or Internet links during deployment.

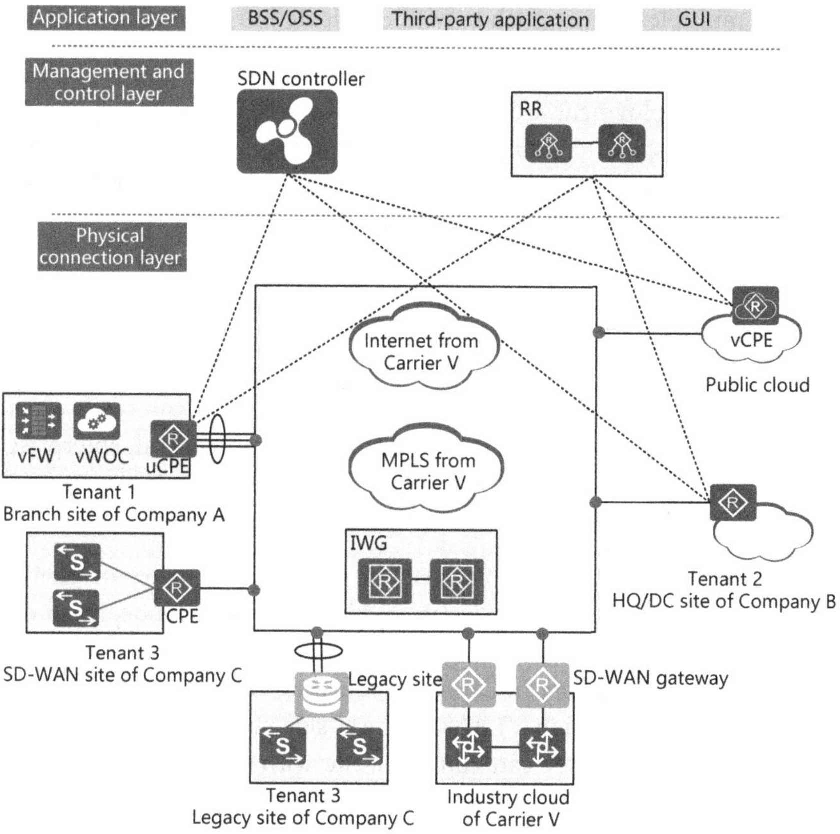

FIGURE 9.4 Carrier V's SD-WAN architecture.

b. Multi-tenant RR and IWG sites can be deployed in the DC as required. They all adopt single-gateway networking to facilitate centralized communication with legacy sites.

c. SD-WAN gateways are deployed on Carrier V's industry cloud, through which enterprise sites can gain access.

d. Enterprise sites can use either single- or dual-gateway networking as required.

e. To facilitate connections between sites of different enterprise tenants, the SD-WAN network should allocate RDs and transport networks to tenants, which correspond to MPLS VPNs and Internet links. Carrier V needs to provide multi-tenant RR/IWG device access services for enterprises and allocate RDs shared by multiple tenants, such as MPLS (oriented to MSPs) and Internet (oriented to MSPs).

f. The deployment sequence is as follows: SDN controller; multi-tenant sites, including RR and IWG sites (these sites are centrally managed by Carrier V); and enterprise sites (enterprise headquarters site preferentially deployed).

2. Site design

Enterprise site

SD-WAN-capable CPEs (CPEs, uCPEs, or vCPEs) are deployed at enterprise network egresses, which Carrier V can resell when building private lines for enterprise users. In general, enterprise users can select one or two SD-WAN CPEs as required.

LAN- and WAN-side routes are designed as follows:

a. **LAN side:** CPEs connect to LAN-side switches or terminals in wired or wireless mode. VRRP can be deployed on a network with two CPEs to improve reliability. If LAN-side switches work in Layer 2 mode, they can communicate with CPEs through VLANs. If LAN-side switches work in Layer 3 mode, they can communicate with CPEs through OSPF, BGP, direct routing protocols, or static routing protocols, in which case the split horizon feature is also enabled to prevent routing loops.

b. **WAN side:** Both CPEs need to connect to MPLS and Internet links. OSPF, EBGP, or static routing protocols are deployed, depending on the routing configuration of the upper-layer network.

Multi-tenant IWG sites

SD-WAN IWGs — which are gateways capable of SD-WAN — are deployed on the private line network of Carrier V to implement communication between MPLS and SD-WAN networks. IWGs can be deployed in active/standby or load balancing mode, thereby ensuring enterprise service reliability.

IWGs support multi-tenant functions, where a single physical IWG can be logically divided into multiple virtual border gateways, and each enterprise connected to the IWG is considered a tenant. Virtual border gateways are invisible to enterprise users and can be managed and controlled by Carrier V's administrator.

LAN- and WAN-side routes are designed as follows:

a. **LAN side:** IWGs connect to ASBR PEs on the legacy MPLS network by using MPLS Option B and exchange route information through MP-EBGP. In this way, SD-WAN overlay access services are available for interworking with legacy MPLS network sites.

b. **WAN side:** IWGs connect to MPLS and Internet links. OSPF, EBGP, or static routing protocols are deployed, depending on the routing configuration of the upper-layer network. MPLS VPNs connected to IWGs are service VPNs shared by tenants. In this way, site routes of different tenants can be imported from tenant VPNs to service VPNs using route leaking on PEs.

Multi-tenant RR sites

In the multi-tenant IWG networking scenario, multi-tenant RR sites are deployed to establish control channels among enterprise sites and between enterprise sites and IWGs. These control channels are centrally managed and maintained by Carrier V.

A multi-tenant RR site operates as a single-gateway (CPE) site. Generally, there is no additional deployment on the LAN side, while two WAN links — one of which is an MPLS VPN link and the other an Internet link — can be deployed on the WAN side.

MPLS VPNs connected to multi-tenant RR sites are service VPNs shared by tenants. OSPF, EBGP, or static routing protocols can be deployed on underlay routes on WAN interfaces, depending on the routing configuration of the upper-layer network. The overlay network exchanges overlay route information with IWGs and enterprise sites through MP-BGP.

3. Networking design

Control plane

Interconnection between enterprise sites depends on overlay routes distributed by RR sites. Generally, an RR site is constructed using either of the following two approaches: (1) deploying an RR site together with the hub site at the enterprise headquarters, or (2) using an MSP multi-tenant RR site assigned by the carrier.

In the IWG interconnection scenario, enterprises should adopt the MSP multi-tenant RR site assigned by the carrier as the unified control plane.

Data plane

According to the SD-WAN solution adopted by Carrier V, the overlay network consists of the following two parts:

a. Overlay interconnection between sites in an enterprise

SD-WAN CPEs deployed at the egresses of enterprise sites can set up an overlay network topology (in hub-spoke, full-mesh, or partial-mesh mode) through the WAN-side underlay network based on the requirements for communication between enterprise sites.

b. Interconnection between enterprise branch sites and IWGs

For enterprises requiring interconnection with the legacy MPLS network, an overlay network can be constructed between enterprise branch sites and Carrier V's IWG sites. Here, enterprise branch sites connect to IWG sites through overlay tunnels to interwork with the conventional MPLS VPNs, communicate with legacy sites, and obtain network services from Carrier V's DC. Depending on the enterprise scale and IWG access quality requirements, the carrier can allow each enterprise to exclusively occupy an IWG site, or multiple enterprises to share the same IWG site.

4. Cloud-network synergy

In addition to granting enterprise customers access to the public cloud, Carrier V also builds an industry cloud that serves enterprise customers, further enhancing customer loyalty and profitability. In order for enterprise customers to flexibly and conveniently access the industry cloud and enjoy services accordingly, Carrier V uses an SD-WAN cloud-network synergy solution.

Figure 9.5 shows the networking of the SD-WAN cloud-network synergy solution for Carrier V.

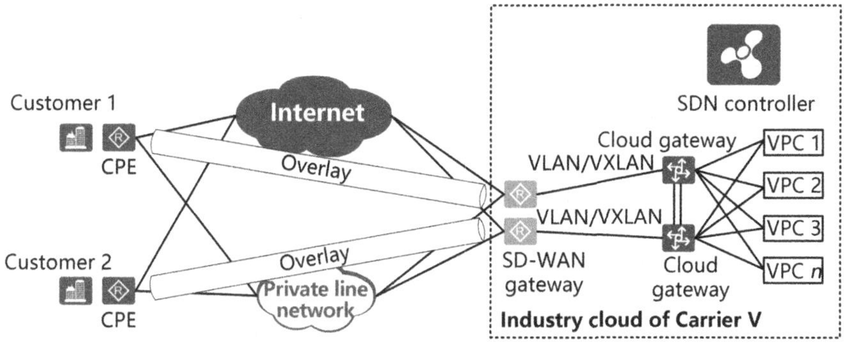

FIGURE 9.5 Carrier V's SD-WAN cloud-network synergy solution.

As illustrated in the figure above, the SDN controller is deployed in Carrier V's industry cloud, which centrally manages CPEs at enterprise branches and SD-WAN gateways.

SD-WAN gateways are deployed at the WAN-side egresses of Carrier V's industry cloud. They establish overlay tunnels with CPEs at enterprise branches by using private lines or the Internet.

High reliability is required for enterprise branch sites to access Carrier V's industry cloud. To this end, multiple SD-WAN gateways are deployed in Carrier V's industry cloud to ensure active/standby backup.

SD-WAN gateways use MPLS Option A VLAN or VXLAN technology to connect to cloud gateways in Carrier V's industry cloud and support multi-tenant functions. This means a single physical SD-WAN gateway can be logically divided into multiple virtual border gateways, and each enterprise connected to the gateway is considered a tenant.

By deploying SD-WAN CPEs, enterprise branch sites can obtain cloud connection services over the Internet or private lines. After Carrier V's SD-WAN O&M professionals enable ZTP, CPEs at enterprise branch sites can automatically go online and establish overlay tunnels with SD-WAN gateways to access the industry cloud. In this case, Carrier V can rate-limit the bandwidth based on service needs.

This SD-WAN cloud-network synergy solution has the following four advantages:

a. Multiple tenants can share the same SD-WAN gateway.

b. SD-WAN gateways support overlay access for CPEs at enterprise branch sites, free from the impact of DC networking.

c. SD-WAN gateways can easily and reliably interwork with cloud gateways.

d. The SDN controller supports multi-tenant management.

5. Service design

Service access

Application identification technology is applied to easily identify the application types on devices at enterprise branch sites. Doing so facilitates the enforcement of related service policies in subsequent service processes, such as intelligent traffic steering.

An enterprise site may have both MPLS and Internet links, where the former are used as the primary link to carry services. Generally, MPLS links carry production services, while Internet links underpin office and video conference services, and allow SaaS traffic to be locally broken out preferentially.

Typically, traffic steering policies are as follows:

a. FTP or ICMP applications preferentially use MPLS links, or Internet links alternatively.

b. HTTP download applications adopt Internet links primarily, and MPLS secondarily.

c. Web applications preferentially use Internet links while adopting MPLS links as backups.

d. During SaaS access, web applications are locally broken out through Internet links based on traffic steering policies.

Simply, Internet and MPLS links back up each other. When the MPLS link is faulty, production services can be switched over to the Internet link. If so, the highest QoS level must be configured in advance.

When the Internet link is faulty, office services can be switched over to the MPLS link. If so, a medium QoS level must be configured. In similar cases, video conference services can be switched over to the MPLS link, and the lowest QoS level should be configured.

When the Internet link is faulty, SaaS traffic that was locally broken out can be switched over to the MPLS link and then centrally routed out through the hub site.

Internet access

Carrier V provides the following three secure Internet access solutions for enterprise customers, depending on Internet link application scenarios:

a. **Solution 1:** The Internet is used only as the underlay transport network for communication between enterprise branch sites. Internet access through local breakout is disabled at enterprise branch sites. Instead, Internet traffic from enterprise branch sites is aggregated to the hub site and then centrally routed out through the firewall at the headquarters. In addition, underlay ACL policies can be configured at enterprise branch sites to improve security.

b. **Solution 2:** Enterprise branch sites adopt a hybrid Internet access approach: Only traffic of specified applications uses local Internet access, while other applications' traffic, by default, is aggregated to the headquarters for centralized Internet access. Enterprise branch sites locally use the FPI function to identify SaaS application traffic types. After that, only specified SaaS application traffic is allowed to be locally broken out to the Internet. Other Internet traffic is concentrated to the headquarters and centrally routed out through the firewall. In addition, underlay ACL policies can be configured, or built-in firewalls can be deployed at enterprise branch sites to ensure branch site security.

c. **Solution 3:** uCPEs are deployed at enterprise branch sites, and vFWs are deployed onto them, which ensures network security. All Internet traffic at enterprise branch sites is locally broken out. If a public network server exists on the LAN side of an enterprise branch site, the underlay NAT function can be enabled on CPEs to map addresses for the LAN-side server. In this way, it is possible to access the public network server on the LAN side of the enterprise branch site from the public network.

Interworking with legacy networks

Carrier V's legacy MPLS networks are not within the management scope of the SDN controller. It is, however, required to support interworking between SD-WAN and legacy MPLS networks. To this end, it is recommended that MPLS Option B be deployed on the SD-WAN IWGs to interwork with legacy MPLS networks. In this way, SD-WAN overlay access services can be provided for enterprises to interwork with legacy MPLS networks.

6. Security design

When deploying the SD-WAN solution, Carrier V must consider the following in order to strengthen security:

a. **Routing policy control:** On the SDN controller, the administrator configures underlay WAN routing policies to filter routes based on whitelists or blacklists. Through this, sites receive routes only from the SDN controller and peer SD-WAN sites.

b. **Access control:** ACL rules are deployed in the inbound direction of the WAN side at enterprise branch sites to filter out malicious packets and restrict access to dangerous ports. This is in line with enterprise customers' network access regulations.

c. **vFWs on uCPEs:** For sites that deploy uCPEs, SFC is enabled on and vFWs are installed onto uCPEs to improve security. vFWs provide security protection functions similar to professional Layer 7 firewalls, such as antivirus.

d. **Overlay tunnel security:** Data transmitted over overlay tunnels based on Internet links is encrypted using IPsec to ensure secure data transmission.

7. uCPE solution

   Enterprises that rent Carrier V's private lines differ greatly from each other; some have only one or more branch sites. For such enterprises, it is recommended that uCPEs be deployed to reduce network construction costs and complexity.

   Third-party VAS software can be deployed onto uCPEs. To do so, third-party VAS software image packages need to be loaded onto the SDN controller. After that, VAS functions such as vFW, software-based load balancing, and software-based WAN acceleration can be deployed once uCPEs go online. Carrier V can periodically update the VAS software version through the SDN controller to ensure ongoing access to the latest version of VAS functions.

8. O&M design

   In an enterprise-built SD-WAN scenario, O&M primarily includes CPE deployment, network service monitoring, and network service maintenance. In a carrier-resale SD-WAN scenario, O&M is more complex, as it also involves multi-tenant management and tenant portal access permission management.

   Multi-tenant management

   In addition to providing all the functions involved in an enterprise-built SD-WAN scenario, the SDN controller also offers rights- and domain-based management, as well as control functions in a carrier-resale SD-WAN scenario. That is, the SDN controller provides differentiated management and control rights for network administrators at Carrier V, lower-level carriers, and enterprises.

   a. The SDN controller involves three levels of administration: system (network administrators of Carrier V), MSP (network administrators of lower-level carriers), and tenant (network administrators of enterprises).

   b. System administrators have the highest level of rights and can create new system, MSP, or tenant administrators by role. When a system administrator adjusts a policy, it affects all tenants.

c. A system administrator can create an MSP account, create a role by using the default administrator of the MSP account, and then create new MSP or tenant administrators based on the role. If there are no lower-level carriers, system administrators can also be MSP administrators.

Tenant portal access permission management

Tenant portal access permission management is provided as a VAS for enterprise customers. Carrier V provides the following three types of accounts for enterprise customers to access the tenant portal:

a. **An account with only monitoring permission:** This account is used to monitor the SD-WAN network status.

b. **An account with monitoring permission and overlay service configuration permission:** This account is used to monitor the SD-WAN network status and configure overlay services, such as intelligent traffic steering, QoS, ACL policies, and multi-VPNs.

c. **An account with monitoring permission, as well as overlay and security service configuration permissions:** This account is used to monitor the SD-WAN network status and configure overlay services (such as intelligent traffic steering, QoS, ACL policies, and multi-VPNs) and security services (e.g., firewall, URL filtering, and IPS).

9. Northbound APIs provided by the SDN controller

The SDN controller provides abundant northbound APIs, through which Carrier V can customize third-party applications as required. Carrier V can also use GUIs provided by the SDN controller as its daily network O&M tool.

Carrier V uses northbound APIs provided by the SDN controller to interwork with third-party applications for the following two reasons:

a. Carrier V's BSS/OSS platform can automatically import SD-WAN CPE deployment parameters into the SDN controller, increasing deployment efficiency and enhancing large-scale SD-WAN site deployment. In addition, the BSS/OSS platform can automatically import existing tenant, site, and device information to the SDN controller. This automation increases service provisioning efficiency, facilitates large-scale deployment, and reduces errors caused by manual information importing.

b. The SDN controller interconnects with Carrier V's BSS/OSS platform through APIs. After that, VASs, which customers subscribe to, can be automatically delivered by invoking the SDN controller through APIs. In this way, services can be easily obtained on demand.

For example, if customers need to expand bandwidth temporarily, they can subscribe to a VAS (doubling the bandwidth for one hour) online. Carrier V's OSS/BSS platform then invokes the APIs to complete related configurations and automatically deliver these configurations. Thereby, the requested VAS can be provisioned quickly and will expire as expected.

9.2.4 SD-WAN in MSPs

The growth of mobile offices, acceleration of video traffic, and migration of services to the cloud mean that enterprises demand larger network bandwidth, and likewise, applications put greater pressure on WAN service quality.

As a result, enterprise O&M teams are struggling to provide QoS assurance for services at DC egresses, particularly for public cloud-based applications. The rapid development of cloud computing, big data, and AI has presented a dilemma for enterprises — a desire for ever-increasing cloud access requirements, yet a lack of sufficient IT capabilities. This opens up huge market space and opportunities for MSPs.

MSPs are an important channel for enterprises that want to obtain network hosting services. By leveraging industry-leading management technologies, MSPs offer systematic network hosting services that enterprises need. The core competitiveness of MSPs lies in their service quality and technical expertise; however, the biggest bottleneck of traditional MSPs is that their technology expertise is insufficient to support large-scale development. As more enterprise services are being migrated to the cloud, MSPs face new challenges in improving their multi-cloud management, enterprise network hosting, and professional service capabilities.

Suppose that MSP Z is a leading MSP in Country C. We will use this case study to illustrate the SD-WAN scenario analysis process and solution design.

9.2.4.1 Scenario Analysis

After a careful analysis of its services and the market, MSP Z has identified the following pain points:

1. "Last-mile" access cost for enterprises is high, and service provisioning is slow.

   MSP Z leases MSTP lines from carriers in Country C. By leveraging MSTP lines, MSP Z constructs an MPLS VPN as the backbone network to provide private line services for enterprises. Through this backbone network, enterprise branches can easily communicate with each other and access the DC.

   If MSP Z cannot provide "last-mile" line facilities for enterprise sites, expensive lines should be routed for enterprise branches to grant them access to the MSP's backbone network. The incurring costs will ultimately be borne by enterprises and will consequently increase the tariffs for purchasing MSP Z's backbone network services. In addition to this, line routing also prolongs the service provisioning time.

2. Private lines for cloud connectivity face long provisioning periods, high line costs, and complex management.

   MSP Z must empower enterprise customers with easy public cloud access. To achieve this, MSP Z needs to bridge the backbone network to enterprise public cloud resources. If traditional private lines are used, MSPs have to pay for both expensive private line routing expenses and cloud private line access services provided by public cloud service providers. This results in high investment for MSPs.

   Furthermore, when different enterprises require access to different public clouds, traditional private lines will suffer from inflexible service provisioning, long service provisioning periods, invisible cloud connection service quality, and complex multi-cloud management.

9.2.4.2 Solution Design

After MSP Z's current situation has been analyzed, an SD-WAN solution provider designs a tailored SD-WAN solution that offers the following features:

- Flexible "last-mile" access accelerates service provisioning while reducing costs.

  Internet links are used to achieve SD-WAN access from enterprise branches to the POP gateways on MSP Z's backbone network, thereby eliminating the need to deploy "last-mile" lines. Additionally, MSP Z adopts ZTP to lower the technical competence for deployment personnel, making it possible to quickly respond to enterprises' demands and accelerate branch service provisioning.

- Cloud access and management functions are provided to meet the demand for enterprise service cloudification.

  MSP Z deploys vCPEs as cloud sites in enterprise VPCs. Doing so can flexibly bridge enterprise public cloud resources to the POPs on MSP Z's backbone network, and also enable enterprise branches' access to the public cloud across POPs. Furthermore, MSP Z can visually manage cloud sites and the link performance of cloud connections.

Figure 9.6 illustrates the SD-WAN architecture for MSP Z.

1. Overall planning

   MSP Z plans to deploy the SDN controller in the DC and introduce multi-tenant RR sites and POP GW sites. It will also reconstruct enterprise customers' legacy sites into SD-WAN sites and help them deploy new SD-WAN sites.

   a. The SDN controller in the DC centrally manages all SD-WAN devices and orchestrates services accordingly. It provides two southbound IP addresses that correspond to private lines and Internet links, respectively, so that SD-WAN sites can register with the SDN controller through private lines or Internet links during deployment.

   b. Multi-tenant RR sites are deployed in the DC as the unified control plane.

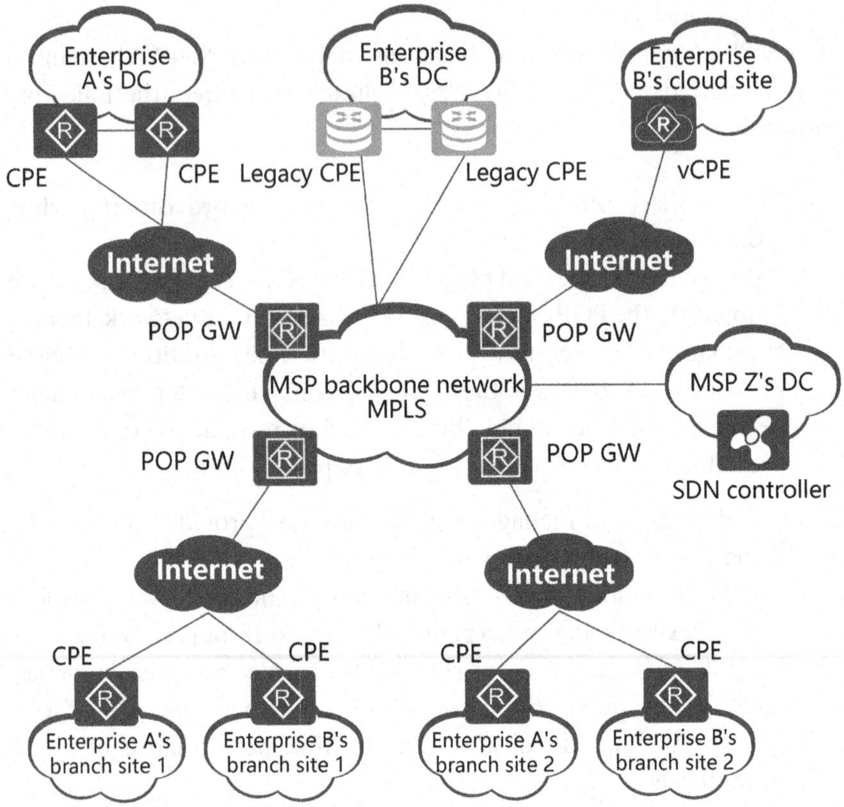

FIGURE 9.6 SD-WAN architecture at MSP Z.

c. Multiple POP GW sites are deployed at the edge of the MSP's backbone network, which function as the SD-WAN access points of the backbone network.

d. Enterprise sites can use either single- or dual-gateway networking as required.

e. MSP Z provides multi-tenant RR and POP GW access services for enterprises. It also plans the overlay network in a unified manner and allocates RDs shared by multiple tenants, such as MPLS (oriented for MSPs) and Internet (oriented for MSPs), to enterprises.

f. MSP Z defines access areas, associates these areas with RR and POP GW sites, and provides backbone network access services for enterprises by access area.

g. The deployment sequence is as follows: SDN controller, multi-tenant RR sites, multi-tenant POP GW sites, and enterprise sites.

2. Site design

Enterprise site

SD-WAN-capable CPEs (CPEs, uCPEs, or vCPEs) are deployed at enterprise network egresses. Enterprise users can select one or two SD-WAN CPEs at an enterprise site as required.

LAN- and WAN-side routes are designed as follows:

a. **LAN side:** CPEs connect to LAN-side switches or terminals in wired or wireless mode. VRRP can be deployed on a network with two CPEs to improve reliability. If LAN-side switches work in Layer 2 mode, they can communicate with CPEs through VLANs. If LAN-side switches work in Layer 3 mode, they can communicate with CPEs through OSPF, BGP, direct routing protocols, or static routing protocols, in which case the split horizon feature is also enabled to prevent routing loops.

b. **WAN side:** CPEs at an enterprise site connect to Internet links for uplink transmission. OSPF, EBGP, or static routing protocols are deployed, depending on the routing configuration of the upper-layer network.

POP GW site

SD-WAN-capable forwarding devices are deployed at the edge of the MPLS VPN, which is provided by MSP Z. These forwarding devices function as the POP GW sites that grant SD-WAN sites access to the MPLS VPN.

The most suitable solution is to regionally deploy POP GW sites (SD-WAN border gateways), which operate as single-gateway CPE sites. POP GW sites are configured in the active and standby regions to ensure the reliability of enterprise services.

POP GW sites support multi-tenant functions. In other words, a single physical POP GW can be logically divided into multiple virtual border gateways, and each enterprise connected to the POP GW is considered as a tenant. Virtual border gateways are invisible to enterprise users, and they can be managed and controlled by MSP Z's administrators.

LAN- and WAN-side routes are designed as follows:

a. **LAN side:** POP gateways connect to ASBR PEs on the legacy MPLS network by using MPLS Option B and exchange route information through MP-EBGP. In this way, SD-WAN overlay access services are available for interworking with legacy MPLS network sites.

b. **WAN side:** POP gateways connect to Internet links. OSPF, EBGP, or static routing protocols are deployed, depending on the routing configuration of the upper-layer network.

Multi-tenant RR sites

In the POP GW networking scenario, multi-tenant RR sites are deployed to establish control channels among enterprise sites and between enterprise sites and POP gateways. These control channels are centrally managed and maintained by MSP Z.

A multi-tenant RR site operates as a single-gateway (CPE) site. In most cases, no additional deployment on the LAN side is necessary, while two WAN links — one of which is a private line link of MSP Z's backbone network and the other an Internet link — can be deployed on the WAN side.

OSPF, EBGP, or static routing protocols can be deployed on WAN interfaces on multi-tenant RR sites, depending on the routing configuration of the upper-layer network. The overlay network exchanges overlay route information with POP gateways and enterprise sites through MP-BGP.

3. Networking design

Control plane

In POP interconnection scenarios, multi-tenant RR sites deployed by MSP Z are used as the control plane.

a. Multi-tenant RR sites are regionally deployed. Overlay control channels are established between the multi-tenant RR sites and POP gateways in the region.

b. An enterprise selects an access area based on its VPN. Overlay control channels are established between enterprise sites and RR sites in the access area. Enterprise sites can be associated with RR sites in the active and standby access areas to enhance reliability of the control plane.

Data plane

MSP Z's SD-WAN overlay network consists of the following two parts:

a. Interconnection between enterprise sites and POP gateways

If an enterprise requires access to the DC or public cloud, or communication between regional branches, enterprise sites can connect to POP gateways through POP interconnection to ensure service experience. Overlay tunnels are established as data channels between enterprise sites and POP gateways.

POP gateways in different access areas communicate with each other through the underlay network of MSP Z's backbone network.

b. Overlay interconnection between enterprise sites

The SD-WAN sites of an enterprise can create a full-mesh topology by using WAN-side Internet links based on the communication requirements between enterprise sites.

4. Service design

Service access

Service bandwidth must be assured for different tenants to access the backbone network. To achieve this, MSP Z configures QoS policies between POP gateways and PEs on the backbone network based on tenant VPNs.

Also, MSP Z limits the bandwidth for traffic accessing the MSP backbone network based on tenant VPNs. All branches of the same tenant VPN share the same bandwidth.

In managed service mode, MSP Z enables the application identification function (specifically, FPI and SA functions) on devices at enterprise branches to identify service applications on the live network. Then, based on the application identification result, MSP Z configures application-based intelligent traffic steering and QoS policies to ensure the bandwidth and priority for key enterprise applications.

If an enterprise branch connects to POP gateways through multiple Internet links, MSP Z can configure a traffic steering policy from the branch to the POP GW to load-balance the service traffic among multiple Internet links. In this way, enterprise customers can be assured of the high-quality experience of key applications while properly utilizing Internet bandwidth.

Internet access

The SD-WAN solution provides different Internet access modes for enterprise customers, including local, centralized, and hybrid Internet access.

Public cloud access

MSP Z deploys vCPEs in enterprises' public cloud VPCs to use enterprise cloud resources as cloud sites. SD-WAN overlay access from cloud sites to POP gateways is achieved based on the Internet or by using the cloud private line underlay network deployed by MSP Z.

SD-WAN-capable enterprise branch sites access public cloud sites through the MSP backbone network in cross-POP GW mode.

Interworking with legacy networks

The legacy enterprise sites on MSP Z's backbone network are not within the management or control scope of the SDN controller. However, the SDN controller is needed to support interworking between SD-WAN sites and legacy MPLS sites. It is recommended that SD-WAN sites connect to POP gateways in order to communicate with legacy sites.

5. Security design

When deploying the SD-WAN solution, MSP Z must consider the following in order to strengthen security:

a. **Routing policy control:** On the SDN controller, the administrator configures underlay WAN routing policies to filter routes based on whitelists or blacklists. Through this, sites receive routes only from the SDN controller and peer SD-WAN sites.

b. **Access control:** ACL rules are deployed in the inbound direction of the WAN side at enterprise branch sites to filter out malicious packets and restrict access to dangerous ports. This is in line with enterprise customers' network access regulations.

c. **vFWs on uCPEs:** For sites that deploy uCPEs, SFC is enabled on and vFWs are installed onto uCPEs to improve security. vFWs provide security protection functions similar to professional Layer 7 firewalls, such as antivirus.

d. **Overlay tunnel security:** Data transmitted over overlay tunnels based on Internet links is encrypted using IPsec to ensure secure data transmission.

6. O&M design

In an enterprise-built SD-WAN scenario, O&M primarily includes CPE deployment, network service monitoring, and network service maintenance. In a carrier-resale SD-WAN scenario, O&M is more complex, as it also involves multi-tenant management and tenant portal access permission management.

MSP Z provides various SD-WAN site deployment modes for enterprise customers, including email-, DHCP-, registration center-, and USB-based deployment.

MSP Z can also take charge of the network O&M for enterprise customers. At the same time, MSP Z can provide tenant accounts with only monitoring rights to enterprise customers, as part of SD-WAN services.

CHAPTER 10

SD-WAN Components

Huawei's SD-WAN Solution consists of two types of key components: NetEngine AR series routers and iMaster NCE-WAN — an SDN controller. This chapter will detail the application scenarios and the main functions of the two components.

10.1 NETENGINE AR SERIES ROUTERS

10.1.1 Overview

Huawei NetEngine AR series routers are next-generation routers that deliver the high performance needed to fully support diversifying and cloud-based enterprise services. Built on Solar AX architecture, NetEngine AR series routers stand out with high-performance forwarding capabilities by using "CPU+NP" heterogeneous forwarding and built-in acceleration engines. They also integrate functions such as SD-WAN, routing, switching, VPN, security, and MPLS.

Flexibly deployed in enterprise headquarters or at branches, NetEngine AR series routers provide the desired egress capabilities for enterprise networks.

Figure 10.1 shows the appearance of NetEngine AR series routers.

Table 10.1 describes the architectural highlights of NetEngine AR series routers.

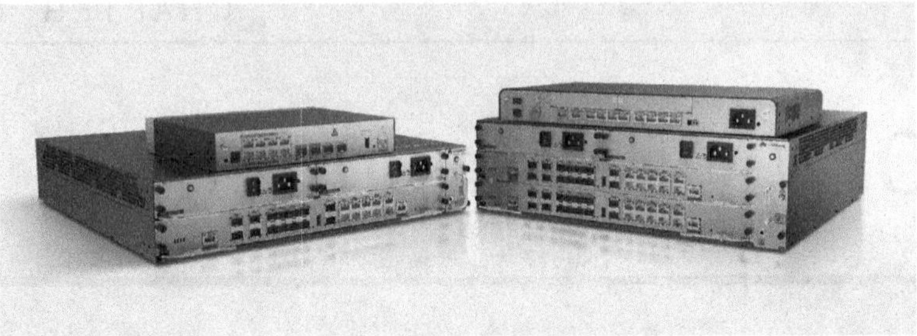

FIGURE 10.1 NetEngine AR series routers.

TABLE 10.1 Architectural Highlights of NetEngine AR Series Routers

Architectural Highlight	Benefits
Multicore processor	High-performance multicore processors deliver high-speed WAN connections, powerful route calculation, and enhanced Layer 4–Layer 7 service processing
Independent NP (SRU-400H/ SRU-600H)	The NP is a powerful hardware forwarding engine that offloads more Layer 1–Layer 4 traffic, greatly enhancing forwarding
Non-blocking switching architecture	High bus bandwidth per slot eliminates service forwarding bottlenecks
High-density service routing unit (SRU)	SRU-400H/SRU-600H provides high-density WAN ports: 14×10GE optical ports (compatible with GE optical) and 10×GE electrical ports. All WAN ports can be configured as LAN ports
Extensive network interfaces	Hot swap is supported Various flexible interface cards, such as LAN, WAN, and LTE interface cards, are available

10.1.2 Application Scenarios

10.1.2.1 Constructing SD-WAN Using Hybrid Links

Huawei NetEngine AR series routers can function as gateway devices at enterprise headquarters or branch sites to build SD-WAN. They support multiple physical links such as MPLS private lines and Internet links, which are centrally and visually managed using the SDN controller.

Figure 10.2 illustrates a typical SD-WAN network. On this network, Huawei NetEngine AR series routers offer abundant SD-WAN features and deliver the ultimate service experience for enterprise users through intelligent application identification, traffic steering, and acceleration features.

10.1.2.2 Building Different Types of VPNs over the Internet

Huawei NetEngine AR series routers provide various secure access functions to facilitate communication among branches and between the headquarters and the branches, where partners can also access enterprise resources. Secure tunnels such as IPsec VPN, GRE VPN, and Layer 2 Tunneling Protocol (L2TP) VPN can be set up for secure data access and transmission.

Figure 10.3 illustrates a typical SD-WAN network. On this network, partners authenticated and authorized by NetEngine AR series routers can remotely access enterprise resources over secure tunnels.

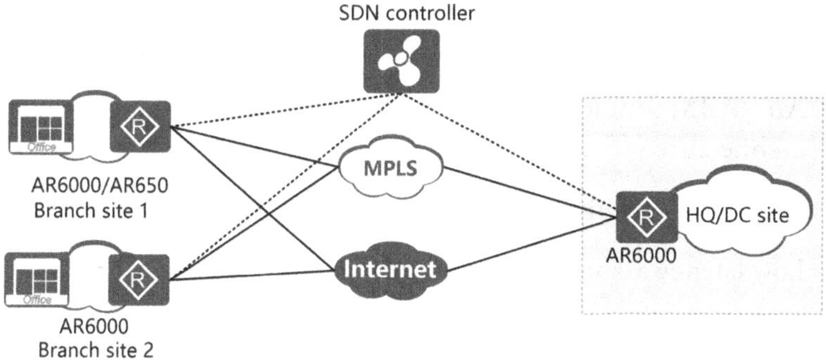

FIGURE 10.2 SD-WAN networking.

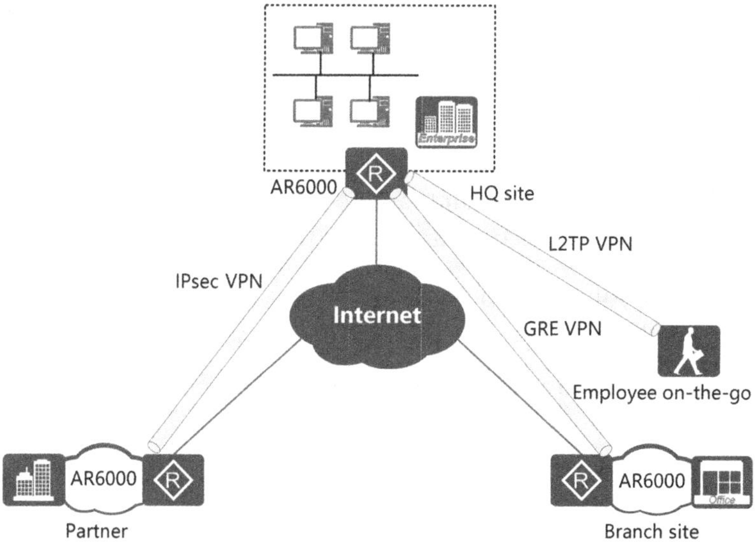

FIGURE 10.3 VPN networking.

10.1.3 Functions and Features

10.1.3.1 High Performance

Huawei NetEngine AR series routers take the lead in introducing innovative "ARM CPU+NP" heterogeneous forwarding into SD-WAN CPE. They are embedded with a rich set of hardware-based intelligent acceleration engines, such as hierarchical quality of service (HQoS), IPsec, ACL, and SA, to implement application-centric service processing at Layer 3–Layer 7. They also fully support VPN, application identification, application monitoring, HQoS, intelligent traffic steering, WAN optimization, security, and other complex services.

NetEngine AR series routers' value-added features include:

- An ARM multicore processor and non-blocking switching architecture
- Three times the industry average performance
- Low latency assurance for mission-critical services
- Unique "CPU+NP" hardware acceleration on SRUs for bottleneck-free service processing.

10.1.3.2 High Reliability

NetEngine AR series routers are designed in compliance with carrier-class standards, and therefore provide reliable and high-quality services for enterprise users. Typical features include:

- Hot-swappable cards are used. Key hardware components, such as SRUs, power supplies, and fan modules, use a redundancy design, ensuring service security and stability.
- Link backup for enterprise services improves service access reliability.
- Fault detection and judgment in milliseconds shortens service downtime.

10.1.3.3 Easy O&M

NetEngine AR series routers can be flexibly managed using the SDN controller, SNMP, or a web system, which simplifies network deployment.

They also support remote maintenance and remote centralized management, which greatly reduces maintenance costs while improving maintenance efficiency.

10.1.3.4 Service Integration

NetEngine AR series routers integrate routing, switching, VPN, security, and WLAN functions to meet continuously diversifying enterprise services at a lower total cost of ownership (TCO).

10.1.3.5 Security

NetEngine AR series routers are designed with built-in firewall, IPS, and URL filtering technologies to deliver comprehensive security defense.

10.1.3.6 SD-WAN Ready

- NetEngine AR series routers work with other SD-WAN components to build cost-effective, commercial-grade Internet connectivity.
- ZTP, including email-, USB-, and DHCP-based deployment, greatly reduces skill requirements for deployment professionals and brings devices online within minutes.
- FPI and SA of applications, as well as intelligent traffic steering based on bandwidth and link quality, ensure high-quality user experience of key applications.

10.2 IMASTER NCE-WAN

10.2.1 Introduction

iMaster NCE-WAN, an SDN controller, is a core component of Huawei's SD-WAN Solution. It manages enterprise interconnection services throughout the entire process; automates deployment of private line services; facilitates configuration of intelligent traffic steering policies; and delivers VAS management, device plug-and-play, and visualized O&M. iMaster NCE-WAN also provides open interfaces for easy interconnection with third-party systems. Figure 10.4 illustrates iMaster NCE-WAN functions.

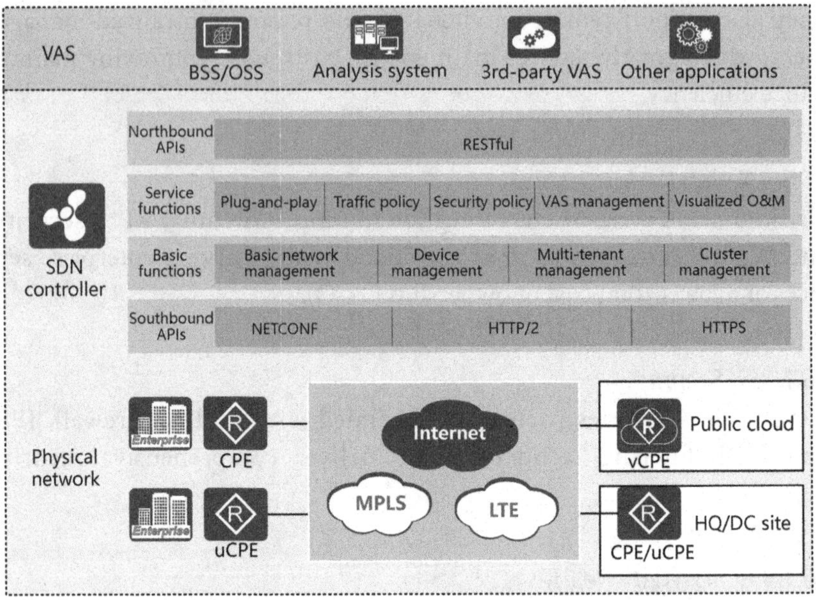

FIGURE 10.4 iMaster NCE-WAN.

iMaster NCE-WAN structure and functions include:

1. **Northbound APIs:** iMaster NCE-WAN provides standard RESTful northbound APIs for easy interconnection and integration with third-party systems, such as NMSs, orchestrators, and analyzers.

2. **Service functions:** iMaster NCE-WAN delivers comprehensive enterprise private line service functions, such as provisioning and deployment, policy control, VAS management, and visualized O&M. These functions help enterprises to simplify the process of private line service management and improve O&M efficiency.

3. **Basic functions:** iMaster NCE-WAN offers basic capabilities, including cluster management, multi-tenant management, basic network management, and device management, to name a few.

4. **Southbound APIs:** iMaster NCE-WAN provides an extensive range of southbound APIs (e.g., NETCONF, HTTP/2, and HTTPS) to connect to CPEs. NETCONF APIs deliver CPE configurations, HTTP/2 APIs transmit CPE performance data, and HTTPS APIs transmit CPE software versions and patches.

iMaster NCE-WAN uses a microservice architecture. Typical microservices include:

- **NE data collection microservice:** collects performance data reported by NEs, then formats and classifies the data, and ultimately inserts it into the database.
- **NE southbound microservice:** provides a unified southbound NE driver for upper-layer services, sorts out and schedules configurations delivered to NEs by model and version, and centrally manages NE-related resources.
- **WAN network microservice:** performs service orchestration on iMaster NCE-WAN, covering overlay network creation, topology orchestration, Internet access, and local breakout.
- **Policy configuration microservice:** performs policy orchestration on iMaster NCE-WAN, covering QoS, ACL, traffic steering, and security policies.
- **Monitoring microservice:** implements monitoring on iMaster NCE-WAN services, including monitoring link quality, traffic data, application quality, and application traffic.
- **O&M microservice:** implements O&M on iMaster NCE-WAN, covering device alarm, device upgrade, fault inspection, accurate application quality monitoring, and device fault information collection.

iMaster NCE-WAN can also be integrated with FusionInsight to implement big data analytics. Specific operations include the analysis of data collected from devices, service visualization, O&M, and monitoring.

10.2.2 Technical Highlights

10.2.2.1 Fast Deployment Accelerates SD-WAN Service Rollouts

iMaster NCE-WAN automates deployment of E2E network services. It also makes CPEs plug-and-play, facilitating quick device provisioning, configuration, and deployment. These capabilities redefine service delivery by reducing the time required to set up private line services from several months to several days. iMaster NCE-WAN also ensures that network services are available no matter the geographical reach of enterprise services.

10.2.2.2 Intelligent Traffic Steering Improves Service Experience

iMaster NCE-WAN makes it easy to configure intelligent traffic steering based on applications (including both predefined and customized applications). It provides differentiated network services for applications based on users' application quality requirements, prioritizing the experience of key applications.

10.2.2.3 On-Demand VASs Speed Up Service Provisioning

iMaster NCE-WAN supports full-lifecycle management of VNFs on uCPEs. This allows enterprises to quickly load VASs.

iMaster NCE-WAN also enables SFC orchestration for VNFs on uCPEs. Therefore, service traffic can pass through multiple VNF nodes in order, meeting various service requirements of enterprises.

10.2.2.4 Visualized O&M Offers Network-Wide Application Traffic Visibility

iMaster NCE-WAN improves O&M efficiency through visualized management of applications and links, network-wide status visibility, and real-time network status control. By monitoring and collecting statistics on real service flows, iMaster NCE-WAN visually displays quality and status trends of applications and links, facilitates quick troubleshooting, and achieves precise fault backtracking.

10.2.3 Functions and Features

10.2.3.1 Plug-and-Play

iMaster NCE-WAN stands out with its plug-and-play function required in SD-WAN scenarios. With email-based deployment or DHCP-based deployment, CPEs can automatically register with iMaster NCE-WAN and then go online after being powered on. Thanks to the plug-and-play function, CPEs at branch sites can quickly go online and provide related services.

10.2.3.2 Tunnel Management

The number of enterprise branches increases, so does the number of communication channels among branches and between the headquarters and the branches. This leads to a more complex network structure. To meet enterprises' requirements for automated networking, iMaster NCE-WAN allows tunnels to be automatically and dynamically established on demand, while ensuring enterprise service security through tunnel encryption.

Featured tunnel management functions include:

- Configuring and orchestrating the VPN network topology
- Supporting full-mesh, hub-spoke, partial-mesh, and hierarchical networking options
- Supporting IPsec encryption

10.2.3.3 Intelligent Traffic Steering

Huawei's SD-WAN solution is capable of transmission over hybrid links. iMaster NCE-WAN delivers application-based intelligent traffic steering to ensure user experience of key applications. Specifically, traffic of quality-demanding applications is transmitted over private lines, while that of quality-insensitive ones is transmitted over Internet links. When the network fails or the quality of a link is unstable, traffic is flexibly switched to qualified links, improving service experience.

Typical intelligent traffic steering features include:

- Identifying predefined and customized applications
- Configuring traffic steering policies based on delay, jitter, packet loss rate, and bandwidth utilization

10.2.3.4 On-Demand VASs

Traditionally, different pieces of hardware provide VASs at enterprise sites. However, fixed hardware functions and complex service deployment and delivery create a disadvantage. To implement quick on-demand VAS deployment, uCPEs now carry virtualized VASs. iMaster NCE-WAN can manage and control VNFs on uCPEs to accelerate VAS deployment on demand. Featured functions include:

- Supporting lifecycle management of VNFs on uCPEs, including deploying, suspending, deleting, stopping, and restarting VNFs
- Remotely logging in to VNFs
- Performing SFC orchestration for VNFs on uCPEs

- Monitoring VNFs on uCPEs, including viewing the management IP address, CPU, random access memory (RAM), and running/operational status of a VNF

10.2.3.5 Visualized O&M

iMaster NCE-WAN enables visualized management of sites, links, and applications, to quickly locate faults and improve O&M efficiency.

Typically, iMaster NCE-WAN can visually display:

- **Network-wide sites:** healthy sites, sites with the worst health score, and site lists
- **A specific site:** average AQM score, bandwidth utilization, throughput trend, and applications with the worst AQM score
- **Network-wide links:** links with the worst link quality measurement (LQM) score, top traffic, and link lists
- **A specific link:** LQM score trend, throughput trend, top *N* applications by traffic, and AQM score distribution of applications
- **Network-wide applications:** AQM score distribution of applications, applications with the worst and highest AQM scores, top *N* applications by traffic, and application lists
- **A specific application:** AQM score trend and throughput trend

10.2.3.6 Performance Monitoring

iMaster NCE-WAN is designed with abundant performance monitoring functions, helping network administrators stay informed of network and service status.

Typical performance monitoring functions include:

- **Site monitoring:** displays network service data by site
- **Inter-site monitoring:** displays network service data by inter-site link
- **Application monitoring:** displays network service data by application

10.2.3.7 Log Management

The pervasive use of networks, the ever-increasing complexity of application environments, and the continuous expansion of the network scale require more intelligent and effective ways to manage networks. In response, iMaster NCE-WAN facilitates network management by providing the following three types of management logs:

1. Security log

   Security logs record user login, creation, logout, modification, and deletion.

   Auditors can export tenants' security logs. System administrators, auditors, and operators can view tenants' security logs.

2. Run log

   Run logs record events that users are interested in, such as patch loading and upgrading, as well as some exceptions.

   Auditors can export tenants' run logs. System administrators can view tenants' run logs.

3. Operation log

   Operation logs record all user operations that involve data changes. Other operations that do not involve data changes, such as query operations, are not recorded into operation logs.

   Auditors can export tenants' operation logs. System administrators, auditors, and operators can view tenants' operation logs.

CHAPTER 11

SD-WAN Outlook

TODAY'S ONGOING TRENDS, SUCH as globalization, digital transformation, and service automation and elasticity, are posing unprecedented requirements for new networks. Emerging cloud-native models, 5G, IoT, smart devices, cybersecurity threats, and immersive applications have had a wide impact on the architecture and operations of IT networks and will continue to do so in the future. We expect that the scale, complexity, and dynamics of these driving forces will soon exceed our capabilities to handle them. For this reason, emerging networks are using disruptive technologies, such as AI, machine learning, and automation, to ensure normal network operations while improving our decision-making abilities.

Technology evolutions and industry trends are driving SD-WAN to move forward, evolve, and improve. This chapter will focus on the perspectives of technology evolutions and industry trends to describe the impact of new technologies and trends on SD-WAN while also depicting what the future holds for SD-WAN.

11.1 TECHNOLOGY EVOLUTION

As SD-WAN gains popularity worldwide, many new network-related technologies are being rapidly developed. This, in turn, is driving SD-WAN technologies forward.

By analyzing the SD-WAN evolution insights of standards organizations such as ONUG and fully leveraging Huawei's years of SD-WAN expertise, we have divided SD-WAN technology evolution into three distinct phases, as illustrated in Figure 11.1.

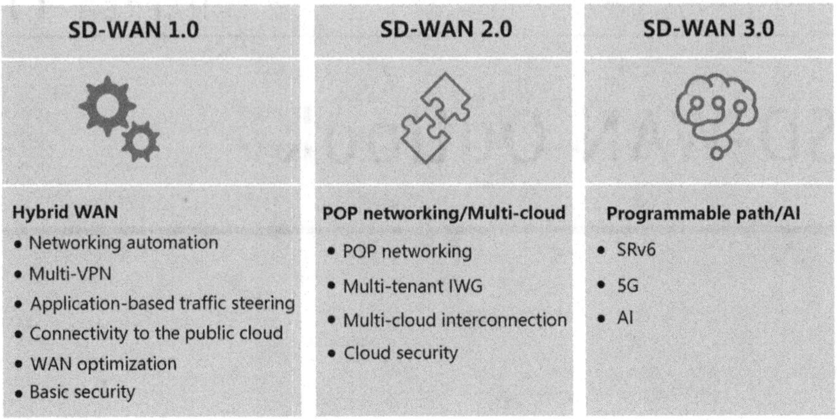

FIGURE 11.1 SD-WAN technology evolution trend.

SD-WAN 1.0 primarily uses a hybrid WAN as the foundation. This phase makes the following possible for enterprises:

- Implementing hybrid networking of multiple WAN links under the centralized management and control of the SDN controller
- Flexibly scheduling and easily configuring multiple WAN links by using application identification and traffic steering
- Preliminarily achieving service-oriented orchestration and automation
- Realizing basic features, such as site-to-Internet, multi-VPN isolation, basic topology planning, basic security, and connectivity to the public cloud

In essence, SD-WAN 1.0 meets enterprises' basic demands for WAN interconnections in the hybrid WAN and cloud era.

At present, we are in the SD-WAN 2.0 phase, which focuses more on the carrier resale scenarios and provides operable capabilities compared with SD-WAN 1.0 that is primarily centered on enterprise-built scenarios. SD-WAN 2.0 further improves enterprise WAN interconnections by enhancing a number of vital carrier-specific SD-WAN features, such as multi-tenant management, IWG, and POP networking, on the basis of SD-WAN 1.0.

ONUG also advocates the expansion of cloud connections to hybrid cloud and multi-cloud connections in the SD-WAN 2.0 phase. From the perspective of carriers and MSPs, SD-WAN 2.0 becomes a new high-quality service in the enterprise 2B market.

Looking ahead to the SD-WAN 3.0 phase, the flexibility, intelligence, and scope of SD-WAN connectivity will be further enhanced by using disruptive technologies such as SRv6, 5G, and AI. Specifically, SRv6-based programmable path technology will greatly improve the flexibility of SD-WAN traffic steering. In addition, 5G will greatly increase the coverage of SD-WAN. And further to this, AI will make SD-WAN more intelligent in terms of application identification and advanced O&M analysis. We can therefore conclude that SD-WAN will enter a wider range of fields such as IoT.

The following sections explore the possible technical trends and solution changes in the future SD-WAN 3.0 phase.

11.1.1 5G

5G — the 5th generation of wireless networks — offers many technical advantages. Three of the key technical differentiators are as follows:

- **Higher speeds:** Compared with 4G technologies, 5G increases the peak rate from 100 Mbit/s to 10–20 Gbit/s.
- **More connections:** As defined by International Telecommunication Union (ITU), 5G will support one million IoT connections per square kilometer, 10 times more than 4G technologies.
- **Lower latency:** The 5G E2E latency defined by ITU is as low as 1 ms, which is 1/10 of the 4G E2E latency.

5G has a wide range of application scenarios and industry values. ITU defines three typical scenarios for 5G: enhanced mobile broadband (eMBB), massive machine-type communications (mMTC), and ultra-reliable low-latency communication (URLLC), as illustrated in Figure 11.2.

5G networks stand out with higher bandwidth, lower latency, greater coverage, and lower costs. As the latest wireless technology, 5G makes "last mile" access more convenient and faster than ever. With these advantages, 5G is ideal for providing WAN links in the SD-WAN solution, with the following three typical benefits:

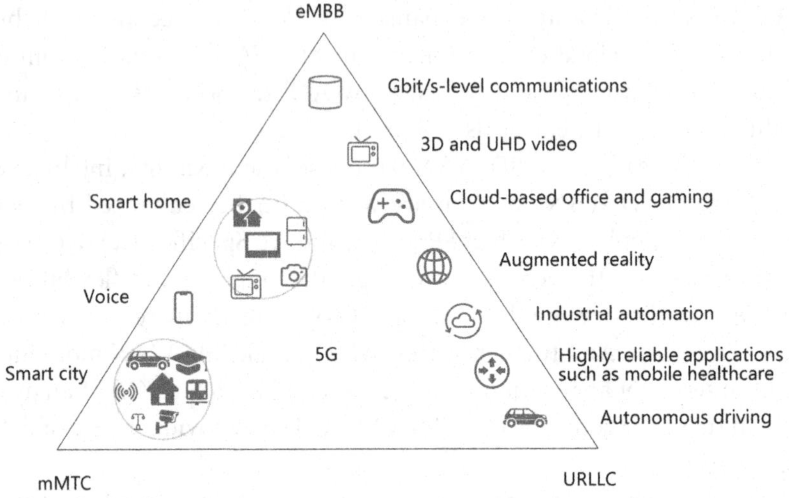

FIGURE 11.2 Typical 5G scenarios and applications defined by ITU.

- 5G delivers high bandwidth and low latency needed for HD video services, so video acceleration functions on SD-WAN, such as FEC, can be fully utilized.
- "5G+MSTP" and access point name (APN)-based network slicing in 5G ensure that SD-WAN can maximize the use of network bandwidth while providing more competitive application-based traffic steering.
- The flexible uplink networking and on-demand network deployment features of 5G make it possible to quickly deploy sites using simplified procedures. This, in turn, improves the competitiveness of the SD-WAN ZTP solution.

The following uses the financial services industry as an example to describe the impact of 5G on the entire industry. 5G has spawned a large number of new financial applications, thereby accelerating the pervasive use of 5G in enterprises, as illustrated in Figure 11.3.

In addition, 5G meets the high bandwidth and low latency requirements of AR/VR applications and therefore supports a large number of new financial applications, such as VR-assisted securities, VR-assisted financial transactions, and VR delivery.

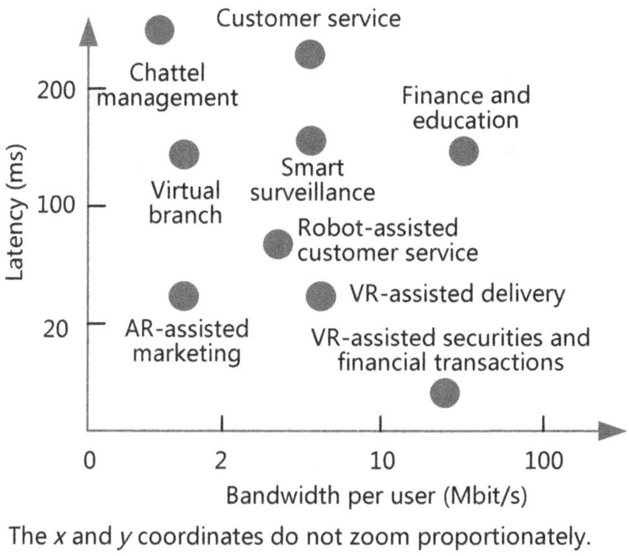

FIGURE 11.3 5G driving new applications in the financial services industry.

5G promotes the prosperity of financial applications, and when paired with SD-WAN, this combination delivers the desired financial solutions that feature high bandwidth, low latency, and optimal experience.

Huawei has taken the lead in proposing a "5G+SD-WAN" solution that is future-proof in multiple financial scenarios.

11.1.1.1 Scenario 1: "5G+SD-WAN" for Unstaffed Bank Branches

"5G+SD-WAN" helps deploy unstaffed bank branches more efficiently. Specifically, multiple types of links, such as 5G and Internet links, are introduced to reduce link costs. Professionals are not needed for onsite operations, simplifying deployment and O&M. All terminals in unstaffed bank branches, including smart home devices and security devices, are managed in a unified manner.

"5G+SD-WAN" unstaffed bank branches deliver tangible benefits typified by high WAN link utilization, low costs, and zero interruption of key services, as illustrated in Figure 11.4.

Currently, some 5G technologies such as network slicing are not sufficiently mature. For this reason, carriers construct Non-Stand Alone (NSA) networks. However, considering network service quality and security, a 5G dedicated base station and an indoor distribution system

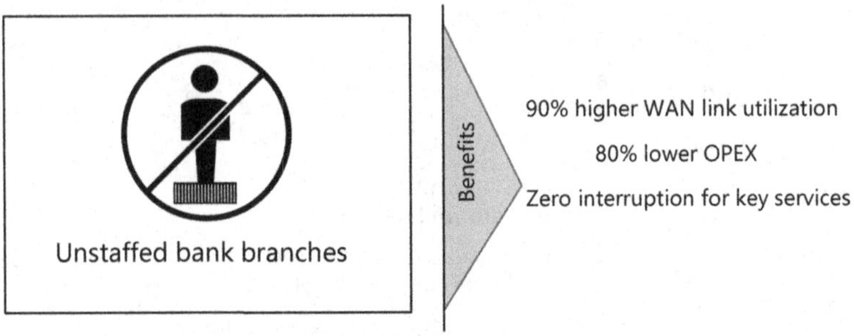

FIGURE 11.4 "5G+SD-WAN" benefits for unstaffed bank branches.

should be deployed in unstaffed bank branches to access a carrier's Stand Alone (SA) network. In this way, frequency resources of a radio can be exclusively occupied, and uplink bandwidth and computing resources of a base station can be exclusively used. Furthermore, a dedicated Virtual Private Dial Network (VPDN) must be constructed for unstaffed bank branches.

At unstaffed bank branches, banking and IoT services are transmitted on the underlay network constructed using 5G and MSTP private lines. On top of the underlay network, the SD-WAN overlay network provides highly reliable and differentiated access networks for services of different levels, implementing intelligent traffic steering based on service types and link quality requirements.

Internet links can also be deployed at unstaffed bank branches to carry Internet traffic and disaster recovery traffic. Different tunnels can be established for 5G, MSTP, and Internet links based on security requirements to ensure data security.

The bank headquarters can establish a 5G access area so that it can directly connect to unstaffed bank branches through 5G. Each unstaffed bank branch then directly connects to the carrier's 5G CPE via a gigabit wired network.

As part of Huawei's "5G+SD-WAN", the high-performance NetEngine AR series is an ideal choice for 5G CPEs. By delivering high-speed 5G and SD-WAN service processing capabilities and implementing intelligent traffic steering, NetEngine AR series ensures zero interruptions of key services such as banking and security services. Additionally, SD-WAN devices are plug-and-play, implementing automatic service provisioning and visualized WAN O&M.

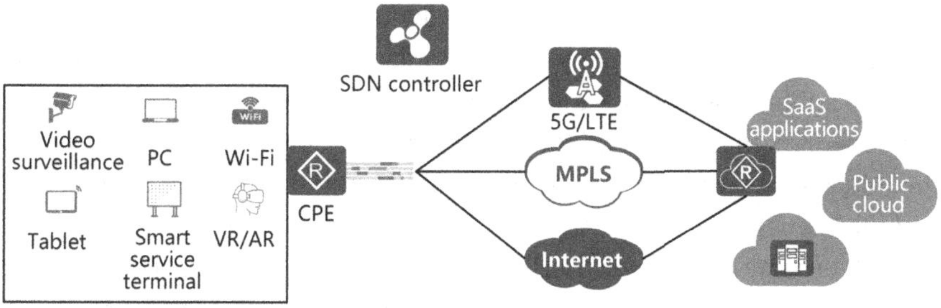

FIGURE 11.5 "5G+SD-WAN" for smart bank branch interconnection.

11.1.1.2 Scenario 2: "5G+SD-WAN" for Smart Bank Branch Interconnection

5G links can be used as WAN links on SD-WAN. Using 5G, bank branches can quickly implement SD-WAN interconnection, as illustrated in Figure 11.5.

In addition to ensuring network quality, "5G+SD-WAN" outperforms traditional private lines when it comes to bank branch interconnection. This is because "5G+SD-WAN" shortens the provisioning period and reduces maintenance costs.

1. SD-WAN high-quality connections powered by 5G

   High-speed 5G access, link resource pooling, and unified scheduling greatly improve bandwidth utilization. Further, application-based intelligent traffic steering and WAN optimization — both unique to SD-WAN — ensure that mission-critical applications are free from interruptions.

2. Unified management and simplified O&M

   LAN and WAN can be centrally managed using one platform, eliminating the need to deploy multiple separate management systems and therefore greatly simplifying O&M. Devices are plug-and-play, services are deployed automatically, and services are rolled out in minutes.

11.1.2 SRv6

11.1.2.1 SR Overview

As we move forward, services become increasingly diversified, posing varied requirements on the network. For example, real-time voice/video applications usually require paths with low latency and small jitter, while big data applications call for high-bandwidth channels with low packet loss rates. If we were to take a conventional approach and enable the network to adapt to services, we would encounter two pressing issues. One is the network's inability to adapt to the rapid development of services, and the other is the growing complexity and difficulty of network deployment and maintenance.

To overcome these issues, we need to introduce a new approach: enable services to drive the network, that is, use services to define network architecture. This means that applications raise specific requirements (e.g., latency, bandwidth, and packet loss rate), with the SDN controller then collecting information such as network topology, bandwidth utilization, and latency, and calculating explicit paths according to services' requirements.

That's where Segment Routing (SR) comes in. With SR, an explicit path can be easily defined. Because nodes on the network only need to maintain SR information, they can efficiently respond to real-time and rapid service development.

SR is a protocol designed to forward data packets on a network based on source routes. It divides a network path into several segments and assigns a segment ID (SID) to each segment and forwarding node. The segments and nodes are sequentially arranged into a segment list to form a forwarding path.

SR encapsulates segment list information that identifies a forwarding path into a data packet header for transmission alongside the data packet. After receiving the data packet, the receive end parses the segment list. If the top SID in the segment list identifies the local node, the node removes the SID and proceeds with the follow-up procedures. Otherwise, the node forwards the packet to a next node.

SR has the following features:

- Supports smooth network evolution by extending existing protocols, such as IGPs
- Supports both SDN controller-based centralized control and forwarder-based distributed control, providing a balance between centralized control and distributed control
- Supports fast interaction between a network and its upper-layer applications

11.1.2.2 SRv6 Overview

Future networks will be 5G oriented, meaning that networks must transform in accordance with the key trends of simplicity, low latency, and SDN/NFV support.

To develop 5G networks, customers hope to use IPv6 addresses so that they are able to implement VPNs more easily. SRv6 was developed as a viable solution to meet this need. It uses the existing IPv6 forwarding technologies and extends the IPv6 header to implement label forwarding-like processing. Some IPv6 addresses are defined as instantiated SIDs, and each SID has its own explicit functions. SIDs are operated to implement simplified VPNs and flexibly plan paths.

An IPv6 packet consists of a standard IPv6 header, extended headers (0, …, *n*), and payload. To implement SRv6 based on the IPv6 forwarding plane, an IPv6 extension header, called segment routing header (SRH), is added.

An SRH specifies an explicit path and stores IPv6 segment list information. A segment list functions the same in both IPv6 SR and IPv4 SR.

The ingress adds an SRH to an IPv6 packet, and each transit node forwards the packet based on path information carried in the SRH.

Figure 11.6 shows the SRH header format.

Figure 11.7 shows the abstract SRH format.

0	7	15	23	31 bit
Next Header	Hdr Ext Len	Routing Type	Segments Left	
Last Entry	Flags	Tag		
Segment List [0] (128 bit IPv6 address)				
......				
Segment List [n] (128 bit IPv6 address)				

FIGURE 11.6 SRH header format.

IPv6 Destination Address=Segment List [N]

SRH(Segment List=n)
<Segment List [0], Segment List [1],
Segment List [2], ..., Segment List [n]>

FIGURE 11.7 Abstract SRH format.

- **IPv6 Destination Address:** Also called IPv6 DA, it is a fixed value in an ordinary IPv6 packet. In SRv6, an IPv6 DA only identifies a next hop of an existing packet and is changeable.
- **<Segment List [0], Segment List [1], Segment List [2],..., Segment List [*n*]>:** SRv6 packet segment list. Similar to an MPLS label stack in SR MPLS, it is generated on the ingress. Segment List [0] is the first segment list to be processed on an SRv6 path, followed by Segment List [1], Segment List [2], and so on. Segment List [*n*] is the $n+1$ segment list to be processed.

An SRv6 SID is an IPv6 address. A wide range of SRv6 SIDs associated with different functions exist, among which an Endpoint SID (also End SID) identifies the prefix of a destination address. After an End SID is generated on a node, the node propagates the SID to all other nodes in the SRv6 domain through an IGP. All nodes in the SRv6 domain know how to implement the instruction bound to the SID.

Figure 11.8 shows the data forwarding process based on an End SID.

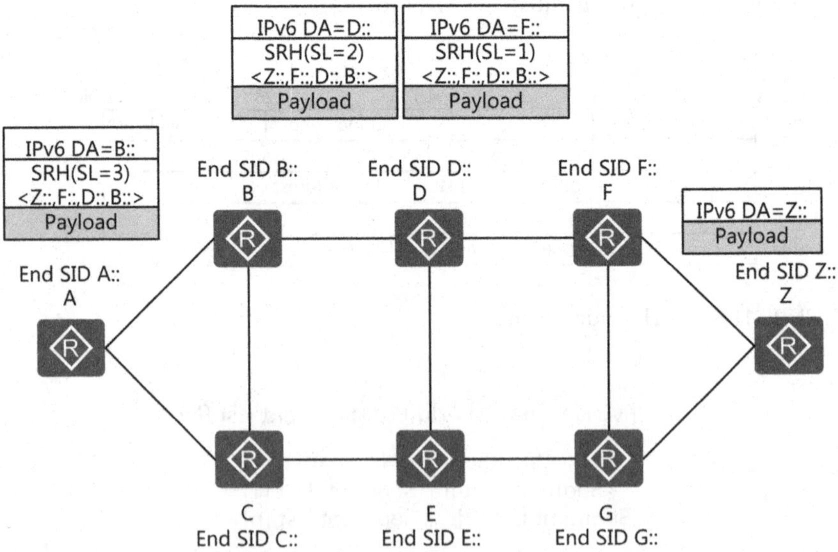

FIGURE 11.8 End SID-based data forwarding process.

The process of forwarding data based on End SIDs from node A to node Z is described as follows:

1. Node A pushes an SRH into a data packet. The path information is <Z::, F::, D::, B::>, and the IPv6 DA is B::.
2. Each time the packet passes through a node, for example, nodes B and D, a node searches the local SID table based on the IPv6 DA, checks the End type, and searches the IPv6 FIB table. After finding a matching outbound interface and next hop, this node then reduces the SL value by one, and changes the IPv6 DA.
3. After the packet arrives at node F, node F searches the local SID table based on the IPv6 DA and checks that the End type is used. It continues to query the IPv6 FIB table and finds the outbound interface. In addition, node F reduces the SL value to 0 and changes the IPv6 DA to Z::. The path information <Z::, F::, D::, B::> becomes meaningless, and therefore node F uses the Penultimate Segment POP (PSP) of the SRH to remove SRH path information and forward the packet to node Z.

The biggest advantage of SRv6 is that it can flexibly define E2E forwarding paths for services. Therefore, SRv6 is especially suitable for combining with SDN to build truly programmable networks.

In the WAN interconnection field, SRv6 and SD-WAN are converged, enabling centralized control through the SDN controller so that E2E service forwarding paths can be specified on the source CPE. This gives rise to a number of high-value scenarios, for example, association between the SD-WAN overlay network and a carrier's underlay network.

SRv6 is usually deployed on a carrier's underlay network (WAN backbone network). Using SRv6, an enterprise's SD-WAN overlay network can be associated with the carrier's underlay network. In this way, the advantages of traffic steering based on application SLA on the carrier's underlay network can be fully leveraged among the enterprise's SD-WAN services, thereby delivering high-quality enterprise application experience.

Figure 11.9 shows this solution for association between an enterprise's SD-WAN overlay network and a carrier's underlay network.

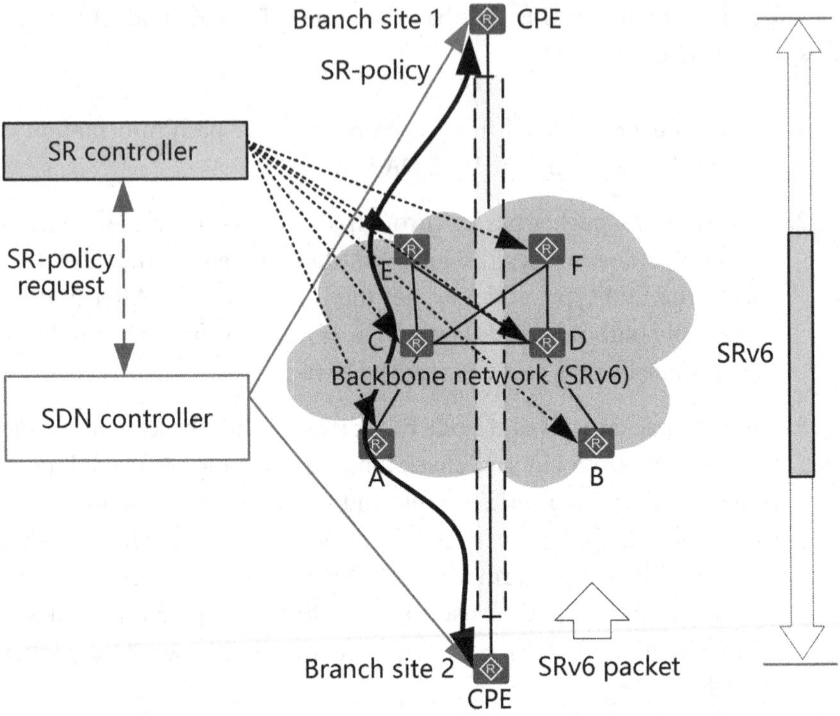

FIGURE 11.9 Association between an enterprise's SD-WAN overlay network and a carrier's underlay network.

This solution decouples a carrier's underlay network from an enterprise's SD-WAN overlay network. Specifically, an SR controller on the carrier's underlay network is responsible for collecting information (such as the topology, link quality, and link bandwidth) and allocating resources to the entire underlay network. The SR controller calculates E2E SR paths that fully meet SLA requirements.

The SDN controller defines SLAs based on application requirements and requests the SR controller to calculate an underlay path. The SR controller then calculates the segment list meeting the SD-WAN SLA requirements and returns the path calculation result to the SDN controller. The result can be identified using SR-Policy.

A CPE at an enterprise branch identifies applications, queries the SID corresponding to SR-Policy from the SDN controller, and then adds the SID to the SD-WAN overlay tunnel packet. The packet is then forwarded to the first-hop device on the carrier's underlay network. This device identifies the SID carried in the packet, adds the SR path calculated by the SR controller into the packet, and continues to forward the packet. Subsequently, the packet is forwarded hop by hop along the SLA-compliant SR path calculated by the SR controller, thereby meeting the forwarding performance requirements of services.

This solution has little impact on the performance of the CPE at the branch. And because the SRv6 header is configured and delivered by the SDN controller, there is no additional burden on the CPE's control plane. The SRv6 data plane still uses IPv6 packet forwarding technology, which is similar to other IPv6 overlay tunnel technologies in terms of the forwarding overhead.

This solution maintains a loose coupling between an enterprise's SD-WAN overlay network and a carrier's underlay network, while effectively implementing service association between the two networks. The resulting costs are low, but the benefits are plentiful.

11.1.3 AI

AI is a new technical science that studies and develops theories, methods, techniques, and application systems for simulating and extending human intelligence. AI involves a wide range of fields, such as machine learning and computer vision. In general, one of the major goals of AI research is to make machines capable of doing complex tasks that used to require humans.

AI theories and technologies have continued to develop ever since AI was conceptualized, and therefore its application scope has been constantly expanding. There is never a better time than now to apply AI to the SD-WAN field.

The following uses application identification as a typical AI use case to predict the future of AI in SD-WAN.

Because traffic steering and visibility, QoS, WAN optimization, and WAN scheduling all depend on application identification results, accurate identification is essential for SD-WAN.

As described earlier, existing application identification technologies on SD-WAN primarily include the following:

- **Protocol identification:** Applications are differentiated based on the IP address and port number of the application server.
- **Signature identification:** Applications are differentiated based on the signatures of service packets.

- **DNS association identification:** Applications are differentiated based on the domain name of the application server.

These legacy technologies face many challenges, such as:

- The IP address and port number of an enterprise's application server change, incurring high costs in maintaining protocol identification entries.
- Application types cannot be identified through signature identification when application traffic is encrypted.
- Not all server IP addresses have corresponding domain names. Therefore, DNS association identification can solve only certain problems.
- A single application consists of multiple services, and the existing application identification methods require human intervention to manually specify the composition relationship.
- Different enterprises have different customized applications. New applications are continuously added within the same enterprise. Identifying each new type of application requires human involvement.

AI can make a difference to application identification. Typical scenarios include the following:

- Application identification based on traffic behavior characteristics

  Service types, such as voice, video, file download, and gaming, are identified. The flow information to be collected includes the packet lengths and intervals of the first M packets of service flows.
- Application identification based on neural networks

  Specific applications, such as instant messaging applications, are identified in a supervised manner. The flow information to be collected includes the first N bytes of service flows.
- Unsupervised application identification

  An enterprise's private applications are identified in zero-touch mode. The flow information to be collected includes the flow creation/termination time, DNS response packets, and URL request packets.

Supervised AI algorithms need to collect data in advance and mark the data. Then, a model is trained based on the data set. However, for an enterprise's private applications, it is difficult to collect and mark data onsite. Furthermore, different enterprises have different private applications, compounding the issue. In this case, it is costly and unfeasible to use supervised AI algorithms. Considering this, SD-WAN can focus on using unsupervised AI algorithms to implement application identification.

In the case of encrypted private application traffic within an enterprise, unsupervised AI algorithms use machine learning to classify and identify network traffic. This approach reduces the reliance on the knowledge of application deployment planning and application flow characteristics while also eliminating the need to mark network traffic, thereby reducing the implementation difficulty.

Unsupervised AI algorithms also make it easy to automatically and continuously identify new applications and recognize inter-application dependency relationships. That is, these algorithms provide visibility of both single applications and inter-application topology relationships.

More importantly, unsupervised AI algorithms can identify applications without user intervention. They use traffic behavior characteristics and rules (topology structure analysis) to distinguish the server from the client. Clustering algorithms and graph analysis algorithms are also used to perform priority-involved, multidimensional, converged clustering analysis from the aspects of time correlation, periodicity, and DNS. The final application identification results are marked and presented using an automatic marking method.

11.2 NEW INDUSTRY CHANGES

As we move into the SD-WAN large-scale deployment phase from the early stages of trial commercial use, the entire SD-WAN industry is undergoing the following new changes, in addition to the technological evolution trends mentioned earlier.

1. Enterprise WAN software and services markets grow rapidly

   A growing number of carriers and MSPs launch a full range of SD-WAN solutions. Driven by this, the role of SD-WAN service providers becomes more prominent than ever. Unlike some large enterprises that may build their own global SD-WAN networks, the majority of small- and medium-sized enterprises tend to obtain

WAN hosting services from service providers, because they cannot efficiently maintain branch networks independently. Given this, service providers need to provide SD-WAN services to stay ahead of the competition.

The past and current SD-WAN developments focus primarily on improving hosting services, for example, improving multi-tenant management and enhancing centralized management. Future SD-WAN technologies will realize joint hosting services across service providers, thereby ushering in global partnerships. It is foreseeable that SD-WAN technologies will also build on open APIs to better integrate and collaborate with other heterogeneous network platforms and service systems.

2. Enterprises focus less on cost saving but more on application experience

   Many enterprise users believe that SD-WAN can reduce enterprises' costs, as it uses low-cost links — such as Internet links — to replace expensive private lines. If an enterprise moves away from expensive MPLS private lines and uses more cost-effective Internet links instead, the enterprise can certainly save costs. This, however, requires the use of network optimization technologies to solve problems caused by the unpredictability of Internet links. And because most enterprises build hybrid networks, they face more bandwidth pressure as the traffic increases. Due to these reasons, enterprise costs are not really reduced as expected.

   This means that enterprises should not aim only to save cost. Instead, they should pay more attention to application experience. Carriers are already aware of this and do not focus heavily on cost saving, because they firmly believe that SD-WAN is more valuable in specific applications, which is more important than merely saving network construction costs.

3. Enterprises focus more on security

   After an enterprise deploys a new SD-WAN or migrates its legacy network to SD-WAN, the network architecture changes and branches can directly access cloud services through local breakout. This change has a profound impact on network security, especially when the number of devices at branch sites that connect to the cloud increases from one to hundreds or even thousands. As such, we should rethink the architecture used in enterprise security systems, paying close attention to the security of clouds, virtual devices, and containerized services.

SD-WAN service providers will strengthen cooperation with security service providers to further integrate security functions, such as next-generation firewalls, unified threat management, and antivirus capabilities. Enterprises can flexibly deploy network security functions at branches, in DCs, or on the cloud.

4. Markets become crowded with players of all types, so market consolidation accelerates

   In Gartner's Magic Quadrant for SD-WAN, there are more than 20 SD-WAN service providers, ranging from startups to small service providers and even large network device vendors. Their businesses span multiple fields, such as network, security, and WAN acceleration. It is expected that more MSPs and edge security vendors will enter the SD-WAN market in the future, providing enterprises with more than 50 options.

   Having a large number of options is beneficial for enterprises. Ultimately, however, this is unfavorable to the market, because supply will gradually exceed demand, which means the SD-WAN market will inevitably be further consolidated. Many vendors who provide only basic SD-WAN functions will be unable to fully meet enterprises' requirements on WANs and therefore be merged or acquired by stronger players. These mergers and acquisitions will optimize the entire SD-WAN market and accelerate the deployment of SD-WAN among global enterprises. As the smaller SD-WAN players decline, the future market will witness more mergers and acquisitions.

11.3 REVIEW AND PROSPECT

SD-WAN is one of the hottest technologies in today's network world and is having a profound impact on enterprise WAN and IT architectures. This is why we have developed this book, so that we can take a dive deep into the essence and core functions of SD-WAN based on Huawei's field-proven practices and explain the main functions, features, implementation principles, and successful use cases of SD-WAN.

Looking ahead, we believe that the market potential of SD-WAN will be huge. According to Gartner's latest report, the global software-defined

WAN infrastructure market is expected to grow by nearly 31% year-on-year by 2023. Similarly, IDC's latest forecast of the SD-WAN market shows that the market will skyrocket to US$5.25 billion in 2023 — a major increase from US$1.4 billion in 2018. Undoubtedly, SD-WAN will have a bright future and the SD-WAN market will offer significant revenue streams.

Let's work together and keep exploring to make SD-WAN an indispensable "key" to the success of enterprises' IT digitization and globalization.

Acronyms and Abbreviations

2B	To Business
ACL	Access Control List
AES	Advanced Encryption Standard
AF	Assured Forwarding
A-FEC	Adaptive-Forward Error Correction
AI	Artificial Intelligence
API	Application Programming Interface
APN	Access Point Name
AQM	Application Quality Measurement
AR	Augmented Reality
ARP	Address Resolution Protocol
AS	Autonomous System
ASBR	Autonomous System Boundary Router
ASIC	Application Specific Integrated Circuit
ASN.1	Abstract Syntax Notation One
AWS	Amazon Web Services
AZ	Availability Zone
BBR	Bottleneck Bandwidth and Round-Trip Propagation Time
BD	Bridge Domain
BE	Best Effort
BGP	Border Gateway Protocol
BGP-LS	Border Gateway Protocol-Link State
BSS	Business Support System
CA	Certificate Authority
CAPEX	Capital Expenditure
CAR	Committed Access Rate

CIFS	Common Internet File System
CLI	Command Line Interface
CMF	Configuration Management Framework
CPCAR	Control Plane Committed Access Rate
CPE	Customer-Premises Equipment
CPU	Central Processing Unit
CRM	Customer Relationship Management
CS	Class Selector
DC	Data Center
DCN	Data Center Network
DH	Diffie–Hellman
DHCP	Dynamic Host Configuration Protocol
DNS	Domain Name System
DS	Differentiated Service
DSCP	Differentiated Services Code Point
DSL	Digital Subscriber Line
DTLS	Datagram Transport Layer Security
E2E	End-to-End
EBGP	External Border Gateway Protocol
ECMP	Equal-Cost Multi-Path
ECS	Elastic Cloud Server
EF	Expedited Forwarding
EIP	Elastic IP Address
eMBB	Enhanced Mobile Broadband
ERP	Enterprise Resource Planning
ESN	Equipment Serial Number
EVI	EVPN Instance
EVPN	Ethernet Virtual Private Network
FEC	Forward Error Correction
FPI	First-Packet Identification
FTP	File Transfer Protocol
GRE	Generic Routing Encapsulation
GUI	Graphical User Interface
HTTP	Hypertext Transfer Protocol
HTTPS	Hypertext Transfer Protocol Secure
IBGP	Internal Border Gateway Protocol
IDEA	Information Digitalization and Experience Assurance
IETF	Internet Engineering Task Force

IGP	Interior Gateway Protocol
IGW	Internet Gateway
IKE	Internet Key Exchange
IM	Instant Messaging
IP FPM	IP Flow Performance Measurement
IPS	Intrusion Prevention System
IPsec	Internet Protocol Security
ISP	Internet Service Provider
ITU	International Telecommunication Union
ITU-T	International Telecommunication Union-Telecommunication Standardization Sector
IWG	Interworking Gateway
IXP	Internet eXchange Provider
L2TP	Layer 2 Tunneling Protocol
L2VPN	Layer 2 Virtual Private Network
L3VPN	Layer 3 Virtual Private Network
LAN	Local Area Network
LLQ	Low-latency Queuing
LQM	Link Quality Measurement
LSO	Lifecycle Service Orchestration
LTE	Long-term Evolution
LZ77	Lempel-Ziv
MAC	Media Access Control
MAN	Metropolitan Area Network
MCE	Multi-VPN-Instance Customer Edge
MEF	Metropolitan Ethernet Forum
MIB	Management Information Base
mMTC	Massive Machine-Type Communications
MP-BGP	Multi-Protocol BGP
MP-EBGP	Multi-Protocol EBGP
MPLS	Multi-Protocol Label Switching
MSP	Managed Service Provider
MSTP	Multi-Service Transport Platform
NAPT	Network Address and Port Translation
NAT	Network Address Translation
NE	Network Element
NETCONF	Network Configuration Protocol
NFV	Network Functions Virtualization

NLRI	Network Layer Reachability Information
NMS	Network Management System
NP	Network Processor
NQA	Network Quality Analysis
NSA	Non-Stand Alone
NTP	Network Time Protocol
NVE	Network Virtualization Edge
NVGRE	Network Virtualization using Generic Routing Encapsulation
NVO3	Network Virtualization over Layer 3
O&M	Operations and Maintenance
OA	Office Automation
ONF	Open Network Foundation
ONUG	Open Networking user Group
OPEX	Operating Expense
OS	Operating System
OSPF	Open Shortest Path First
OSS	Operational Support System
OTN	Optical Transport Network
P2P	Peer-to-Peer
PA	Provider Address
PACK	Periodic-Acknowledge
PCEP	Path Computation Element Communication Protocol
PKI	Public Key Infrastructure
POC	Proof of Concept
POP	Point of Presence
PSN	Packet Switched Network
PSP	Penultimate Segment POP
QoS	Quality of Service
RAM	Random Access Memory
RD	Routing Domain
RD	Route Distinguisher
REST	Representational State Transfer
RFC	Request for Comments
RLE	Run Length Encoding
RPC	Remote Procedure Call
RR	Route Reflector
RS	Reed–Solomon

RSA	Rivest–Shamir–Adleman
RT	Route Target
RTT	Round-Trip Time
SA	Security Association
SA	Service Awareness
SD	Software Defined
SDH	Synchronous Digital Hierarchy
SDN	Software-Defined Networking
SD-WAN	Software-Defined Wide Area Network
SF	Service Function
SFC	Service Function Chain
SHA	Secure Hash Algorithm
SID	Segment ID
SLA	Service Level Agreement
SMB	Server Message Block
SMI	Structure of Management Information
SNMP	Simple Network Management Protocol
SRv6	Segment Routing over IPv6
SRH	Segment Routing Header
SRU	Service Routing Unit
SSH	Secure Shell
SSL	Secure Socket Layer
STUN	Session Traversal Utilities for NAT
SWVC	SD-WAN Virtual Connection
SWVC EP	SD-WAN Virtual Connection End Point
TCO	Total Cost of Ownership
TCP	Transmission Control Protocol
TDM	Time Division Multiplexing
TLS	Transport Layer Security
TN	Transport Network
TNP	Transport Network Port
TVC	Tunnel Virtual Connection
UCS	Underlay Connectivity Service
UDP	User Datagram Protocol
UI	User Interface
UNI	User Network Interface
URI	Uniform Resource Identifier
URL	Uniform Resource Locator

URLLC	Ultra-Reliable Low-Latency Communications
USB	Universal Serial Bus
VAS	Value-Added Service
VGW	Virtual Gateway
VIF	Virtual Interface
VLAN	Virtual Local Area Network
VM	Virtual Machine
VN	Virtual Network
VNF	Virtual Network Function
VNI	VXLAN Network Identifier
VoD	Video on Demand
VPC	Virtual Private Cloud
VPDN	Virtual Private Dial Network
VPN	Virtual Private Network
VR	Virtual Reality
VRF	Virtual Routing and Forwarding
VRRP	Virtual Router Redundancy Protocol
VSID	Virtual Subnet ID
VTEP	VXLAN Tunnel Endpoint
VXLAN	Virtual eXtensible Local Area Network
WAN	Wide Area Network
WDM	Wavelength Division Multiplexing
WFQ	Weighted Fair Queue
WLAN	Wireless Local Area Network
WOC	WAN Optimization Controller
XML	Extensible Markup Language
YANG	Yet Another Next Generation
ZTP	Zero-Touch Provisioning

References

1. ONUG. Software-Defined WAN Use Case. [R/OL] 2014-10.
2. MEF. Understanding SD-WAN Managed Services, Service Components, MEF LSO Reference Architecture and Use Cases. [R/OL] 2017-07.
3. MEF. MEF 70, SD-WAN Service Attributes and Services. [R/OL] 2019-07.

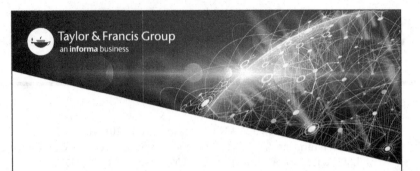

Printed in the United States
by Baker & Taylor Publisher Services